Latin American Viticulture Adaptation to Climate Change

Gastón Gutiérrez Gamboa • Mercedes Fourment
Editors

Latin American Viticulture Adaptation to Climate Change

Perspectives and Challenges of Viticulture Facing up Global Warming

Springer

Editors
Gastón Gutiérrez Gamboa
Instituto de Investigaciones Agropecuarias
INIA Carillanca
Vilcún, Chile

Mercedes Fourment
University of the Republic
Faculty of Agronomy
Montevideo, Uruguay

ISBN 978-3-031-51324-4 ISBN 978-3-031-51325-1 (eBook)
https://doi.org/10.1007/978-3-031-51325-1

This Springer imprint is published by the registered company Springer Nature Switzerland AG
The registered company address is: Gewerbestrasse 11, 6330 Cham, Switzerland

Paper in this product is recyclable.

Contents

Chapter 1
Opening Remarks and General Overview of the Current Scientific Scenario of Latin American Vitiviniculture: A Critical View

Gastón Gutiérrez Gamboa, Philippo Pszczólkowski, and Mercedes Fourment

1.1 Background of the Current Science Scenario of the Latino American Vitiviniculture

The growing season temperature is increasing as a result of climate change (IPCC, 2023; Verdugo-Vásquez et al., 2023). This increase in temperature has considerably affected the viticulture developed under Mediterranean climates, leading to production of wines with high pH and alcohol content and low levels of acidity and phenolic content (Gutiérrez-Gamboa et al., 2021). An excessive accumulation of soluble solids in the grapes has a negative impact on the alcoholic fermentation process (Gutiérrez-Gamboa et al., 2020a, b), resulting in wines with undesirable aromas, high alcohol content, and potential stuck and sluggish fermentations, affecting the winery's economic sustainability (Bell & Henschke, 2005; Gutiérrez-Gamboa et al., 2020a). High levels of alcohol in wines give heavier and warmer sensations, such as high astringency and green tannins if they are not adequately compensated by other sensory components of a balanced wine (Mira de Orduña, 2010). In particular, several studies indicate that the increase in temperature during berry development has

G. Gutiérrez Gamboa (✉)
Instituto de Investigaciones Agropecuarias, INIA Carillanca, Vilcún, Chile

Escuela de Agronomía, Facultad de Ciencias, Ingeniería y Tecnología, Universidad Mayor, Temuco, Chile
e-mail: gaston.gutierrez@inia.cl

P. Pszczólkowski
Pontificia Universidad Católica de Chile, Facultad de Agronomía e Ingeniería Forestal, Departamento de Fruticultura y Enología, Santiago, Chile

M. Fourment
Facultad de Agronomía, Universidad de la República, Montevideo, Uruguay

G. Gutiérrez Gamboa, M. Fourment (eds.), *Latin American Viticulture Adaptation to Climate Change*, https://doi.org/10.1007/978-3-031-51325-1_1

a negative effect on berry composition and wine quality (Martínez-Lüscher et al., 2017; Villalobos-Soublett et al., 2021).

Scientific literature has reported that most of the South American viticultural valleys have been considerably affected by climate change favoring the viticultural aptitude of the Southern valleys (Mills-Novoa et al., 2016; Solman et al., 2018; Straffelini et al., 2023; Verdugo-Vásquez et al., 2023). Contrary to this, the most famous viticultural valleys of Chile and Argentina that are located in the central valleys have lost their aptitude for the cultivation of short-cycle grapevine varieties (Mills-Novoa et al., 2016; Solman et al., 2018; Straffelini et al., 2023). There is little research on the interaction between climate change and viticulture in the Southern Hemisphere compared to the Northern Hemisphere (Straffelini et al., 2023). Some examples include climatic trends and variability in the Chilean viticultural regions (Montes et al., 2012; Verdugo-Vásquez et al., 2023), climatic trends and impacts of spring frost in Brazil (Bardin-Camparotto et al., 2014; Campos et al., 2017), climatic trends and effects on interannual yield variability, heatwaves, and extreme rainfall in Argentina (Agosta et al., 2012; Straffelini et al., 2023), increase in maximum temperature in Ica, Perú (Yzarra et al., 2015), and climate trends and variability effects on local perceptions, vulnerability, and adaptative responses in Uruguay (Fourment et al., 2013, 2020).

Despite that there is an interesting development of viticultural production in México, Bolivia, Colombia, Venezuela, Paraguay, Costa Rica, Dominican Republic and Haiti with an important historical context (Fig. 1.1), there is a scarce development in the sector mostly due to little support from the state to its producers and for research (Corzo, 1987; Morales-Payan & Morales-Payan, 2004; Aleixandre et al., 2013; Valenzuela Solano et al., 2014; Lacoste, 2015; Castillo et al., 2018; Salas Arreaga, 2018; García-Rodea et al., 2022; Santander Racines et al., 2022).

The Latin American wine industry is backed by different organizations, such as technical and scientific institutes and departments of viticulture and enology in public and private universities, which train technicians in the management of vineyards and wineries (Aleixandre et al., 2013). This support gives some Latin American countries a great prestige within the group of New World winemaking countries, in particular, Argentina, Chile, Brazil and Uruguay, who produce close to 90% of the wines in the area (Aleixandre et al., 2013). The Organisation for Economic Co-operation and Development (OECD) points out that Latin American countries spend the least on research and development in relation to their Gross Domestic Product (GDP). Notwithstanding, Brazil, Chile, and Argentina reached the 7th, 12th and 19th positions among the countries that most publications have in the vitiviniculture area (Cimini & Moresi, 2022). Two Latino American institutions (Consejo Nacional de Investigaciones Científicas y Técnicas in Argentina and Empresa Brasileira de Pesquisa Agropecuária – Embrapa in Brazil) are positioned in the top-twenty place among research institutions that develop science in the vitiviniculture sector (Cimini & Moresi, 2022). Only one Latin American researcher from a Spanish institution appears in the top ten of those who publish the most in the area related to grape production (Cimini & Moresi, 2022). Interestingly, Castillo et al. (2018) reported that Mexican scientific output relied to a large extent (36%) on

Fig. 1.1 Vineyards established in Ecuador (**a**), Venezuela (**b**), and Perú (Cañete, Lunahuaná (**c**) and Ica (**d**))

collaborations with research groups from developed countries, while collaborative papers with Latin American countries were comparatively rare (7.9%). This is probably due to the greater investment in equipment and personnel in more developed countries that are capable of sustaining high scientific productivity and solving technical problems in the viticultural area compared to Latin American countries. In Chile, although significant progress has been made in the last decades with regard to research financing, there is still a reduced number of researchers and insufficient investment to face the urgent challenges of the agricultural sector (del Pozo et al., 2021). The lack of investment in research conditions the development of the wine industry in less developed countries as in Latino America. It should be noted that the term map of the most cited papers in the vitiviniculture area coincided with the gene, chromatography, disease, anthocyanin, expression, grape seed, regulation, antioxidant activity, fermentation, red wine, among others words (Cimini & Moresi, 2022). Some of these words describe techniques or variables that must be analyzed using equipment of high economic value, which is why many research projects in Latin American countries are supported by foreign scientists and exchange programs.

Several governmental initiatives have been taken to encourage development (Bontis, 2004; Helleiner, 2009; Macekura, 2013; Bebbington & Unerman, 2018).

The most acclaimed is the one initiated with President Truman's speech in 1949 announcing the doctrine of "fair treatment" for countries of the so-called third world (Latin America, Asia, and Africa) that from that very moment became "underdeveloped areas" (Escobar, 1995). The development seeks to produce important qualitative changes between marginalized areas and those with development potential (de Janvry & Sadoulet, 2004; Schejtman & Berdegue, 2008), but in practice, this longed-for transformation has only increased the socioeconomic problems present in these areas (Escobar, 1995). The 2030 Agenda for Sustainable Development, adopted by all the United Nations Member States in 2015, provides a shared blueprint for peace and prosperity for people and the planet, now and into the future (United Nations, 2023). The member established 17 Sustainable Development Goals (SDGs), which are an urgent call for action by all countries, developed and developing, in a global partnership (United Nations, 2023). The goals recognize that ending poverty and other deprivations must go hand-in-hand with strategies that improve health and education, reduce inequality, and spur economic growth, all while tackling climate change and working to preserve our oceans and forest (United Nations, 2023). SDGs encourage fairness, equity, consistency, and clarity in the use and reporting of race and ethnicity in all areas of development and this is a current controversial subject in medical and science journals (Flanagin et al., 2021). Liu et al. (2017) used a dataset of 1,000,000 papers from six publishers over the past two decades, reporting fewer non-white editors than would be expected based on their share of authorship. In addition, these authors evidenced that non-white scientists endure longer waiting times between the submission and acceptance of their manuscripts, and upon publication, their papers receive fewer citations than would be expected based on textual similarity. These findings highlight ways through which non-white scientists suffer from inequalities, potentially hindering their academic careers. Journals related to vine and wine sciences encourage similar discriminations. Figure 1.2 shows the number of editors of Latino American origin in journals

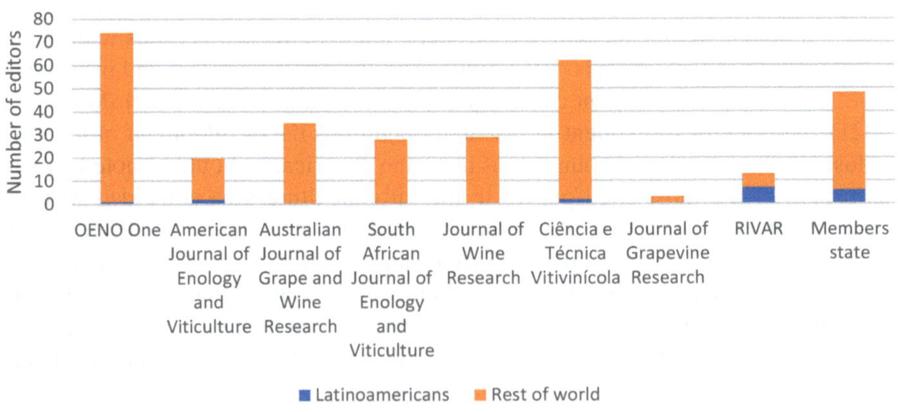

Fig. 1.2 Number of editors of Latino American origin in journals related to vine and wine sciences in 2022

related to vine and wine sciences in 2022. Figure 1.2 was made by counting the editors and their affiliations that are displayed in the editorial board member platform of each journal, including editors from the member states and observers of the International Organisation of Vine and Wine. The Revista Iberoamericana de Viticultura, Agroindustria y Ruralidad (RIVAR) belonging to the Universidad de Santiago de Chile presented the highest diversity in the number of editors reaching 53.8% of Latino Americans and 46.2% of the rest of the world members. In addition, the editorial board of RIVAR presents an equitable incorporation of gender, reaching the same number of male and female editors. Notably, in the first year of metric, RIVAR reached Q1 in Cultural Sciences and History, Q2 in Archeology, and Q3 in Agronomy and Crop Sciences, Food Sciences, and Horticulture by Scimago. Journals related to the vitiviniculture from France, the United States, Australia, South Africa, Germany, and Canada reached the lowest diversity that varied from 0.0% to 10.0% of Latino American editors, contributing to inequality. *OENO One* (France) and *Australian Journal of Grape and Wine Science* (Australia) reached the highest impact factor of the subject and between the two journals, there is only one male editor of Latin American origin. These journals cover specific areas related to Food Science and Horticulture linked to solve technical problems, which makes it difficult for authors from less developed countries to publish who valorize technocracy and mostly heritage, society, and history. Kreimer and Vessuri (2018) reported that researchers belonging to Latin America are in a good position to critically analyze the current relations between science and society, to assist decision-makers and help the public understand the implications of present-day technoscientific change, as well as to support the development of fairer, more equitable solutions to combat the challenges of today's changing world. It goes without saying that, far from having reached maturity, this is a space in a permanent state of construction.

Latino America vitiviniculture has particular traits that provide a more sustainable development. Cotagaita and Cintis valleys, belonging to the Bolivian viticulture, develop an activity that implements management tending to conserve their activity sustainably, particularly by cultivating the vines in living tutors of molle trees (*Schinus molle* L.) and chañar (*Geoffroea decorticans*) (Castro et al., 2022) as agroforestry system, which is a current research topic in Europe. Similar management is evidenced in Codpa valley, Chile (Gutiérrez-Gamboa et al., 2023), which is carried out by Aymara descendants. These services are expected to protect trees against climate hazards and flooding, disease control and maintenance of soil fertility, improving environmental sustainability (Oliva Oller et al., 2022). Notably, the scientific research carried out in Bolivia was supported by foreign scientists belonging to Georgetown University, L'institut Agro Montpellier, and the French National Research Institute for Agriculture, Food and Environment (INRAE) institutions.

Contrary to the general trend of producing wine from the most famous grapevine varieties associated with the French paradigm, such as Cabernet Sauvignon, Merlot, Pinot Noir, Syrah, Sauvignon Blanc, and Chardonnay, there is a current tendency to revalorize and preserve minority or autochthonous grapevine varieties worldwide, mostly in Latin America (Gutiérrez-Gamboa et al., 2020b). Some of these varieties and unknown vine genotypes are under study by some Latino American research groups, mainly from the Instituto de Investigaciones Agropecuarias (INIA) in Chile and the Instituto Nacional de Tecnología Agropecuaria (INTA) in Argentina (Pérez et al., 2018; Fanzone et al., 2019; Mendoza et al., 2022; Pszczólkowski et al., 2022; Torres et al., 2022). These authors aim to provide plant material adapted to global warming, revalorizing local genetic resources. In addition, the Empresa Brasileira de Pesquisa Agropecuária (EMBRAPA) has been strongly working for more than two decades in a genetic improvement plan with public and private funds, which has allowed it to provide table grape and grape juice producers with new varieties and interspecific hybrids adapted to the subtropical conditions of Brazil, mainly seedless plant material adapted to *Plasmopara viticola* (Ahmed et al., 2019; Koyama et al., 2020).

Different authors have promoted double cropping in vines as a novel bud-forcing strategy to face the negative effects of global warming in European warm climate vineyards (Poni et al., 2020, 2021; Martínez de Toda, 2021). The principle of this strategy is to maintain the primary crop and obtain a second, late-ripening crop through release of dormancy of the axillary buds during the current season (Poni et al., 2021). Double or even triple cropping system strategy in subtropical and tropical areas is a reasonably consolidated approach for table grape production (Bo et al., 2016; Ahmed et al., 2019; Koyama et al., 2020). To our knowledge, Brazilian researchers were the first that published research on double-cropping in the literature (Kishino, 1981). These researchers developed an experiment carried out in the County of São Miguel Arcanjo, State of São Paulo, Brazil, to delay the vegetative cycle of Italia grapevines through double-cropping strategies. The double cropping strategy delayed the grape ripening by 44 days. In this study, removing the shoots at 20 cm longitude did not affect the yield. However, removing shoots when they were 50 and 100 cm caused a considerable decrease in productivity.

Latino American vine and wine sciences provides to the world different research opportunities for the mitigation of negative effects of global warming in viticulture, favoring adaptability, sustainability, and social equity that the international community should consider to provide programs that promote equity and access to science. Based on this, initiatives promoted by Germany, France, and the United States have helped to improve these gaps through programs such as BAFAGRI in Brazil, CHILFAGRI in Chile, ARFAGRI in Argentina, German Academic Exchange Services (DAAD) and US Fulbright Program (Lally, 2022).

1.2 Latino American Vitiviniculture

1.2.1 Argentinian Vitiviniculture

Argentina is the most important wine producer in Latin America, being the fifth-largest wine-producing country in the world (OIV, 2022). In 2022, the vineyards occupied 207.047 hectares (INV, 2023). In the same year, grape production was 19.368.031 kg, and wine production was 11.456.000 hl. In Argentina, most grapes are used for winemaking (mainly for wine production and must production), with lower volumes in the production of dry grapes (*uvas pasa*) and for fresh consumption (Hernández, 2021). Argentinian emblematic variety is Malbec because of its enological personality. However, grapevine genotypic diversity in Argentina, and probably also in South America, is higher than previously thought. In that sense, there is a current effort to revalue the vegetable heritage that exists with the *criollas* varieties.

Cirivini and Mazzini (2012) identify three stages in the Argentinian's viticulture: (1) Proto-industrial: Since the sixteenth century, with the Spanish colonization and the arrival of the first religious orders, viticulture—as an economic activity and cultural practice—was transferred to the region and adapted to local conditions. Production tended to satisfy the local market and the needs of the scarce inhabitants of the cities of present-day Argentina. (2) Industrialization and first modernization: This period comprises the first moment of intense growth of the winemaking activity between 1885 and 1930. It was characterized by a vertiginous growth driven by liberal economic policies, large-scale irrigation works, the importation of technology, the prominence of large-scale European immigration, and new communications, particularly the railway. The second stage, between 1930 and 1990, corresponds to the expansion (territorial and economic) of the production model and its crisis. It is important to note that the crisis of the model began around 1970 and was finally triggered in 1990. (3) Crisis and second modernization: The wine industry manifested a sustained critical situation between 1970 and 1990 as a consequence of overproduction of low-quality wines, the narrowness of markets, and the collapse of large establishments of traditional family businesses. From the last decade of the twentieth century onwards, a new stage of modernization took place (which is continuing), characterized by a marked orientation towards high-quality wines aimed at the national and international market and exalting its territorial identity and varietal production.

Due to climatic conditions and historical and cultural factors, the provinces with the largest area under vine are Mendoza, where in the last six decades, approximately 70% of the hectares have been located, and San Juan, with around 20% of the hectares. The rest of the production is distributed among La Rioja, Río Negro, Catamarca, Salta, Neuquén and others (Hernández, 2021). Despite that, Argentina has 110 geographic indications (IG) and 2 recognized and protected designations of origin (DOC as Luján de Cuyo and San Rafael, both in Mendoza) (INV, 2023). Based on a national law, the *Vitis vinifera* varieties that are recognized as suitable for the production of quality wines with GI and DOC have been determined (INV, 2023).

1.2.2 Brazilian Vitiviniculture

Brazilian vitiviniculture has been characterized by historical and cultural aspects from its beginnings to the present day (García-Rodea et al., 2022). The first cuttings cultivated in Brazil were brought by the Portuguese colonizers around 1532 and were established along the coast of the country (Falcade, 2011; Wurz et al., 2017). The vines were cultivated in the states of Sao Paulo, Pernambuco, and Bahia and on the continental side, by the Spaniards, who began their plantations in Paraná and Rio Grande do Sul (García-Rodea et al., 2022). Subsequently, Italian immigrants arrived in the nineteenth century in Rio Grande do Sul, who brought with them the grape culture and the habit of wine consumption, forming a cultural heritage in this region (García-Rodea et al., 2022). The most important grape variety introduced by the Italians was a hybrid of *Vitis labrusca* and *V. vinifera* called Isabel (Wurz et al., 2017).

Similar to those observed in Chile with the País variety, Isabel grape cultivation was affected due to the rapid replacement of the national vineyard by European grapevine varieties, which was promoted by local laws (García-Rodea et al., 2022). The aim was to establish the basis for the development of the region by the production of commercial wines with more competitive varieties (García-Rodea et al., 2022) without taking into account the particular identity of this variety and the contribution made by the Italian colony to Brazilian viticulture.

Brazilian wine production is the third most important in the South American region, after Argentina and Chile and before Uruguay, accounting for an estimated wine production volume of 3.2 mhl in 2022 (OIV, 2022). In 2018, Brazilian wine production accounted for 3.1 mhl and the surface of vineyards reached 82,000 ha (Castro et al., 2019). Rio Grande do Sul produces about 90% of the national wine, which is reflected in the fact that approximately 57% of the area planted with vines is located in this state (Castro et al., 2019). Vale dos Vinhedos is the main viticultural region for wine production and is located in the Serra Gaucha, which territorially encompasses the municipalities of Bento Gonçalves, Monte Belo do Sul, and Garibaldi (Sousa et al., 2013; de Alcântara Bittencourt César, 2019).

Tropical viticulture is mostly concentrated in the Northeast regions of Brazil, while subtropical viticulture is found towards the Southeast regions of the country, especially under high-altitude mesoclimates (Camargo et al., 2008). American (*V. labrusca*) and hybrid grapevine varieties are mostly cultivated in the south to produce grape juices and wines. In contrast, in the other regions, American, hybrids, and European (*V. vinifera*) table grapes are cultivated both for the domestic market and for export (Gazzola et al., 2020). Brazil has developed a serious breeding program to provide the industry plant material adapted to subtropical and tropical viticulture, and the future perspective is to continue its breeding program to provide seedless table grapes of high quality for the international market (Gazzola et al., 2020). The most important vitiviniculture products of Brazil are table grapes, especially the exports of fresh grapes from the Pernambuco and Bahia states (Submédio São Francisco Valley), as well as wines, sparkling wines, and grape juices, which are mostly produced in Rio Grande do Sul, which also registered an interesting increase in exports (Kist et al., 2022).

1.2.3 Bolivian Vitiviniculture

Bolivian viticulture seems to have begun in the La Paz river (1550) below 2600 meters above sea level (Buitrago Soliz, 2014). However, a legend mentions that Vicchoca locality (Cotagaita valley, Potosí) was the first place where the vines were planted. The criolla variety Vicchoqueña, which probably originated as a natural cross breeding between the Criolla Negra and the Moscatel de Alejandría, was introduced in this place during the Spanish conquest. In colonial times, the training system used was the Mollar, in which the molle (*Schinus molle*) or chañar (*Geoffroea decorticans*) trees were used as a natural support for vine development. Mollar training systems can still be observed in vineyards in Cotagaita (Potosí) and de los Cintis (Chuquisaca) valleys (Castro et al., 2022; Oliva Oller et al., 2022). Bolivian viticulture played a key role in the development of other South American vitiviniculture, particularly those of Perú, Chile, and Argentina, as a result of the high demand for wine and distillates generated by the mining exploitation of the Cerro Rico de Potosí (Basadre, 1884; Billinghurst, 1893). In 1650, this city had a population of 160,000 habitants, a determining factor for the productive and commercial success of the time (Buitrago Soliz, 2014).

Bolivian vitiviniculture development remained stagnant for four centuries and since 1960 some improvements in viticulture and enology have been incorporated to the enhancement of the value chain of the sector (Buitrago Soliz, 2014). This improvement began with promoting the European paradigm, in which some varieties were selected over others as the most suitable for wine production, leading to a new development of Bolivian vitiviniculture. Extensive viticulture began between 1976 and 1982, particularly in the Central Valley of Tarija, which today concentrates 65% of the national surface area (Buitrago Soliz, 2014). However, this period was followed by a short time of stagnation and the national vitiviniculture reached a new significance in the 1990s which continues today. During the last decades, new viticultural regions such as Samaipata (Mesothermic valleys of Santa Cruz) were developed, and an incipient wine export process began with international recognition, particularly for wines produced with the Tannat variety promoted by "Wines of Bolivia" (Buitrago Soliz, 2014). Singani, a distillate obtained from Muscat of the Alexandria variety and which has been a Denomination of Origin since 1992, should be highlighted in terms of production, trade, and culture for Bolivia. There is also an interesting development of table grapes, with varieties such as Red Globe and Italia Piróvano 65, grown in the same areas where viticulture is developed and others (Gutiérrez-Gamboa et al., 2020b) such as the Chaco Tarijeño (400 m.a.s.l.). Currently, Bolivia accounts for close to 3000 ha of vineyards located in Chuquisaca (11%), Potosí (10%), Santa Cruz (7%), Cochabamba (4%), La Paz (3%), and Tarija (65%).

1.2.4 Chilean Vitiviniculture

Chile holds as of 2021 approximately 139,000 ha of vineyards, decreasing plant surface in the last decade by around of 50,000 ha (SAG, 2021). Close to 93% of the national vine surface is destined for wine production and the rest for Pisco elaboration (SAG, 2021). Pisco was the first wine product protected by the Chilean state under a Denomination of Origin (1931). Pisco was able to consolidate over time, becoming the main D.O. in South America and an emblematic product of both Chile and Peru (Lacoste et al., 2013), while some other D.O. in Chile as "asoleado" and "pajarete" did not reach this success (Castro et al., 2016). The most important grapevine varieties planted in Chile are Cabernet Sauvignon (11.01%), Sauvignon Blanc (8.32%), Merlot (8.32%), País (cv. Listán Prieto) (8.04%), Chardonnay (7.95%), and Carmenère (7.93%) (SAG, 2021).

Chile is the largest exporter of fresh grapes and the fourth largest exporter of wine after Spain, France, and Italy (OIV, 2023). Currently, the wine development model is considered beneficial for the country because it has a number of advantages compared to the commodity development model, as it is compatible with small property, industrialization, intensive labor, capital and technology investment, and creates a range of secondary activities (Lacoste, 2005). The great wine development achieved in recent years by the national industry has brought a series of negative consequences, such as the increase in the grape price, mainly affecting the viticulturists from Itata, Biobio, and Maule valleys (Erices Castro & Escalona Ulloa, 2022). This situation encouraged the change or abandonment of this activity, which led to a decrease in the vineyard surface area in these above-mentioned valleys (Erices Castro & Escalona Ulloa, 2022). Unfortunately, many of these producers had old vineyards with genetic material that was not studied and for economic reasons they converted to forestry producers. However, to date, there are still producers who seek to maintain the old traditions, investing in their own wineries and establishing themselves through guilds.

The historical discursive construction of wine production in the last decades has a clear positivist vision in economic terms, which downplays the importance of the grape varieties and wines produced not only from Biobío and Itata valleys (Erices Castro & Escalona Ulloa, 2022), but also from Codpa, Pica, Maule, Malleco, Cautín, and the rest of the Southern valleys. This phenomenon marginalizes them from development and positions them as a secondary or tertiary actor in the vitiviniculture sector (Erices Castro & Escalona Ulloa, 2022). To modernize the country, public institutions imposed French practices as the only valid form of wine production, which meant in the long term a devaluation of criolla wines and grape varieties, confining them to certain geographical areas and finally making them invisible in the national wine industry (Lacoste, 2021; Erices Castro & Escalona Ulloa, 2022; Jerković et al., 2022). Based on this, Chilean viticultural valleys located in the Atacama Desert are not recognized by the actual Decree-Law that establishes the Chilean viticultural zoning (BCN, 2020), probably due to the scarcity of the surface planted vines (SAG, 2021), despite their great historical importance. In addition, the natural or artificial crossbreeding of these varieties gave rise to others that have great agronomical and enological potential (Gutiérrez-Gamboa et al., 2020b). Still, the wine they produce cannot be sold because the national legislation does not cover them.

1.2.5 Uruguayan Vitiviniculture

Uruguay has an area of vineyards covering 5849 hectares, 91.8% of which are in the south of the country (Canelones, Montevideo, Colonia, and San José Departments) (INAVI, 2023). In 2022, the grape production was 106,672,752 kg of grapes; 97% of the grapes were destined for wine production.

Small and medium-sized family producers characterize the Uruguayan wine sector with a strong immigrant identity. Nearly 71.7% of the producers have less than 20 hectares.

The most popular planted varieties are Tannat, Merlot, Moscatel de Hamburgo, and Ugni blanc. Tannat is considered the flagship *cépage* because of its adaptability.

Winegrowing in Uruguay is an immigrant sector (Fourment et al., 2015). To this day, there is a strong emphasis on cultural traditions in how wine is produced and sold. The winegrowing tradition has a cultural and identity-related dimension, as explored by Vitale (2003). The author notes how the sector's cultural heritage is strongly linked to the immigrant communities who helped to bring it into being and spurred its development. Vitale (2003) recognises the importance of the role played by immigrants at the end of the nineteenth century in the modernization of Uruguayan agriculture and the role that they and their descendants continue to play in several agricultural production sectors, particularly winegrowing, where in most cases they are the majority players. Reliving work habits and valuing activity and cultural conceptions of work can be recognized and interpreted as part of the traditions of the immigrant community.

In the last decades, the sector is suffering a retraction of the surface area. At the same time, national production is globally maintained thanks to the technologies applied in the vineyard and winery. The current challenges in the sector revolve around preventing small family producers from disappearing, the competitiveness of products at national (in terms of profitability) and international levels, and the sustainability of production systems (economically, socially and environmentally).

1.2.6 Caribbean Vitiviniculture

Haïtienne vitiviniculture is mostly developed in Chardonnières (18.2747° Lat., and −74.1644° Lon), which is bordered by the Massif de la Hotte Mountains on the north and the Caribbean Sea lying in the south. Based on Köppen classification, Chardonnières presents a tropical dry climate (Aw). Since scarce chilling hours accumulation for vines is reached in Chardonnières, grapevine planting should be moved towards Paillant (18.41841° Lat., and −73.143° Lon). The vineyard soils of Chardonnieres are calcareous (Fig. 1.3a) so there are usually iron deficiencies (Fig. 1.3b, c). The vines are trained randomly, forming an overhead system that provides shade (Fig. 1.3d). The main vine diseases in Chardonnières are downy mildew and powdery mildew, which are sometimes managed using magic (Fig. 1.3e, f).

Fig. 1.3 Some particularities of the Chardonnierès vitiviniculture: calcareous soil (**a**); leaves with iron deficiencies (**b**, **c**); vines in an overhead system (**d**); pest and disease management using magic (**e**, **f**)

Fig. 1.4 Vitiviniculture in the Dominican Republic. Industria Vinícola de Neyba (**a, b**); grapevines and cuttings produced in Neyba (**c, d**)

The first cuttings were introduced in the 1490s and early 1500s in the Dominican Republic (Morales-Payan & Morales-Payan, 2004). Since the 1980s, the grape and wine consumption in the Dominican Republic has been increasing. The national vineyard surface for grape production is about 300 ha, mostly concentrated in the Neyba Mountain Range. Most of the wine consumed in the Dominican Republic is imported. Still, local wine production could be enhanced by the adoption of varieties more adequate for wine production, increasing quality control in the winemaking process, and modernizing winery facilities (Morales-Payan & Morales-Payan, 2004). Currently, the most important viticultural product of the Dominican Republic is table grape, whose production is based mostly on Red Globe. Some wines are also produced in the Dominican Republic, using European grapevine varieties, especially Tempranillo and Syrah, which has been promoted by the Industria Vinícola de Neiba (Fig. 1.4).

1.3 Concluding Remarks

The vitiviniculture in Latin America faces significant challenges and could provide opportunities for research for solving the current problems in the context of climate change and global wine production. Climate change has led to an increase in temperatures, affecting grapevine development and wine quality. Most of the wines produced in the region are characterized by high levels of pH and alcohol content and low acidity and phenolic content, impacting the sensory attributes of wines. Despite these challenges, the Latin American wine industry has made significant progress in recent years. In this sense, Argentina, Chile, Brazil, and Uruguay are gaining recognition in the world, producing characterized products mostly from European grapevine varieties, such as Cabernet Sauvignon, Chardonnay, Malbec, Pinot Noir, Sauvignon Blanc, Syrah, and Tannat. However, from a scientific point of view, there are disparities in research and development support among the Latin American countries, with some of them accounting for limited resources for scientific research in this sector. By leveraging the region's unique characteristics and promoting research collaboration, Latin American vitiviniculture can continue to evolve and contribute to the global wine market while preserving its cultural and historical heritage.

References

Agosta, E., Canziani, P., & Cavagnaro, M. (2012). Regional climate variability impacts on the annual grape yield in Mendoza, Argentina. *Journal of Applied Meteorology and Climatology, 51*(6), 993–1009.

Ahmed, S., Roberto, S. R., Shahab, M., Colombo, R. C., Silvestre, J. P., Koyama, R., & de Souza, R. T. (2019). Proposal of double-cropping system for 'BRS Isis' seedless grape grown in subtropical area. *Scientia Horticulturae, 251*, 118–126.

Aleixandre, J. L., Bordeu, E., Aleixandre-Tudó, J. L., Bolaños, M., & Aleixandre-Benavent, R. (2013). Scientific productivity and collaboration in viticulture and enology in Latin American countries. *Ciencia e Investigación Agraria, 40*(2), 429–443.

Bardin-Camparotto, L., Constantino Blain, G., Júnior, M. J. P., Hernandes, J. L., & Cia, P. (2014). Climate trends in a non-traditional high quality wine producing region. *Bragantia, 73*(3), 327–334.

Basadre, M. (1884). *Riquezas peruanas: colección de artículos descriptivos escritos para "La Tribuna"*. Imprenta de La Tribuna.

BCN. (2020). *Decreto 464. Establece zonificación vitícola y fija normas para su utilización.* Ministerio de Agricultura. Biblioteca del Congreso Nacional.

Bebbington, J., & Unerman, J. (2018). Achieving the United Nations sustainable development goals: An enabling role for accounting research. *Accounting, Auditing & Accountability Journal, 31*(1), 2–24.

Bell, S., & Henschke, P. (2005). Implications on nitrogen nutrition for grapes, fermentation and wine. *Australian Journal of Grape and Wine Research, 11*, 242–295.

Billinghurst, G. E. (1893). *La irrigación en Tarapacá.* Patrimonio Cultural Común, Imprenta y Librería Ercilla.

Bo, W., Yang, B., XianJin, B., Ying, Z., TaiLi, X., JinBiao, L., AiJun, C., BingHai, L., JianJun, H., Ling, L., YongMei, Z., & XiongJun, C. (2016). Introduction performance and double-harvest-a-year cultivation technique of 'Shine Muscat' grape in Nanning, Guangxi. *Journal of Southern Agriculture, 47*, 975–979.

Bontis, N. (2004). National intellectual capital index: A United Nations initiative for the Arab region. *Journal of Intellectual Capital, 5*(1), 13–39.

Buitrago Soliz, A. (2014). The emerging viticulture of Bolivia and opportunities for Singani. *Revista Iberoamericana de Viticultura, Agroindustria y Ruralidad, 1*(2), 90–102.

Camargo, U. A., Protas, J. F. S., & Mello, L. M. R. (2008). Grape growing and processing in Brazil. *Acta Horticulturae, 785*, 51–58.

Campos, C. G. C., Malinovski, L. I., Marengo, J. A., Oliveira, L. V., Vieira, H. J., & Silva, A. L. (2017). The impact of climate projections when analyzing the risk of frost to viticulture in the southern region of Brazil. *Acta Horticulturae, 1188*, 165–172.

Castillo, G., Castro-López, L., & Méndez, S. (2018). Contribution of Mexican scholars to viticultural and oenological research: Where do we stand? *OENO one, 25*(4), 1–6.

Castro, A., Pszczólkowski, P., Mujica, F., Lacoste, P., Núñez, E., Cofré, C., Adunka, M. L., & Soto, N. (2016). El pajarete de Huasco y Elqui (Chile): Historia de un vino escogido. *Idesia, 34*(4), 23–35.

Castro, A., Villena, W., & Pszczólkowski, P. (2022). Cotagaita and Cintis: Patrimonialize the spirit of the place? *Revista Iberoamericana de Viticultura, Agroindustria y Ruralidad, 9*(25), 153–172.

Cimini, A., & Moresi, M. (2022). Research trends in the oenological and viticulture sectors. *Australian Journal of Grape and Wine Research, 28*, 475–491.

Cirvini, S. A., & Manzini, L. (2012). El paisaje vitivinícola: Identificación y caracterización. Mendoza, Argentina. *Revista de Historia Americana y Argentina, 47*(1), 101–131.

Corzo, P. (1987). Tropical viticulture in Venezuela. *Acta Horticulturae, 199*, 27–29.

de Alcântara Bittencourt César, P. (2019). Procesos de transformaciones territoriales del Vale dos Vinhedos – Bento-Gonçalves – Brasil. *Estudios y Perspectivas en Turismo, 28*(2), 354–371.

de Janvry, A., & Sadoulet, E. (2004). Estrategias de ingresos de los hogares rurales de México: el papel de las actividades desarrolladas fuera del predio agrícola. Seminarios y Conferencias 6717, Naciones Unidas Comisión Económica para América Latina y el Caribe (CEPAL).

de Oliveira Castro, Á. L., de Barros Vilas Boas, L. H., Carvalho de Rezende, D., & Pimenta, M. L. (2019). Behind the wine glass: Values that guide consumption in Minas Gerais – Brazil. *British Food Journal, 121*(10), 2477–2489.

del Pozo, A., Engler, A., & Meza, F. (2021). Agricultural sciences in Chile: Institutions, human resources, investment and scientific productivity. *Chilean Journal of Agricultural Research, 81*(4), 664–673.

Erices Castro, J., & Escalona Ulloa, M. (2022). Power disputes in Chilean viticulture: Valleys of Biobio under the shadows of development. *Economía, Sociedad y Territorio, 22*(69), 309–337.

Escobar, A. (1995). Encountering Development: The Making and Unmaking of the Third World. STU-Student edition, Princeton University Press, pp 312.

Falcade, I. (2011). A paisagem como representação espacial: a paisagem vitícola como símbolo das indicações de procedência de vinhos das regiões Vale dos Vinhedos, Pinto Bandeira e Monte Belo do Sul (Brasil). Doctoral Thesis. Porto Alegre, Universidade Federal do Rio Grande do Sul.

Fanzone, M. L., Griguol, R., Mastropietro, M., Sari, S., Pérez, D., Catania, A., Mussato, E., Salafia, A., Assof, M., & Jofré, V. (2019). Chemical and sensory profile of Torrontés riojano wines from different geographical zones of Argentina. *Investigación, Ciencia y Universidad, 3*(4), 22–29.

Flanagin, A., Frey, T., & Christiansen, S. L. (2021). Manual of style committee. Updated guidance on the reporting of race and ethnicity in medical and science journals. *Journal of the American Medical Association, 326*(7), 621–627.

Fourment, M., Ferrer, M., & Quénol, H. (2013). *Vitis vinifera* L. cv. Tannat: respuesta a la variabilidad climática. *Agrociencia, 17*(2), 45–54.

Fourment, M., Pérard, P., & Beretta, A. (2015). Uruguay: un vignoble d'inmigration. Rencontres de Clos Vougeot "Le vin en héritage: anciens vignobles, nouveaux vignobles". Chaire UNESCO "Culture et traditions du vin". Dijon, France. 325p.

Fourment, M., Ferrer, M., Barbeau, G., & Quénol, H. (2020). Local perceptions, vulnerability and adaptive responses to climate change and variability in a winegrowing region in Uruguay. *Environmental Management, 66*, 590–599.

García-Rodea, L. F., Thomé-Ortiz, H., Espinoza-Ortega, A., & Bittencourt-César, P. A. (2022). Viniculture and tourism in the New World of wine: A literature review from the American continent. *Wine Economics and Policy, 11*(1), 127–140.

Gazzola, R., Gründling, R. D. P., & Aragao, A. A. (2020). Production and international trade of grapes. *Revista Brasileira de Agrotecnologia, 10*(3), 68–74.

Gutiérrez-Gamboa, G., Alañón-Sanchez, N., Mateluna-Cuadra, R., & Verdugo-Vásquez, N. (2020a). An overview about the impacts of agricultural practices on grape nitrogen composition: Current research approaches. *Food Research International, 139*, 109946.

Gutiérrez-Gamboa, G., Liu, S. Y., & Pszczólkowski, P. (2020b). Resurgence of minority and autochthonous grapevine varieties in South America: A review of their oenological potential. *Journal of the Science of Food and Agriculture, 100*(2), 465–482.

Gutiérrez-Gamboa, G., Zheng, W., & Martínez de Toda, F. (2021). Current viticultural techniques to mitigate the effects of global warming on grape and wine quality: A comprehensive review. *Food Research International, 139*, 109946.

Gutiérrez-Gamboa, G., Contreras Cortez, V., Jara, S., Verdugo-Vásquez, N., Pszczólkowski, P. (2023). Andean Viticulture: Edaphoclimatic Characterization of Vineyards Located in the Codpa Valley. *Revista Iberoamericana de Viticultura, Agroindustria y Ruralidad, 10*(28), 26–40.

Helleiner, E. (2009). The development mandate of international institutions: Where did it come from? *Studies in Comparative International Development, 44*, 189–211.

Hernández, J. J. (2021). La promoción de la vitivinicultura Argentina: seis décadas, una función, múltiples actores. *Mundo Agrario, 22*(50), 1–17.

INAVI. (2023). Estadisticas de viñedos. Retrieved from: https://www.inavi.com.uy/uploads/vinedo/e1141691ff8dd5bd2a83547b5a8c60636eb4aebcc.pdf

INV. (2023). Principales datos vitivinícolas. Retrieved from: https://www.argentina.gob.ar/inv/vinos/principales-datos-vitivinicolas

IPCC. (2023). Ara Begum, R., Lempert, R., Ali, E., Benjaminsen, T.A., Bernauer, T., Cramer, W., Cui, W., Mach, K., Nagy, G., Stenseth, N.C., Sukumar, R., & Wester, P. 2022. Point of departure and key concepts. In H.-O. Pörtner, D. C. Roberts, M. Tignor, E. C. Poloczanska, K. Mintenbeck, A. Alegría, M. Craig, S. Langsdorf, S. Löschke, V. Möller, A. Okem, & B. Rama (Eds.), *Climate change 2022: Impacts, adaptation and vulnerability. Contribution of Working Group II to the sixth assessment report of the intergovernmental panel on climate change* (pp. 121–196). Cambridge University Press. https://doi.org/10.1017/9781009325844.003

Jerković, M., Skewes, J. C., Duhart, F., Mujica, F., & Lacoste, P. (2022). Viñas patrimoniales y vinos naturales de Chile: el legado mestizo góticomapuche. *Revista Austral De Ciencias Sociales, 42*, 349–371.

Kishino, A. Y. (1981). Videira Itália (*Vitis vinifera* L.) – Produção tardia da uva com variações no sistema e na época de poda. Master Thesis, Escola Superior de Agricultura Luiz de Queiroz.

Kist, B. B., Carvalho, C., de Beling, R. R., & Machado, M. O. (2022). *Anuário Brasileiro de Horti & Fruti 2022* (pp. 1–96). Gazeta Santa Cruz.

Koyama, R., Borges, W. F. S., Colombo, R. C., Hussain, I., de Souza, R. T., & Roberto, S. R. (2020). Phenology and yield of the hybrid seedless grape 'BRS Melodia' grown in an annual double cropping system in a subtropical area. *Horticulturae, 6*(1), 3.

Kreimer, P., & Vessuri, H. (2018). Latin American science, technology, and society: A historical and reflexive approach. *Tapuya: Latin American Science, Technology and Society, 1*(1), 17–37.

Lacoste, P. (2005). El vino y la nueva identidad de Chile. *Universum, 20*(2), 24–33.

Lacoste, P. (2015). The rise and secularization of viticulture in Mendoza: The Godoy family contribution, 1700–1831. *The Americas, 63*(3), 385–407.

Lacoste, P. (2021). Rise, fall and rebirth of Listán Prieto in the Southern Cone of America. *Idesia, 39*(2), 75–84.

Lacoste, P., Jiménez, D., Castro, A., Rendón, B., & Soto, N. (2013). A binational appellation of origin: Pisco in Chile and Peru. *Chilean Journal of Agricultural Research, 73*(4), 107–114.

Lally, M. (2022). Understanding the experiences of Fulbright visiting scholars—A qualitative systematic review. *Education Sciences, 12*(2), 90.

Liu, F., Rahwan, T., & AlShebli, B. (2017). Non-White scientists appear on fewer editorial boards, spend more time under review, and receive fewer citations. *Proceedings of the National Academy of Sciences, 120*(13), e2215324120.

Macekura, S. (2013). The point four program and U.S. international development policy. *Political Science Quarterly, 128*(1), 127–160.
Martínez de Toda, F. (2021). Grapevine double cropping: A reality, not a myth. *IVES Technical Reviews, 1*, 1–2.
Martínez-Lüscher, J., Chen, C. C. L., Brillante, L., & Kurtural, S. K. (2017). Partial solar radiation exclusion with color shade nets reduces the degradation of organic acids and flavonoids of grape berry (*Vitis vinifera* L.). *Journal of Agricultural and Food Chemistry, 65*(49), 10693–10702.
Mendoza, L., Aliquó, G., Prieto, J., Flores, J., Casas, A., Grados, M., Aybar, L., & Torres, M. R. (2022). Prospection and identification of traditional-heritage Peruvian grapevine cultivars (*Vitis vinifera* L.) from Ica and Cañete valleys. *Vitis, 61*(2), 47–51.
Mills-Novoa, M., Pszczólkowski, P., & Meza, F. (2016). The impact of climate change on the viticultural suitability of Maipo Valley, Chile. *The Professional Geographer, 68*(4), 561–573.
Mira de Orduña, R. (2010). Climate change associated effects on grape and wine quality and production. *Food Research International, 43*(7), 1844–1855.
Montes, C., Perez-Quezada, J. F., Peña-Neira, A., & Tonietto, J. (2012). Climatic potential for viticulture in Central Chile. *Australian Journal of Grape and Wine Research, 18*, 20–28.
Morales-Payan, J. P., & Morales-Payan, M. O. (2004). Viticulture and enology in the Dominican Republic: Situation, limitations, and possibilities. *Acta Horticulturae, 640*, 369–374.
OIV. (2022). *World wine production outlook. OIV first estimates*. International Organisation of Vine and Wine.
OIV. (2023). *State of the world vine and wine sector in 2022 April 2023*. International Organisation of Vine and Wine.
Oliva Oller, P., Notaro, M., Langer, E., & Gary, C. (2022). Structure and management of traditional agroforestry vineyards in the high valleys of southern Bolivia. *Agroforestry Systems, 96*, 375–386.
Pérez, D., Assof, M., Bolcato, E., Sari, S., & Fanzone, M. (2018). Combined effect of temperature and ammonium addition on fermentation profile and volatile aroma composition of Torrontés Riojano wines. *LWT, 87*, 488–497.
Poni, S., Gatti, M., Tombesi, S., Squeri, C., Sabbatini, P., Lavado Rodas, N., & Frioni, T. (2020). Double cropping in *Vitis vinifera* L. Pinot Noir: Myth or reality? *Agronomy, 10*(6), 799.
Poni, S., Del Zozzo, F., Santelli, S., Gatti, M., Magnanini, E., Sabbatini, P., & Frioni, T. (2021). Double cropping in *Vitis vinifera* L. cv. Pinot Noir: Agronomical and physiological validation. *Australian Journal of Grape and Wine Research, 27*, 508–518.
Pszczólkowski, P., Ganga, M. A., Veliz, L., Castro, M. H., Meneses, M., Castro, A., Cañón, P., & Hinrichsen, P. (2022). Molecular and ampelographic characterization of genotypes used for Pajarete wine, an old Denomination of Origin from Huasco and Elqui Valleys in northern Chile. *Vitis, 61*(1), 37–44.
SAG. (2021). Catastro Vitivinícola 2021. Servicio Agrícola y Ganadero. Retrieved from: https://www.sag.gob.cl/
Salas Arreaga, C. D. (2018). Wine culture and its development in Ecuadorian society. *Centro Sur, 2*(2), 28–36.
Santander Racines, A. B., Rodríguez Santos, E. M., Toapanta Custode, C. D., & Suárez Carrillo, R. A. (2022). *Vitis vinifera*, a case of study at Chaupi Estancia, Pichincha province. *Siembra, 9*(2), e3731.
Schejtman, A., & Berdegue, J. (2008). La desigualdad y la pobreza como desafíos para el desarrollo territorial rural. *Revista Española de Estudios Agrosociales y Pesqueros, 218*, 99–121.
Solman, S., Cabré, M. F., González, M. H., & Núñez, M. N. (2018). Bioclimatic zoning of Argentinian Malbec grape productivity regions by means of a unique combined index. *Climate Research, 74*, 185–199.
Sousa, C., Mesquita, D., de Souza, F., Moraes, J., & Corrêa, J. (2013). Processo de Desenvolvimento do Vale dos Vinhedos. *Gestão e Tecnologia para a Competitividade*, 1–15.
Straffelini, E., Carillo, N., Schilardi, C., Aguilera, R., Estrella Orrego, M. J., & Tarolli, P. (2023). Viticulture in Argentina under extreme weather scenarios: Actual challenges, future perspectives. *Geography and Sustainability, 4*(2), 161–169.

Torres, R., Aliquó, G., Toro, A., Fernández, F., Tornello, S., Palazzo, E., Sari, S., Fanzone, M., De Biazi, F., Oviedo, H. J., Segura, R., Laucou, V., Lacombe, T., & Prieto, J. A. (2022). Identification and recovery of local *Vitis vinifera* L. cultivars collected in ancient vineyards in different locations of Argentina. *Australian Journal of Grape and Wine Research, 28*, 581–589.

United Nations. (2023). *Progress towards the Sustainable Development Goals: Towards a Rescue Plan for People and Planet*. General Assembly Economic and Social Council.

Valenzuela Solano, C., Ruiz Corral, J. A., Ramírez Ojeda, G., & Hernández Martínez, R. (2014). Effects of climate change on the wine production potential of Baja California, Mexico. *Revista Mexicana de Ciencias Agrícolas, 5*(10), 2047–2059.

Verdugo-Vásquez, N., Orrego, R., Gutiérrez-Gamboa, G., Reyes, M., Zurita Silva, A., Balbontín, C., Gaete, N., & Salazar-Parra, C. (2023). Climate trends and variability in the Chilean viticultural production zones during 1985–2015. *OENO one, 57*(1), 345–362.

Villalobos-Soublett, E., Gutiérrez-Gamboa, G., Balbontín, C., Zurita-Silva, A., Ibacache, A., & Verdugo-Vásquez, N. (2021). Effect of shading nets on yield, leaf biomass and petiole nutrients of a Muscat of Alexandria vineyard growing under hyper-arid conditions. *Horticulturae, 7*, 445.

Vitale, A. (2003). Tradición y saberes en la cultura de la vid y del vino. In *Actas del 2do Congreso de Historia Vitivinícola. Uruguay en el contexto regional (1870–1950)*. Montevideo, Uruguay. 809 p.

Wurz, D., Pereira, B., Allebrandt, R., Bonin, B., Dalmolin, L., Canossa, A., Rufato, L., & Kretzschmar, A. (2017). New wine-growing regions of Brazil and their importance in the evolution of Brazilian wine. *Bio Web of Conferences, 9*, 01025.

Yzarra, W., Sanabria, J., Cáceres, H., Solis, O., & Lhomme, J.-P. (2015). Impact of climate change on some grapevine varieties grown in Peru for Pisco production. *OENO one, 49*(2), 103–112.

Chapter 2
The History of Winemaking in Latin America and New Trends: Identity, Market, and Consumption

Amalia Castro San Carlos, Fernando Mujica, and Pablo Lacoste

2.1 Introduction

The arrival of Spanish culture in America implied a series of changes in the lifestyles of the first inhabitants of the continent. Among many of these changes, one stands out in particular for this study: the introduction of wine varieties and the knowledge around wine production. However, before the Spanish arrival, local communities produced fermented products known today under the generic name of "chicha". This beverage is not only linked to American identity but also to the feminine art of its elaboration and its use in ceremonies of high religious and political importance.

Since the introduction of vines by the Spaniards in what is known as the American colonial period (which begins with the arrival of Columbus in 1492 and ends with the independence of the colonies and their conversion into independent republics during the early decades of the nineteenth century), viticulture in Latin America has followed a diverse process: the historical thread that links the first traditional viticultural practices can be traced through the French wine paradigm to the Anglo-Saxon one with the consolidation of the New Wine World. In less than 200 years, the wine industry went from the pre-industrial to the industrial model and later to the technological, scientific model. These wine revolutions, carried out under the idea of progress and development, marginalized the knowledge acquired during the first stage of the history of Latin American wine. However, this legacy has not been forgotten by

A. Castro San Carlos (✉)
Centro de Investigación en Artes y Humanidades, Universidad Mayor, Santiago, Chile
e-mail: amalia.castro@umayor.cl

F. Mujica
Escuela de Sommeliers de Chile, Santiago, Chile

P. Lacoste
Instituto de Estudios Avanzados, Universidad de Santiago de Chile, Santiago, Chile

G. Gutiérrez Gamboa, M. Fourment (eds.), *Latin American Viticulture Adaptation to Climate Change*, https://doi.org/10.1007/978-3-031-51325-1_2

the local laborers, who still maintain resilient mestizo viticulture that stands out as an attractive alternative in the wine market precisely because its history and knowledge distinguish it from the viticulture of the larger industry.

In the present study, a general historical overview of viticulture in Latin America is developed with special emphasis on the South American Southern Cone. We will explore the development of viticulture in Latin America through the three paradigms that prevailed throughout this development: the colonial paradigm, the French wine paradigm, and the Anglo-Saxon wine paradigm. This essay will emphasize the rise and fall of wine production centers within the markets with the distinction of origin and the revaluation of mestizo wines thanks to the reevaluation of the historic vines of the territories and local laborers' knowledge and innovation in their exploration.

2.2 Material and Methods

For this study, the heuristic-critical method, typical of history, is used in reviewing existing literature of interest on the history of viticulture in Latin America.

2.3 Results and Discussion

2.3.1 *Colonial Paradigm: Traditional Viticulture (Sixteenth, Seventeenth, Eighteenth, and Nineteenth Centuries)*[1]

Traditional viticulture was developed in the sixteenth, seventeenth, eighteenth, and nineteenth centuries. Colonial Paradigm history was described based on different references (Corona Páez, 2011; Chamot, 2021; Del Pozo, 1998; Giglo & Morello, 1980; Pardo & Pizarro, 2016; Lacoste, 2005, 2019; Lacoste et al., 2010; Lacoste & Castro, 2013; Unwin, 2001).

Colonial viticulture in Latin America developed in a global context of enormous technological, social, and economic innovations between 1500 and 1750. The growing population of the eighteenth century and the recovery of land productivity, thanks to new industrial techniques applied to agriculture, produced an expansion of the productive economy within Europe and an opening to the world. These shifts were largely facilitated by the new discoveries of land, which shaped a global economy, in which trade was becoming increasingly specialized. Among the consequences of increased traffic of goods, people and ideas on world markets, these changes impacted how viticulture was organized and traded.

[1] This section is based on data from: Corona Páez (2011), Chamot (2021), Del Pozo (1998), Giglo and Morello (1980), Pardo and Pizarro (2016), Lacoste (2005, 2019), Lacoste et al. (2010), Lacoste and Castro (2013), Unwin (2001).

It is known that the rapid expansion of vine cultivation in Latin America resulted from the Spanish conquistadors' fondness for wine. Spanish wine, along with other elements of Iberian food culture, such as ham and olive oil, was included in the provisions of all the expeditions that arrived in the Americas. However, European taste alone is insufficient to explain in depth the accelerated dissemination of wine.

Although there were native vines in North America before the arrival of European settlers, there are no indications of wine-making by the native peoples living in that area. However, they did make other fermented beverages, such as pulque, tesgüino, pozol, balché, tascalate and colonche. The basis of most of these fermented products was corn, tree bark, prickly pear fruits, and cocoa. Such traditional elaborations have been in the hands of women since pre-hispanic times, and these drinks have been used ancestrally to enhance social cohesion, in ritual contexts, and as food and medicine.

The first vines (*criolla* or missionary grape cuttings) were brought into New Spain (México) and were imported by Cortés during the early 1520s. In 1524, it was legislated that the new European landowners should plant 1000 vines for every 100 indigenous inhabitants at their service *(encomendados)*. From Mexico, viticulture expanded rapidly throughout Latin America, following in the footsteps of the conquistadors. Between 1531 and 1545, the current territories of Peru, Bolivia, Colombia, and Chile fell under Spanish territory. Around 1556, the grape cuttings were introduced to Argentina by a Jesuit priest. In this way, viticulture expanded throughout Latin America over 30 years from Mexico southward along the entire western coast of Latin America to Concepción (Chile).

Due to the economic characteristics of the commercial monopoly that the Spanish empire imposed in Latin America, the complaints of the Spanish wine producers soon made themselves felt, faced with the increase in wine production in the colonies. Although the market of New Spain was very large and exceeded the production capacity of the Iberian Peninsula, the pressure that the Spanish industry exerted upon the American market had two great consequences. The first of these was the prohibition of winemaking in New Spain, except for Santa María de Parras on the northern border. Secondly, the settlers of South America, especially in Peru and Chile, insisted on obtaining permits to cultivate vineyards, since the total American demand for wine could not be met by the Iberian Peninsula. In this way, the colonial map of wine production in America was outlined.

The wine-growing nodes that led the production, market, and consumption were those of the Southern Cone, since this was the most important region in the American continent. Until the seventeenth century, the Viceroyalty of Peru was the main wine-growing region of America. The valleys of Guayuri, Ica, and Pisco were the largest Peruvian producers. Critical to this region's momentum, this industry was strongly influenced by the market at Lima—the largest capital of South America—and by Potosi—the world's primary center for mining during this era.

Peru led in the number of vineyards and in wine production until 1700 when its decline began. This trend was accentuated in the eighteenth and nineteenth centuries when Chile took the lead and consolidated itself as the main wine-growing center of South America. This situation remained until the twentieth century. The wine crisis that occurred in Peru between the eighteenth and nineteenth centuries had several reasons. Earthquakes and volcanic eruptions, the sugar cane boom (for manufacturing cane brandy), the white gold (cotton) rush (which caused the uprooting of vineyards), and the expulsion of the Jesuits, along with the decline of Potosi mining at the end of the eighteenth century, caused Peru to abandon the leadership of the continental wine industry. When this happened, Chile arose as the main wine-growing producer of America between the eighteenth and nineteenth centuries along the Pacific coast.

On the Atlantic coast, Paraguay dominated the Río de la Plata basin thanks to the quality of its wine and the commercial circuits from Asunción del Paraguay to Santa Fe and from there to the provinces of Plata. However, Paraguay's boom cycle was relatively short, declining towards the eighteenth century as its productive interest turned towards a process of specialization of labor. Such labor division led to its regional exportation of yerba and tobacco, sustained by the markets of Potosi and the army that was waging the so-called Arauco war in Southern Chile. Over two centuries, the wine industry leadership moved along the Atlantic axis from Paraguay to Mendoza. The provinces of Cuyo and the Kingdom of Chile share a common past in wine production, a process that began in the sixteenth century with the founding of the city of Santiago in 1541. Mendoza and San Juan were founded twenty years later, in 1561 and 1562, respectively. Despite being on different sides of the Andean Mountain Range, they formed a political unit until 1776, when the Viceroyalty of the Río de la Plata was created. This political formation was integrated into the current Argentine Republic in 1810 due to the creation of the Viceroyalty of the Rio de la Plata, Cuyo. However, the cultural unity remained, an identity forged around wine production.

Between the sixteenth and seventeenth centuries, the desert in which Mendoza, capital of the province of Cuyo, is located, was almost uninhabited. It was only during the first half of the seventeenth century that wineries were opened and, thanks to establishing commercial routes and markets for San Juan and Mendoza, 250,000 liters of wine and brandy were exported to the Río de la Plata in that century. By the eighteenth century, the population had doubled, with 150 families among its leaders. Of these families, 105 owned vineyards, including some of the richest owners of the time. Between secular and religious producers, a total of 650,000 vines were put into production under a model of small intensive agricultural ownership. By 1780, Mendoza exported 1 million liters of wine and brandy.

Between the sixteenth and nineteenth centuries, Chile was quietly building its own wine identity. While it was recognized as a country of major military actions, it had space to develop a wine culture imbricated with the territory and its particularities. The traditional viticulture in Chile extended for 300 years until the mid-nineteenth century, differentiating seven productive poles: Concepción, with 15,500 ha (represented more than 50% of the total), Aconcagua valley, 5000 ha, Cauquenes, 4500 ha, Santiago, 2000 ha, Coquimbo, 1600 ha, Colchagua, 1240 ha and Talca, 700 ha. In total, the Central Valley of Chile had 30,000 ha of vineyards in production, which consolidated it as the main wine-growing pole of Latin America at the time.

2.3.2 *Mestizo Viticulture Landscape: Between Wine and Chicha*[2]

Mestizo viticulture history was described according to different references (Corona Páez, 2011; Chamot, 2021; Del Pozo, 1998; Giglo & Morello, 1980; Pardo & Pizarro, 2005, 2016; Lacoste, 2005, 2019; Lacoste et al., 2010; Lacoste & Castro, 2013; Castro et al., 2016a, b, Castro, 2020; Pszczólkowski et al., 2022; Pszczólkowski, 2015; Jerkovic et al., 2022; Soto, 2023).

The vineyard landscape in the southern cone of Latin America is a part of a production rooted in a mestizo knowledge. In this landscape, the vineyard cultivation is combined with other crops, such as fruit trees and cereals, in addition to maintaining small amounts of domestic animals. This form of mix farming protects biodiversity and takes advantage of the synergies between plants, animals, soil, woods, and light to ensure the good health of the orchard. This tradition comes from the native American peoples from the Mexican milpa in the north toward the Mapuche crops in Southern Chile.

The colonial landscape of the Kingdom of Chile emerged between 1550 and 1850 (considering the cultural maintenance of colonial socio-economic characteristics). Adopted from a Moorish heritage that travelled with the conquistadors, it combined cool adobe houses and cellars with interior gardens to alleviate the heat. The cellars, within their thick walls, maintained the perfect temperature for winemaking. The vineyards were also protected with perimeter fences made of raw earth, with a stone foundation, tapia walls, and a tile fence. Raw earth as a building material is a tetra continental heritage, as it is present in Asia (Middle East), Africa (North), Europe (South) and pre-Columbian America.

Although Chile led the continental viticulture there was no monoculture or plantation culture. Instead, properties supported between 500 and 3500 vines per property in combination with fruit orchards and cereal and legume production. Between 1550 and 1850, the vines were grown '*en cabeza*' with a tutor or "*rodrigón*" and "*bracero*" as a way of the trellis system. There were also embedded grapevines ("*parrones encatrados*"): the small ones were next to the house as part of an outdoor living space, and the large ones were installed in the patios, to protect from the sun and provide shade.

In Latin America, there are ways of trellis systems that indicate much older practices, which link the oldest vineyard productions in the world, such as Egypt, with the continent. In Bolivia's Cotagaita and Cintis valleys, the practice of having the vine driven by fruit trees present in the territory is continued even today. Classical authors describe these systems as predating the use of wooden poles or stakes. Bolivian cultivators have had great success with this system, turning disadvantages

[2] This section is based on data from: Corona Páez (2011), Chamot (2021), Del Pozo (1998), Giglo and Morello (1980), Pardo and Pizarro (2005, 2016), Lacoste (2005, 2019), Lacoste et al. (2010), Lacoste and Castro (2013), Castro et al. (2016a, b, 2020), Pszczólkowski et al. (2022), Pszczólkowski (2015), Jerkovic et al. (2022), Soto (2023).

into advantages thanks to the treatment of the territory as a cultural unit. Agriculture in the valleys is carried out on small plots located along the Cotagaita River, where vines are grown with other fruit trees and combined with horticulture to provide food for local families. The post-harvest dehydration of crops for winter consumption, and the traditional method of cultivating, with plows pulled by animals, also accounts for the maintenance of a mestizo and colonial production system until the present day.

As for the varieties, the first grape to arrive in America was the Listán Prieto, known in Chile as the country grape "*uva país*", in Mexico as "*Misión*" or "*Misionera*", in the USA as "*Mission*", in Peru as "*Negra Criolla*" or "*Negra Corriente*", as "*Misionera*" in Bolivia and in Argentina as "*Criolla Chica*". This grape was most frequently used in the production of colonial wine. Originally from the Canary Islands, it arrived in America around 1550 and was the hegemonic variety until 1850. In second place in importance was the "*Muscat of Alexandria*", also known as "*Italy*" in Peru and in some parts of Chile. It arrived in Mendoza in 1669, in La Serena in 1720 and in Colchagua in 1760. These two cultivars are the mothers of American viticulture. They are even called Spanish grapes or cultivars. The "*Mollar*" grape has been referenced in documents since 1750. Its origin has not yet been identified, so there is no certainty whether it is a "criolla" grape, that is, originating from the natural genetic crossing of the first two vines or a third variety introduced by the Spaniards. From these three varieties, in the multivarietal vineyards of Latin America, particularly the Southern Cone, a process of natural genetic exchange began which gave rise to new cultivars known, for being daughters of Spanish grapes born in America, as criolla cultivars. A cross between Listán Prieto and Mollar is Quebranta, a grape famous for making Pisco in Peru. In the same way, other criollas varieties appear in the area, such as Torontel (syn. Torrontés Riojano or Yellow Muscat), Moscatel Rosada (syn. Rosa Pastilla or Pastilla), Moscatel de Austria (syn. Torrontés Sanjuanino), Pedro Giménez, among others. The set of these varieties, both Spanish and Criollas, can also be referred to as colonial cultivars.

Having noted this, the intention of the present text is to bring to discussion Spanish, criolla, and mestizo cultivars concepts and to propose new ones. It turns out that the concept criolla was the best way to designate the children of Europeans, primarily Spaniards, born in America. However, when reviewing the history of the first two varieties that arrived in America and pursuing their origin, it is discovered that the Listán Prieto or Listán Negro grape originates from Spain (Europe). At the same time, the Muscat of Alexandria comes from Egypt (Africa). Based on this, it is possible to group both varieties and call them Mediterranean cultivars. However, the natural genetic crossing that occurred in the American vineyards between both varieties gave rise to grapes misnamed criollas. The biological exchange between African and European vines should be considered as mixed cultivars, "mestizo" grapes. In this sense, the cultivars born in America, whose genes have part of these Mediterranean grapes and another part, probably, of the new cultivars (thorough genetic studies have not yet been carried out to know exactly how to recognize the type of DNA of each of these new varieties which have been described) correspond to mestizo grapes.

The productive equipment of its vineyards, with wineries and wine jars made of clay, along with earth bottles and jars, are also part of the traditional way of making wines. The wine press house and wine press of traditional viticulture have been investigated in depth for the Kingdom of Chile, detecting 682 wine presses between the sixteenth and mid-nineteenth centuries. These were pieces of various materials, costs, and sizes. The construction of a certain type of winepress depended on both the available materials and the resources of the winemaker and shows the social and economic diversity around winemaking, especially the possibilities of social mobility for poor workers.

The wine presses could be permanent or perishable. The permanent ones could be made of lime, brick and clay, stone, or a combination of these materials. Among the perishable presses, we find them constructed from leather, wood, and basketry. The permanent winepresses and those made of wood and basketry were the most frequently used in Latin America, from Mexico to Peru, and had links with ancient Oriental and European traditions in the way of making wine. These wineries also served as installations for tanning leather, dyeing cloth, and salting meat in such a way that viticulture contributed to a wide productive spectrum in the colonial economy.

However, the leather winepress was the great innovation of traditional South American viticulture. Derived from the Peruvian canvas and the Mapuche *tracal*, they appeared in the 1740s. They were not used in Mexico or Peru, but their use spread from the southernmost vineyards of present-day Chile and Argentina. Its diffusion began from the Maule region, Parral, Talca, Cauquenes, and the Itata Valley, areas associated with the small regional mestizo viticulture. It was associated with livestock farming and made it possible for poor winegrowers to have their own facilities for stomping on grapes and producing wine.

The process defined for the elaboration of the wine, landscape, and equipment outlines the figure of a mestizo producer, characterized by a mix of native American people culture and European culture. The first relevant aspect is the verification of knowledge before the Spanish arrival regarding the elaboration of fermented products. These fermented products, which have received various names throughout America, and which for the purposes of this study we will include under the generic term 'chicha', have a long history. Although it is not possible to define a date of origin, the culture of making fermented products is as old as the world. Since there has been fruit, its natural fermentation has produced sugar and alcohol, something that insects and animals detected early.

The observation and understanding of the environment are essential features for explaining the creation of fermented beverages: in Panama, the mantled howler monkey celebrates the fallen fruit of the Astrocaryum, getting drunk with the fermented fruit (Forsyth, 2019). In Colombia, the corn chicha has a revealing origin. According to a Muisca legend, the chicha was discovered by an indigenous woman who, slighted by the punishment she received for being unfaithful to her husband, fled to the Guatavita lagoon. In this remote place, her only consolation was the corn ferment that she found on the slopes of the pond.

Beyond its place in creation myths, chicha was a feminine product that played a central role in the construction of pre-Hispanic societies. Its consumption, which could be daily, was widespread throughout America and was often ceremonial: the chicha was present in all the rites of passage, in welcoming a foreigner, in the political decisions of establishing alliances, and as an offering to the gods. It was also intimately linked to the agricultural cycles of sowing and harvesting and in religious ceremonies for promoting fertility.

When the conquistadors arrived on the continent, they were surprised at the immense quantity and variety of fermented beverages that were offered to them. Yet they did not manage to grasp its symbolic value, and very soon, restrictive laws surfaced around the consumption of chicha by the indigenous people to favor the social European order. At the same time, new fruits were planted for fermentation. Female-made chicha, the central element of the American religion, was displaced by male-led wine production central to the Western Christian tradition. In the ensuing unequal power struggle, the culture of wine was imposed upon the culture of chicha. However, this confrontation also nourished the traditional mestizo viticulture of Latin America, which resulted in knowledge transmission, austerity linked to the land, respect for the environment, and a culture of work, effort, and innovation.

2.3.3 French Wine Paradigm. Introduction of Modern European Strains and Change of Productive Paradigm (Nineteenth to Twentieth Centuries)[3]

French wine paradigm history was described using different references (Del Pozo 1998; Unwin 2001; Baptista, 2008; Beretta, 2015, 2016; Borcosque, 2011; Castro, 2014, 2016; Coira, 2010; Cueto 2009; Giamportone, 2014; Paredes, 2004; Lacoste, 2005, 2006, 2019; Lacoste et al., 2014; Riffo 2007).

In the middle of the nineteenth century, in the era of industrialization and enlightened knowledge, led by the overseas commercial networks of the British Empire, America suffered a French cultural conquest. Western European fashion seduced the whole world, particularly the political and social elites, in terms of architecture, costumes, decoration, art, music, language, production, industry and gastronomy, among others. In this scenario, the French paradigm of wine was established as a new colonialism that took place in American lands.

New techniques related to wine production transformed the landscape of mestizo wine culture. The French trellis system was introduced, which led to the implementation of irrigation systems in order to ensure high levels of grape and wine production. Along with that, different types of machinery related to wine production were

[3] This section is based on data from: Del Pozo (1998), Unwin (2001), Baptista (2008), Beretta (2015, 2016), Borcosque (2011), Castro (2014, 2016), Coira (2010), Cueto (2009), Giamportone (2014), Paredes (2004), Lacoste (2005, 2006, 2019), Lacoste et al. (2014), Riffo (2007).

imported from Europe. The introduction of this machinery intended to automate winemaking, leaving aside the manual labor of mestizo winegrowers. This new paradigm of wine production altered not only the mestizo wine landscape but the entire social organization in colonial winemaking.

The mid-nineteenth century saw the introduction of French European vines in Latin America, considered to be of high oenological value. The Cabernet Sauvignon, Merlot, Carmenère, Sauvignon Blanc, and Semillon vines were successfully introduced in these territories and, along with them, an intense process of transformation and change in the landscape and the viticultural culture. In this process, there were also other cultivars, considered of lower oenological value, such as Carignan and Cinsault, called "tintoreras".

The Chilean state supported this wine-making modernization and hired specialists skilled in the art of modern winemaking. Unfortunately, the technocrats were blind to the valuation of traditional viticulture and the products they produced, so the mestizo producers were dismissed by associating them with a Hispanic paradigm, from which the new republics wanted to distance themselves. In this way, enlightened elites imposed their liberal ideas in order to modernize the young republics. Along with important actions, such as abolishing slavery and manorial privileges, the technocrats also discredited the original creations of the mestizo producers and stigmatized them as part of a repudiable Spanish past.

The cultural atmosphere of Latin American countries was one of admiration for Europe from the second half of the nineteenth century onward. This was evident in the case of the rentier economies of the tropical Caribbean and northern South America, which emerged as a result of the plantation economy. These large estates, controlled by a few people, ostentatiously displayed their wealth through differentiated consumption, which privileged the import of European products to show off on their bodies, houses and dining tables, distancing themselves as much as possible from their cultural reality. In countries such as Chile, these behaviors were decanted as a result of territorial expansion processes, which allowed higher revenues to the treasury and lower tariff rates. For example, after the Pacific War, imports of bottled white wines from Europe increased (Couyoumdjian, 2006), with which the elites joined the fashion of identity alteration that prevailed in Latin America.

Wine was a central factor in this stage of change and differentiated two worlds that ended up separating almost completely: the farmer's world and the lower social strata, who continued to consume traditional colonial wines because they were part of their identity, accompanying productive hours, moments of leisure, and during civil and religious celebrations. On the other hand, the middle and upper strata differentiated themselves by embracing European wine consumption, especially prestigious French but also Spanish and Portuguese, such as Champagne, Bordeaux, Sherry, Port, and Cognac. Towards the end of the nineteenth and the beginning of the twentieth century, this process was in full swing and would soon culminate with the transition to the viticulture of the New Wine World.

In this context, between 1850 and 1900, the viticulture of Chile and Argentina competed as the great Latin American productive centers. However, Chile developed earlier due to the introduction of French vines, the incorporation of technology, the arrival of European winemakers, and the creation of institutions of singular importance, such as the National Society of Agriculture and Quinta Normal de Agricultura. In Argentina, immigration, the modernization of agronomy from 1850, and the arrival of the railway linking Buenos Aires to Mendoza all produced a boom in the Cuyo wine industry, which by 1885 reached its Chilean competitor in volume. In the words of Lacoste (2004) "*At the beginning of the twentieth century, both countries produced about 3,000,000 hl per year, which positioned them as the two main producing powers of America: Argentina and Chile, together, produced more than 70% of the total wine produced throughout the continent*". Argentina became the largest wine producer in Latin America. Chile and Argentina, as part of their commercial model, began with the specialization in varietal wines such as Cabernet Sauvignon, Malbec, and Torrontés. These cultivars will become the icon-based wines of Argentina and Chile in the twentieth century.

In the same period, new participants appeared on the Latin American wine scene, such as Brazil. Here in 1840, vines were brought from the USA and adapted, and the new world winemaking rapidly developed, establishing this nation at the beginning of the twentieth century among the main Latin American producers, although below Argentina and Chile.

2.3.4 Anglo-Saxon Paradigm: Modern Industry and the Survival of Mestizo Viticulture (Twentieth to Twenty-first Centuries)[4]

Anglo-saxon paradigm history was described using different references Buitrago (2014), Castro (2014, 2020), Castro et al. (2015, 2016a, b, 2018, 2022), Giglo & Morello (1980), Hirschegger (2012), Olguín (2019), Pszczólkowski et al. (2022), Pszczólkowski (2015), Jerkovic et al. (2022), Lacoste et al. (2013, 2015a, b, c, 2016), Lacoste (2016, 2019, 2022), Mateu (2014), Medina et al. (2014), Mujica & Castro (2021), Riffo (2007), Serrano (2010), Vargas (2017).

Until the 1970s, despite the paradigm changes, a wine culture remained closely linked to the mestizo farmers world, especially in Chile, which still maintained 63% of "*rulo*" (without irrigation) vineyards, with an older profile than Argentine viticulture, for example, which in Mendoza deploys irrigated vineyards in their entirety. This occurred because its production was directed toward the domestic market with a high consumption of wine per capita, which reached 90 liters per capita in the 1930s in Chile and in the 1960s in Argentina.

[4] This section is based on data from: Buitrago (2014), Castro (2014, 2020), Castro et al. (2015, 2016a, b, 2018, 2022), Giglo and Morello (1980), Hirschegger (2012), Olguín (2019), Pszczólkowski et al. (2022), Pszczólkowski (2015), Jerkovic et al. (2022), Lacoste et al. (2013, 2015a, b, c, 2016), Lacoste (2016, 2019, 2022), Mateu (2014), Medina et al. (2014), Mujica and Castro (2021), Riffo (2007), Serrano (2010), Vargas (2017).

From the mid-twentieth century onward, Argentina took the lead as the main wine producing center, thanks to the size of its internal market and the contributions of European immigrants. Immigrants strengthened Argentine viticulture by revolutionizing cultivation, the wine making process, and marketing it. In this swing, Argentina maintained its predominance until the last quarter of the twentieth century.

The transformation of colonial mestizo viticulture in Latin America into the modern New Wine World was the outcome of a major historical political-economic context, which was responsible for the implementation of accelerated and, to a large extent, countercultural changes. A series of coups d'état in Latin America spurred a process of large-scale militarization of the continent between the 1960s and 1970s. The result of these movements was the abolition of the traditional idea of the state and the centrality that public institutions had in the articulation of political life in society (Serrano, 2010). The traditional Latin American state, which shaped developmentalist and/or populist projects, lost centrality in political and economic decisions to favor neoliberalist ideas that would unite Latin America to a world capitalist system.

The imposition of this new model of state is associated with the so-called Cold War. The interference that the United States had in the series of coups d'état that took place in Argentina, Bolivia, Brazil, Chile, Costa Rica, Dominican Republic, El Salvador, Guatemala, Haiti, Nicaragua, Panama, Paraguay, Perú, and Uruguay. This marked an inflexion point in the wine-producing countries in Latin America.

In 1970, there was a boom in the export of wines from the USA and Australia, countries that correspond to the New Wine World, with a young tradition in making wines, which began in the mid-nineteenth century directly associated with the French wine paradigm and produced in the context of the Anglo-Saxon productive revolution.

The new economic model aimed at export, so the wine had to be produced under these criteria. As a result, countries such as Chile and Argentina uprooted their traditional "criolla" vines to replace them with new varieties, genetically improved, with row driving, and with single-variety production criteria. The wineries were also being modernized, implementing the use of stainless steel, cooperation to change barrels for sanitary measures, and incorporation of harvesting machines instead of manual harvesting, among others.

In Chile, entrepreneurs like Miguel Torres and companies like Cánepa were the first to embrace this model. The changes they implemented tended to make the traditional wine culture invisible, ignoring its contributions and intrinsic value. A new, simpler way of drinking wine was imposed, which left out the old wine consumer.

In this process of wine globalization or "oenological revolution" (Medina et al., 2014), the New Wine World shows a competitive organization, managed by large business firms since 1990. In general terms, this meant an increase in the area of the vineyard, implementation of single variety with higher unit yields, and a consequent increase in the supply of wines. The industry shapes the taste of new consumers, homogenizing it. The differentiation of the product is linked to brands, like AO, IG, varieties, etc.) There was an increase in world wine exports: the New Wine World went from exporting 2% in the 1980s to 34% in the twenty-first century. In this context, although Latin American wines gained prominence, Argentina and Chile stood out in international markets, with a greater acceptance of their products. In this model, business networks were developed with corporate strategies of vertical integration. Companies now participate in the entire production process, from the vineyard to the bottling and storage, and also part of the distribution and marketing process to the main markets.

Chile and Argentina are under this model, which gives a growing competitiveness to the wine-growing countries of the southern hemisphere. Large companies in the industry compete in different segments due to their wide diversification of markets and varieties of strains, their high negotiating capacity with suppliers and the fact that they have their own marketing and distribution networks in dynamic and expanding markets (Medina). Other Latin American countries, such as Mexico, Peru, Uruguay and Paraguay, while they have embraced the Anglo-Saxon model, have not managed to achieve the levels of production and export of their wines at the same level as Argentina and Chile. These two countries are the only Latin American countries that are consistently among the top 10 largest global wine producers. Brazil occupies the 22nd place in this table, Mexico is at number 32, and Peru is at 34th place, at least until 2018.

The emergence of Appellation of Origin and Geographical Identification after 1990 did not follow the same path as the traditional European indications, although the model was copied following a commercial rather than a cultural strategy. The oldest AO in Latin America is the case of the AO Pisco in Chile (1931). A good example of this is what happened in Chile. At the beginning of the export period, the Chilean industry discovered the European AOS and the legislation that protected them with less taxes. Therefore, it was decided to take advantage of a cultural fact as an economic advantage, and in 1994, Decree 464 of the Chilean Wine-Growing Zoning was published, by means of which the Chilean AO were delimited. The problem is that this rule was imposed from "above", without recognition of a wine and its cultural particularity, its unique link with a territory and knowledge that results in a distinguished product. The criterion that was imposed was political, which reproduced the political-administrative jurisdictions of the country. Similarly, other Latin American countries, such as Argentina, followed similar steps in this direction.

In any case, the same industry introduced legitimate valuations of certain products, which led, first of all, to the valuation of distinguished and distinctive cultivars in different regions of the southern cone: Malbec in Mendoza, Torrontés in Cafayate, Carmenère and Cabernet Sauvignon in Chile, Tannat in Uruguay. This led, in a second instance, to the valuation of unique, peasant wines, a process that can be observed more clearly since 2010. Gradually, natural wines are being valued, after the large industry generated organic and biodynamic wines looking for cleaner and environmentally friendly processes.

The earth, as a complex production system, is made up of a network of production processes that interact and regulate each other. One of the principles to understand this dynamic is that of impermanence: the trajectory of this complex system is directed towards transformation. However, the discourse that globalization and the green revolution brought with it was the permanence of things. Because of that we think, since things will not change, we can always continue to do the same things without any prejudice. The main consequence of this misconception is the actual environmental crisis. The socio-economic system also suffered in the globalization process. In Chile, the modern wine industry detached mestizo society from their traditional wine beverages: more than 80% of wine production is exported. Traditional wine consumers were forgotten in this process.

In this context, new trends are perceived in the market of valuation of the mestizo viticulture, through the inclusion of the landscapes of the traditional vineyard in tourist routes that highlight the mestizo knowledge and the wine harvests festivals, in which the traditional practices of the mestizo wines can be seen, such as the treading of the grapes and the valuation of criolla and mestizo vines as of oenological value to produce quality wines.

2.4 Conclusions

In the history of wine in Latin America, the knowledge of native societies, prior to the Spanish arrival, on the elaboration of fermented wines has been systematically dismissed. This could be explained by two central reasons: firstly, it was a knowledge associated with the feminine. Beyond the differences between the role of women in American and European culture, in the rigors of the conquest, women suffered a process of invisibility and dispossession of knowledge that was important to pre-hispanic societies. Secondly, the ceremonial and political American fermented beverage had to be replaced because a new religion and a new politics were imposed, in which wine functioned as a ceremonial drink.

The map of wine production in Latin America is configured from the first stage indicated in colonial times. As a result of the policies imposed by the Spanish crown, the main production centers in South America were formed. Perú led in the number of vineyards and wine production until 1700. Subsequently, Chile achieved American leadership in wine production during the eighteenth and nineteenth centuries. Finally, Argentina led these processes from the late nineteenth century to the twentieth century. From the 1970s onwards, Chile and Argentina became the largest producers and exporters of wine on the continent, competing for the first positions on the international stage. The imposition of the Anglo-Saxon paradigm, and the consequent industrialization in winemaking, has led countries, such as Chile, to unprecedented figures in wine production and export

Modern wine production in Chile had various effects on society: economically, they are profitable industries for the country, but the cost has been the loss of identity. This phenomenon is replicated in the different wine-producing countries of Latin America. The traditional vineyards failed to maintain themselves. The traditional vines were replaced by more productive varieties for industry and commerce in a type of plantation that relies on irrigation, monoculture, and intensive use of agrochemicals.

Although the industry has detected the heritage of mestizo wine in the line of natural wines as a new creation, this type of wine remained protected by mestizo winemakers in rural areas with no other denomination other than wine. It is necessary to reflect on their chances of staying. Countries such as Peru and Bolivia keep their traditional mestizo drinks at the center of their identity. However, Pisco and Singani are distilled and the wines they produce also belong to the Anglo-Saxon paradigm. However, they have maintained small wine valleys that protect the traditional colonial knowledge in the mestizo landscape and the trellis systems, with the emblematic case of Bolivia's Cotagaita and Cintis.

Chile is the only Latin American country that has preserved an identity treasure. Heritage vineyards are spread over 15,000 ha, planted with Listán Prieto, Moscatel de Alejandría, Torontel, and Cariñena, among others. The peasants of valleys such as Itata, Maule, Concepción, Cauquenes, Codpa, and Pica reflect a five-century tradition linked to the production of mestizo wine. Although mestizo winegrowers are resilient, the phenomenon of a decrease in the amount of heritage vineyard hectares is worrisome. A century ago, Chile had 60,000 ha of heritage vineyards. This happened because the mestizo winemakers have a bond of love with their culture, their land, their vines, and their wines. The heritage vineyards of a mestizo nature are not only oriented to the market, they are places of reunion with a more life-friendly past, the environment and a sense of transcendence. They are festivals that celebrate the harvest and winemaking, cultural tourism, and territorial integration.

The decline of this heritage is alarming, even though there are two clear ways of action that have the potential ability to stabilize and ensure the permanence of this mestizo wine culture. The first is to advance in the process of patrimonialization, nationally and internationally, of traditional vineyards and their wines. Second is to involve consumers in the discovery of mestizo wines and vines, for example, Chacolí, Chicha, Pajarete, Asoleados, Pintatani, Pipeño, Pisco, Aguardiente, Country grape wines, Moscatel de Alejandría, Cinsault, Carignan, and Semillon, among others. Focusing the domestic consumption of countries towards these mestizo wines can ensure them, to a large extent, a long and healthy life.

Acknowledgement Project ATE220008 "Mestizo Heritage and Appreciation of the Local Culture. Forgotten Lessons from the Cold War".

References

Baptista, B. (2008). La temprana vitivinicultura en Uruguay: surgimiento y consolidación (1870–1930). *América Latina en la Historia Económica, 29*, 101–129.

Beretta Curi, A. (2015). *Historia de la viña y el vino de Uruguay. El viñedo y su gente (1870–1930)*. Universidad de la República, Uruguay. Comisión Sectorial de Investigación Científica.

Beretta Curi, A. (2016). *Historia de la viña y el vino de Uruguay. El vino uruguayo y sus espacios, imagen y consumo (1870–1930)*. Universidad de la República, Uruguay. Comisión Sectorial de Investigación Científica.

Borcosque, L. (2011). Desarrollo y consolidación de la vitivinicultura sanjuanina (1870–1915). *Revista Digital de la Escuela de Historia, 4*, 103–133.

Buitrago, A. (2014). The emerging viticulture of Bolivia and opportunities for Singani. *Revista Iberoamericana de Viticultura, Agroindustria y Ruralidad, 1*(2), 90–102.

Castro, A. (2014). Healing grief, cheer the soul. Alcohol and wine in chilean farmers stories. *Revista Iberoamericana de Viticultura, Agroindustria y Ruralidad, 1*(1), 57–74.

Castro, A. (2016). Chicha and cider apple in Chile (1870–1930): Apples with origin identification. *Revista Iberoamericana de Viticultura, Agroindustria y Ruralidad, 9*, 4–25.

Castro, A. (2020). The Society of the Chichera apple. Chiloé, 1849–1924. *Revista Iberoamericana de Viticultura, Agroindustria y Ruralidad, 7*(20), 145–162.

Castro, A., Mujica, F., & Fabiola, A. (2015). Between Pintatani and Codpa. Landscape and typical products in peasants stories, 1847–2013. *Revista Iberoamericana de Viticultura, Agroindustria y Ruralidad, 3*(6), 70–86.

Castro, A., León, A., Cussen, F., & Lacoste, P. (2016a). ¡Viva la chicha nueva! La chicha en la vida popular y campesina chilena. *Idesia, 34*(1), 77–83.

Castro, A., Pszczólkowski, P., Mujica, F., Lacoste, P., Núñez, E., Cofré, C., Adunka, M., & Soto, N. (2016b). El pajarete de Huasco y Elqui (Chile). Historia de un vino escogido. *Idesia, 34*(4), 23–35.

Castro, A., Mujica, F., & Urriola, D. (2018). Chiloé apple Chicha: Epic story of a typical product, 19th–20th centuries. *Hemispheric & Polar Studies Journal, 9*(2), 1–21.

Castro, A., Villena, W., & Pszcolkowski, P. (2022). Cotagaita and Cintis: Patrimonialize the Spirit of the Place? *Revista Iberoamericana de Viticultura, Agroindustria y Ruralidad, 9*(25), 153–172.

Coira, L. (2010). El boom vitivinícola de Mendoza (Argentina) y su impacto espacial interno (1883–1914). *Estudios Avanzados, 14*, 145–173.

Corona Páez, S. (2011). *El País de La Laguna*. Universidad Iberoamericana Laguna.

Couyoumdjian, J. (2006). Vinos en Chile desde la Independencia hasta el fin de la Belle Époque. *Historia, 39*(1), 23–64.

Cueto, A. (2009). *Historia de la vitivinicultura a través de sus protagonistas*. Universidad Nacional de Cuyo.

Dargent Chamot, E. (2021). *El vino peruano. Patrimonio e historia. Serie cuadernos para el Bicentenario 4*. Perú.

Del Pozo, J. (1998). *Historia del vino chileno*. Editorial Universitaria.

Forsyth, M. (2019). *Una breve historia de la borrachera*. Planeta.

Giamportone, T. (2014). Government and the wine industry in Mendoza (Argentina) through the Album Argentino Gloriandus. *Revista Iberoamericana de Viticultura, Agroindustria y Ruralidad, 1*(2), 103–125.

Giglo, N., & Morello, J. (1980). Notas sobre la historia ecológica de América Latina. *Estudios Internacionales, 13*(49), 112–148.

Hirschegger, I. (2012). Políticas estatales y su repercusión en la vitivinicultura de la Provincia de Mendoza y del Municipio de San Rafael, 1946–1955. *América Latina en la Historia Económica, 19*(3), 57–97.

Jerkovic, M., Skewes, J., Duhart, F., Mujica, F., & Lacoste, P. (2022). Chilean heritage vineyards and natural wines: The Mestizo Gothic-Mapuche legacy. *Revista Austral de Ciencias Sociales, 42*, 349–371.

Lacoste, P. (2004). La vid y el vino en América del Sur: el desplazamiento de los polos vitivinícolas (siglos XVI al XX). *Universum, 19*(2), 62–93.

Lacoste, P. (2005). El vino y la nueva identidad de Chile. *Universum, 20*(2), 24–33.

Lacoste, P. (2006). Viticultura y política internacional: el intento de reincorporar a Mendoza y San Juan a Chile (1820–1835). *Historia, 39*(1), 155–176.

Lacoste, P. (2016). *El Pisco nació en Chile*. Génesis de la primera Denominación de Origen de América.

Lacoste, P. (2019). *La vid y el vino en el Cono Sur de América. Argentina y Chile (1545–2019)*. Inca Editorial.

Lacoste, P. (2022). Wine, party and vintage in Chile (1930–1970). *Cuaderno de Historia, 57*, 185–212.

Lacoste, P., & Castro, A. (2013). Vines and social mobility in the Kingdom of Chile (1636–1814). *Estudos-Iberoamericanos, 39*(1), 8–18.
Lacoste, P., Yuri, J. A., Aranda, M., Castro, A., Quinteros, K., Solar, M., Soto, N., Gaete, J., & Rivas, J. (2010). Variedades de uva en Chile y Argentina (1550–1850): genealogía del torrontés. *Mundo Agrario, 20*(1), 1–36.
Lacoste, P., Jiménez, D., Castro, A., Rendón, B., & Soto, N. (2013). A bi national appellation of origin: Pisco in Chile and Peru. *Chilean Journal of Agricultural Research, 73*(4), 424–429.
Lacoste, P., Jiménez, D., Briones, F., Castro, A., Rendón, B., & Jeffs, J. (2014). Bordeaux of Talca and Champagne of Mendoza: Appellations of Origin and identity contamination of wines in Argentina and Chile. *Mundo Agrario, 15*(29).
Lacoste, P., Muñoz, J., & Castro, A. (2015a). Brandy and Vineyards in Chile: Quillota, Colchagua y Cauquenes from 1777 census. *Universum, 30*(2), 105–127.
Lacoste, P., Castro, A., Briones, F., Cussen, F., Soto, N., Rendón, B., Fernando Mujica, F., Aguilera, P., Adunka, M., Núñez, E., & Cofré, C. (2015b). Typical Wines of Chile: the rise and decline of Chacolí (1810–2015). *Idesia, 33*(3), 97–108.
Lacoste, P., Castro, A., Briones, F., & Mujica, F. (2015c). The Pipeño: Story of a typical wine Southern Central Valley of Chile. *Idesia, 33*(3), 87–96.
Lacoste, P., Castro, A., Rendón, B., Pszczólkowski, P., Soto, N., Adunka, M., Salas, A., Cofré, C., & Núñez, E. (2016). "Asoleado" ("Sunny") of Cauquenes and Conception: The Rise and Fall of a Chilean wine with Appelation of Origin. *Idesia, 34*(1), 85–99.
Mateu, A. (2014). Las postulaciones oscilantes de una centenaria corporación vitivinícola frente a la regulación de la industria, Mendoza, 1930–1955. *América Latina en la Historia Económica, 21*(1), 134–173.
Medina-Albadejo, F., Martínez-Carrión, J., & Ramon-Muñoz, J. (2014). El mercado mundial de vino y la competitividad de los países del hemisferio sur, 1961–2019. *América Latina Historia Económica, 21*(2), 40–83.
Mujica, F., & Castro, A. (2021). The Chacolí de Doñihue Festival: historical and cultural analysis of a celebration (1975–2020). *Revista Iberoamericana de Viticultura, Agroindustria y Ruralidad, 8*(23), 183–202.
Olguín, P. (2019). Empresas públicas y proyectos de desarrollo económico: los planes gubernamentales para la empresa Bodegas y Viñedos Giol (Mendoza, Argentina, 1954–1964). *América Latina en la Historia Económica, 26*(1), 1–25.
Pardo, O., & Pizarro, J. L. (2005). *La Chicha en el Chile Precolombino*. Mare Nostrum.
Pardo, O., & Pizarro, J. L. (2016). *Chile, bebidas fermentadas prehispánicas*. Mare Nostrum.
Paredes, A. (2004). Industria, política y relaciones internacionales en la frontera: el vino de Chile hecho en Mendoza. *Universum, 19*(2), 138–149.
Pszczólkowski, P. (2015). Sauvignon Blanc, Cabernet-Sauvignon and Carmenère: Three key grape varieties in current viticulture of Chile. *Revista Iberoamericana de Viticultura, Agroindustria y Ruralidad, 2*(4), 1–16.
Pszczólkowski, P., Ganga, M. A., Veliz, L., Castro, M. H., Meneses, M., Castro, A., Cañón, P., & Hinrichsen, P. (2022). Molecular and ampelographic characterization of genotypes used for Pajarete wine, and old Denomination of Origin from Huasco and Elqui Valleys in northern Chile. *Vitis, 61*, 37–44.
Riffo, M. (2007). Impactos Espaciales y Socioeconómicos de la Vitivinicultura en Chile y Argentina. *Revista Geográfica, 143*, 163–210.
Serrano, F. (2010). Estado, golpes de Estado y Militarización en América Latina: una reflexión histórico política. *Nueva Época, 23*(64), 175–193.
Soto, N. (2023). Agricultura y Vitivinicultura en Chile colonial. In M. Llorca-Jaña & J. J. Martínez-Barraza (Eds.), *Historia económica de Chile colonial* (pp. 59–87). FCE.
Unwin, T. (2001). *El vino y la viña. Geografía histórica de la viticultura y el comercio del vino*. Tusquets Editores.
Vargas, L. (2017). *Pintatani*. Arica.

Chapter 3
Sustainability of Latin American Viticultural Firms. Sustainability Frameworks Development in a Context of Global Challenges

Francisco J. Fernández and José Machado

3.1 Introduction

Like many other agricultural industries, the wine industry faces significant challenges due to climate change. However, what sets the wine industry apart from other agricultural industries is its reliance on particular climatic conditions and the constrained climate range in which grapevines can grow (Mozell & Thach, 2014). As global temperatures rise and weather patterns become more unpredictable, vineyards confront a range of challenges, including altered growing seasons, increased frequency of extreme weather events, and shifts in water availability (Sabir, 2018; Schultze & Sabbatini, 2019). These changes significantly impact the quantity and quality of grape yields and the overall ecological balance of vineyard ecosystems. All these factors make the wine industry particularly susceptible to climate change's short- and long-term effects (Jones & Webb, 2010; Mozell & Thach, 2014).

On the other hand, the wine industry has been recognized for its significant environmental impact. The environmental impacts of wine production include the use of energy-intensive pesticides and fertilizers (Jourdaine et al., 2020), water and carbon footprints (Rinaldi et al., 2016), and waste management (Soceanu et al., 2021), among others. In the same context, consumers are increasingly aware of these impacts associated with wine production and are increasingly interested in

F. J. Fernández (✉)
Centro de Economía para el Desarrollo Sostenible (CEDES), Facultad de Economía y Gobierno, Universidad San Sebastián, Santiago, Chile

Center of Applied Ecology and Sustainability (CAPES), Santiago, Chile
e-mail: francisco.fernandez@uss.cl

J. Machado
Independent Researcher, São Paulo, Brazil

G. Gutiérrez Gamboa, M. Fourment (eds.), *Latin American Viticulture Adaptation to Climate Change*, https://doi.org/10.1007/978-3-031-51325-1_3

sustainability (McMillan, 2023; Schäufele & Hamm, 2017). The demand for sustainable products has increased among younger generations, who are willing to pay higher prices to ensure that their purchasing decisions support environmentally friendly industries (Schäufele & Hamm, 2017).

Because of the above, there is a consensus that dealing with sustainability is a necessary challenge imposed on the wine industry, considering climate change, environmental degradation, and its social consequences that affect even the economic dimension (Flores, 2018). On the other hand, sustainability can also be a competitive factor in this industry, a driving market strategy, and a key to innovation (Fiore et al., 2017; Nidumolu et al., 2009).

To face this challenge, wine regions are developing sustainability frameworks, presented as national (or regional) programs, to make sustainability work in their areas and address local issues (Flores, 2018; Moscovici & Reed, 2018). These frameworks respond to customers and markets and are a way to systematize sustainable actions or an effort to improve wine management.

Although there is no common sustainability framework for the entire wine industry, many sustainable certifications are available in the market, each with its characteristics and particularities. However, although the topic of "Sustainability in viticulture and winemaking" has received increased attention from academic research in recent decades (Costa et al., 2022), different authors sustain that the wine industry is in the nascent stages of sustainability certification (Moscovici & Reed, 2018). Moreover, from an academic point of view, even with the interdisciplinary literature, there is still a gap in sustainable wine certification research (Moscovici & Reed, 2018).

As an attempt to contribute to filling this gap, this chapter aims to present an overview of sustainable wine certification, focused on major producers from Latin America. Latin America now produces an important share of the world's wine thanks to recent rapid growth in the region's wine industry. Moreover, the wine industry in Latin America is an important driver of economic growth, generating jobs and income for rural communities and contributing to the development of the local economy.

This chapter is structured as follows. The next section presents an overview of the major Latin American wine producers, comparing surface, production, and export trends with major European wine producers. Section 3.3 covers a global overview of sustainability in the wine industry, reviewing the major milestones and describing how sustainability has been established through sustainability frameworks in wine producers' regions. The development of sustainability of major Latin American wine industries and a description of their sustainability frameworks is discussed in Sect. 3.4. Section 3.5 discusses major challenges and opportunities for sustainability in the Latin American wine industry. Finally, Sect. 3.6 concludes the chapter.

3.2 Latin American Role in the Global Wine Industry

3.2.1 Surface

Statistics provided by the International Organization of Vine and Wine (OIV) (2023) indicate that the total world area under grapevines in 2022 reached 7.3 million hectares (including vines for all purposes and not yet productive young vines). Among them, the major vine-growing countries like Spain, France, and Italy cover an area that represents more than 34% of the world under grapevines.

In South America, the major vineyards are in Argentina, Chile, and Brazil, with 207 kha, 196 kha, and 81 kha, respectively, representing 6.6% of the total world area under grapevines. Argentina's vineyard surface has been on a decline since 2015 and will reach a reduction of 4 kha (2%) in 2022 compared to 2021. Argentina's reduction in its vineyard surfaces can be explained by climatic factors such as water scarcity, rising temperatures, and drought-like conditions. On the other hand, the size of the vineyard in Chile in 2022 remained almost unchanged compared to 2021. Finally, after a continuous decline of eight years, Brazil increased the size of its vineyard in 2022 by 0.8%. Figure 3.1 shows a comparison of the development of the vineyard surface of the major global producers vs. major producers from Latin America

3.2.2 Production and Exports

The patterns of vinified production are similar to those related to the surface under grapevines. Italy, France, and Spain accounted for 51% of the world's wine production in 2022, with 49.8 mhl, 45.6 mhl, and 35.7 mhl, respectively (OIV, 2023).

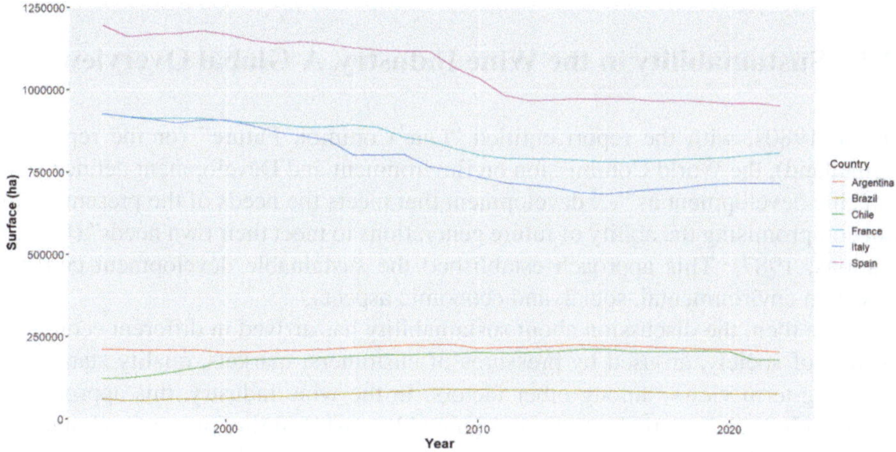

Fig. 3.1 Vineyard surface trends from major global producers vs. major Latin American producers

Among these top three wine producers, Italy is relatively stable in wine production, with −1% compared to 2021 and +2% concerning its last five-year average. On the other hand, France recorded an increase in wine production not only compared to the low volume of 2021 (+21%) but also with respect to its last five-year average (+7%). Notwithstanding drought and limited access to water in many regions, Spain's 2022 wine production levels are at +1% compared to 2021, but is 5% below its last five-year average.

In Latin America, Chile is the largest producer, with wine production close to 12.4 mhl, followed by Argentina with 11.5 mhl in 2022 (OIV, 2023). Brazil experienced an increase in its wine production in 2022, with a level of 3.2 mhl. This represents an increase of 14% in comparison to its last five-year average.

Regarding exports, in 2022, global wine exports amounted to 107 mhl, a 5% decrease compared to the historically high of 2021. Italy was the largest exporter in 2022, with 21.9 mhl, accounting for 20% of the global exports. The international trade of wine is dominated by three EU countries – Italy, Spain, and France – which together exported 57 mhl in 2022, accounting for 53% of the world's wine exports. In terms of volume, these three countries have all declined compared with 2021. However, with different degrees, Italy exported 21.9 mhl (only −0.6% compared to 2021), Spain, with 21.2 mhl exported, saw the most significant decrease with respect to 2021 (−11%), and France, with 14.0 mhl, recorded −5% with respect to 2021. These three countries account for 61% of the global exports in value.

Chile is the leading exporter in South America and the fourth exporter at the global level. In 2022, Chile exported 8.3 million hl and reached a value of 1.7 bn EUR. On the other hand, Argentina exported 2.7 million hl in terms of volume, reaching a value in exports of 752 m EUR. On the other hand, Brazil's wine exports are marginal, reaching just 78,000 hl during 2022. Figure 3.2 shows the 2022 wine production and exports of the three major global producers and exporters versus the three major Latin American major producers.

3.3 Sustainability in the Wine Industry. A Global Overview

In the 1980s, with the report entitled "Our Common Future" (or the report of Brundland), the World Commission on Environment and Development defined sustainable development as "… development that meets the needs of the present without compromising the ability of future generations to meet their own needs" (United Nations, 1987). This approach established the sustainable development concept based on environmental, social, and economic aspects.

Since then, the discussion about sustainability has arrived in different economic sectors of society, aroused by pressures of customers, markets, quality standards, and long-term views, among other factors. In the wine industry, this approach is essential because it is tied to activities guided by sustainability processes (Lamastra et al., 2016). In this sense, wine producers and vine growers have been increasingly engaged in it (Costa et al., 2022; Flores, 2018).

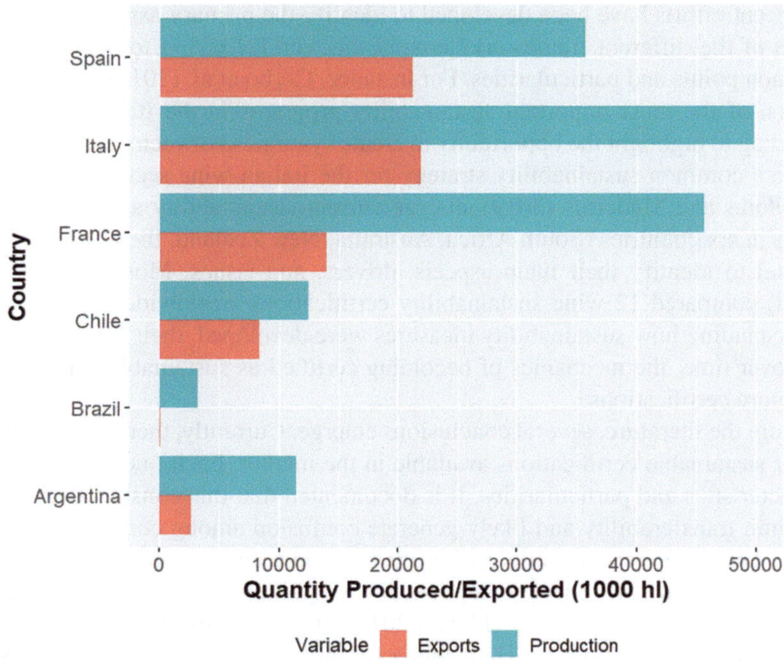

Fig. 3.2 Trends in quantity produced and exported by major global and Latin American wine producers

The importance of sustainability in the wine industry is reflected in that the notion of sustainable viticulture is supported by official documents from the International Organization of Vine and Wine (OIV), which include definition (OIV, 2004), guidelines (OIV, 2008), and general principles (OIV, 2016). In parallel, wine regions have responded to the challenge of sustainability through the development of their own national (or regional) sustainability frameworks (most of them translated into wine sustainability certifications). Each one of them aim to adapt sustainability in their contexts and face their particular environmental, social, and economic challenges (Barbosa et al., 2018). The existing certification approaches followed in different countries worldwide reaffirm the sustainability commitment of the industry.

In this context, although the topic of "Sustainability in viticulture and winemaking" has received increased attention from academic research in recent decades (Costa et al., 2022), different authors sustain that the wine industry is in the nascent stages of sustainability certification (Moscovici & Reed, 2018). Moreover, even with the interdisciplinary literature, there is still a gap in sustainable wine certification research (Moscovici & Reed, 2018). This is particularly interesting given the wide range of frameworks covering different regions and production profiles and that the records for the first sustainable frameworks are from the 1990s (e.g., Sustainable Winegrowing New Zealand was initiated in 1994).

Recent efforts have been developed to identify the primary aspects, drivers, and issues of the different frameworks around the world, looking to understand their common points and particularities. For instance, Corbo et al. (2014) present a comparison of the most important sustainability programs in the Italian wine sector, intending to highlight the opportunity to create synergies between the initiatives and define a common sustainability strategy for the Italian wine sector. Flores (2018) and Flores and Medeiros (2019) analyze current sustainability assessment frameworks in six countries (South Africa, Australia, New Zealand, the USA, Chile, and France) to identify their main aspects, drivers, and issues. Moscovici and Reed (2018) compared 12 wine sustainability certifications worldwide with the aim of understanding how sustainability measures were developed, their level of membership over time, the mechanics of becoming certified as sustainable, and their plan for future certifications.

From the literature, several conclusions emerge. Currently, there are a high number of sustainable certifications available in the market. Each one of them with its characteristics and particularities. It is documented that this considerable variation can limit transferability and likely generate confusion among consumers (Gerling, 2015). Second, the frameworks are certifications in voluntary programs, using qualitative approaches to check compliance with guidelines without proposing the improvement of sustainability (Flores, 2018). It is also observed that countries in the "New World" (Southern Hemispheres wine countries and the United States) have been pioneers in introducing sustainability in the wine industry (e.g., USA and NZ) (Corbo et al., 2014). However, among them, Latin American countries with a presence in the wine industry, have presented diverse realities, with sustainable frameworks highly developed (as in the case of Chile), processes that currently are in their nascent stage (Argentina), and frameworks that they haven't even started (Brazil).

In the following section, we will discuss the main characteristics regarding the sustainability of these three major viticultural countries of Latin America.

3.4 Latin American Sustainability in Viticultural Firms

As mentioned in the above sections, Latin American wine producers are far from the major European producers in terms of area cultivated, production, and exports. However, the development of Latin American producers has presented rapid growth in the wine industry. Figure 3.3 shows the trends in production and exports of 6 selected Latin American countries.

As presented in Fig. 3.3, on a detailed level, this growth has been mainly caused by two Latin American countries, Chile and Argentina, while other Latin American regions, such as the South of Brazil, although without an important role in wine exports, have presented important developments in production and vine surface.

Based on the above, we describe the development of sustainability frameworks of these three major Latin American producers.

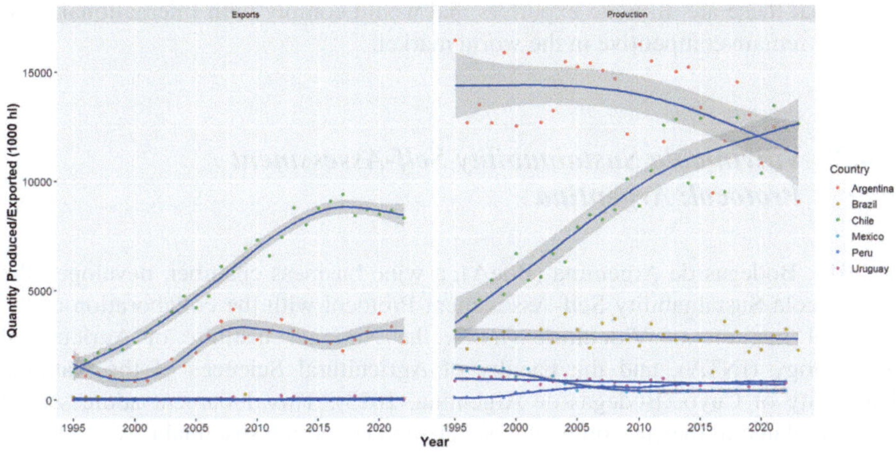

Fig. 3.3 Exports and production of wine of selected Latin American countries

3.4.1 National Code of Sustainability for the Chilean Wine Industry

The National Code of Sustainability for the Chilean wine industry was developed by the Wines of Chile consortium in 2007 and formally launched in 2011 (Flores & Medeiros, 2019). The program certifies vineyards, wineries, and management practices and is compatible with international standards, such as the Global Reporting Initiative (GRI[1]). The Sustainability Code is based on requirements in four complementary areas: green (vineyards), red (process), orange (social), and purple (tourism) (Sustentavid, 2023). The Code allows wine companies to be certified in the Green, Red, and Orange Areas and for individual grape producers to be certified in the Green Area and wineries in the Red Area. The purple standard complements the Sustainability Code in its three areas. Companies that request certification in this area must be certified in the Code, or if it is their first audit, they can request for their incorporation voluntarily. Wines that meet the standards can have the Wines of Chile-Sustainable label on their bottles. Today, 84 vineyards are certified, representing about 80% of bottled wine for export. To participate, it is necessary to register through a one-time fee which gives the user the right to receive Sustainability Code information such as the standard, checklists, and other basic documents as well as training activities on sustainability. Moscovici and Reed (2018), indicate that the Chilean certification is one of the most expensive internationally—approximately US $500. On the other hand, Valenzuela and Maturana (2016) indicate that the justification for registering and working for the certification lies on strategic reasons,

[1] A framework guiding the construction of sustainability reports, and establishing the principles and indicators that businesses can use to measure and publicise their economic, environmental, and social performance in the pursuit of sustainability.

given that these are all wine exporters that would comply with international standards to remain competitive in the world market.

3.4.2 *Vitivinícola Sustainability Self-Assessment Protocol: Argentina*

In 2011, Bodegas de Argentina (BdeA), a wine business chamber, developed the Vitivinícola Sustainability Self-Assessment Protocol with the collaboration of the National Institute of Viticulture (INV), the National Institute of Agricultural Technology (INTA), and the Faculty of Agricultural Sciences of the National University of Cuyo (Bodegas de Argentina, 2023). This document addresses all aspects related to the environmental, social, and economic sustainability of viticulture, which aims to make viticulture and conventional winemaking more sustainable, improving the competitiveness of Argentine wines. The stages that wineries must complete to certify the Protocol are self-assessment, implementation, and verification audit by an authorized company and certification by BdA. To date, 171 units (farms and warehouses) have been certified in Mendoza, San Juan, San Rafael, Salta, and Río Negro (Bodegas de Argentina, 2023).

In 2018, the opportunity arose to develop a new broader protocol, which would serve as a non-certifiable methodological guide. This Guide is a supplementary document to the Protocol. Just as the latter is designed for companies with a high level of implementation of management systems that export to demanding markets, the Guide is aimed at small wineries and vineyards with little evolution in their management systems so that in this way, with less demand, they can access and implement sustainable practices.

Despite the Protocol beginning in 2011, a comparison made by Salas and Farreras (2022) indicates that the level of adherence of the Argentine wine industry to the BdeA Protocol is far below the levels of adherence of other wine industries. While in Argentina, adherence to its protocol is below 1%, among its main competitors, the adherence exceeds, in some cases, 95% (Salas & Farreras, 2022).

3.4.3 *Integrated Grape Production for Processing*

Despite the emerging growth that Brazil has had in the wine sector (especially in the southern regions) (Barbosa et al., 2018), recent literature indicates that, in terms of sustainable viticulture, Brazil has just isolated and not systematized initiatives, and there is a lack of a common framework (Flores & Medeiros, 2019). However, the recent implementation of the Normative Instruction MAPA N°. 21, of June 2, 2022 (which amends MAPA Normative Instruction No. 42, of November 9, 2016), called Integrated Grape Production for Processing (PIUP), could be the starting point for future sustainability frameworks in the country. Such normative instruction refers to the "Farm" and "Industry" stages of the PIUP, covering all the processes conducted in agricultural production and processing. For both, the "Farm" and Industry stages, there are different thematic areas that correspond to mandatory norms to be followed, recommended and prohibited, when applicable.

To verify the conformity of the items mentioned in the previous tables, the Brazilian Agricultural Research Corporation (EMBRAPA) provides checklists, and field booklets, in addition to a series of Technical Manuals for the Integrated Production of Grapes for Processing—Wine and Juice (Technical Manuals of PIUP).

Having checked the standards established by the Ministry of Agriculture, Livestock and Food Supply on the property, the producer must validate the conformity assessment (certification). This process is carried out by certifiers accredited by the National Institute of Metrology, Quality and Technology (INMETRO). Table 3.1, shows the scope and coverage of each of the frameworks assessed.

3.5 Challenges and Opportunities

This chapter presents a general overview of how Latin American viticultural firms are considering and developing sustainability in the wine industry. From the literature reviewed, several challenges and opportunities were detected.

In the first place, compared with the first sustainability frameworks that started in the 90s, the Latin American wine industry is still in the nascent stages of sustainability certification. In this context, the Chilean wine industry has developed a successful framework, where certified vineyards represent more than 80% of bottled wine production for exports. However, the high cost of the process can delay its implementation by smaller businesses. In the case of Argentina, the level of adherence to the sustainability protocol has low adherence rates compared to their neighbours. This suggests a need for greater engagement and implementation of sustainable practices.

In the second place, some opportunities arise from this review. As was observed, different efforts have been made to identify common aspects, drivers, and issues among different sustainability frameworks worldwide. Emerging initiatives, such as the Integrated Grape Production for Processing (PIUP) in Brazil, provide an opportunity to follow successful examples from neighbouring countries, which can help create synergies between initiatives and define common sustainability strategies.

3.6 Conclusions

The wine industry in Latin America has experienced significant growth, primarily driven by Chile and Argentina. These two countries have played a major role in increasing the production and export of wine in the region. Overall, Chile and

Table 3.1 Scope and coverage of sustainability frameworks for Chile, Argentina, and Brazil

Country	Framework	Scope	How much does it cover
Chile	National code of sustainability for the Chilean wine industry	Developed by the consortium Wines of Chile in 2007, this program was launched in 2011 and certified vineyards, wineries, and management practices. The main guidelines are agricultural management, chemicals, and water resources protection to vineyards; energy, water, and waste management to wineries; ethics, environment, quality of life at work, community and marketing, and customer compromise in management practices	Currently, there are 84 certified vineyards, which represent more than 80% of the bottled wine production for export.
Argentina	Bodegas de Argentina's sustainability protocol (Bodegas de Argentina sustainability code)	The BdA code, certified by specialized agencies that audit each producer, certifies wineries with sustainable practices in viticulture, soil management, energy efficiency, and water. conservation, solid waste management, air quality, human resources, and community.	To date, 171 units (farms and warehouses) have been certified in Mendoza, San Juan, San Rafael, Salta and Río Negro
Brazil	Produção Integrada de Uva para Processamento-PIUI (Integrated grape production for processing)	Certification system that follows the norms of the Ministry of Agriculture. Requires audit, carried out by an independent certifier accredited to the Instituto Nacional de Metrologia, Qualidade e Tecnologia-INMETRO (National Institute of Metrology, Quality and Technology). Once approved, the producer receives the Brazil Certified stamp to be printed on the label. The System prioritizes food safety through the application of safe methods with a focus on sustainability. It has thematic areas divided into "Farm" and "Industry", which cover all processes carried out in agricultural production and processing, respectively	The practical application of the system can be considered recent. According to the Empresa Brasileira de Pesquisa Agropecuária- EMBRAPA (Brazilian Agricultural Research Corporation), the first wines certified for PIUP are from the 2017 to 2018 harvest

Argentina have made significant progress in developing sustainability frameworks for their wine industries, being the first recognized and studied framework in mainstream literature. For the Argentinean protocol, on the other hand, despite starting at a similar time to the Chilean code, the level of progress has not been at the same pace, having a low level of adherence compared to other wine industries. Finally, Brazil is still in the early stages of establishing a comprehensive framework. However, the recent implementation of Integrated Grape Production for Processing (PIUP) could be the initial step for a future common sustainability framework in the country.

Acknowledgments The authors would like to thank ANID PIA/BASAL FB0002 and ANID/FONDAP/15130015.

References

Barbosa, F. S., Scavarda, A. J., Sellitto, M. A., & Lopes Marques, D. I. (2018). Sustainability in the winemaking industry: An analysis of Southern Brazilian companies based on a literature review. *Journal of Cleaner Production, 192*, 80–87. https://doi.org/10.1016/j.jclepro.2018.04.253

Bodegas de Argentina. (2023). Bodegas de Argentina. Retrieved from: https://www.bodegasdeargentina.org/protocolo-sustentabilidad-2/

Corbo, C., Lamastra, L., & Capri, E. (2014). From environmental to sustainability programs: A review of sustainability initiatives in the Italian wine sector. *Sustainability, 6*(4), 4. https://doi.org/10.3390/su6042133

Costa, J. M., Catarino, S., Escalona, J. M., & Comuzzo, P. (2022). Achieving a more sustainable wine supply chain—Environmental and socioeconomic issues of the industry. In J. M. Costa, A. Catarino, J. M. Escalona, & P. Comuzzo (Eds.), *Improving sustainable viticulture and winemaking practices* (pp. 1–24). Academic. https://doi.org/10.1016/B978-0-323-85150-3.00009-8

Fiore, M., Silvestri, R., Contò, F., & Pellegrini, G. (2017). Understanding the relationship between green approach and marketing innovations tools in the wine sector. *Journal of Cleaner Production, 142*, 4085–4091. https://doi.org/10.1016/j.jclepro.2016.10.026

Flores, S. S. (2018). What is sustainability in the wine world? A cross-country analysis of wine sustainability frameworks. *Journal of Cleaner Production, 172*, 2301–2312. https://doi.org/10.1016/j.jclepro.2017.11.181

Flores, S. S., & Medeiros, R. M. V. (2019). A framework proposal for sustainability management in wine industry. *International Journal of Innovation and Sustainable Development, 13*(3-4), 348–375. https://doi.org/10.1504/IJISD.2019.100422

Gerling, C. (2015). *Environmentally sustainable viticulture: Practices and practicality*. CRC Press.

Jones, G. V., & Webb, L. B. (2010). Climate change, viticulture, and wine: Challenges and opportunities. *Journal of Wine Research, 21*(2-3), 103–106. https://doi.org/10.1080/09571264.2010.530091

Jourdaine, M., Loubet, P., Trebucq, S., & Sonnemann, G. (2020). A detailed quantitative comparison of the life cycle assessment of bottled wines using an original harmonization procedure. *Journal of Cleaner Production, 250*, 119472. https://doi.org/10.1016/j.jclepro.2019.119472

Lamastra, L., Balderacchi, M., Di Guardo, A., Monchiero, M., & Trevisan, M. (2016). A novel fuzzy expert system to assess the sustainability of the viticulture at the wine-estate scale. *Science of the Total Environment, 572*, 724–733. https://doi.org/10.1016/j.scitotenv.2016.07.043

McMillan, R. (2023). Wine Industry Trends and Report 2023 | Silicon Valley Bank. Retrieved from: https://www.svb.com/trends-insights/reports/wine-report

Moscovici, D., & Reed, A. (2018). Comparing wine sustainability certifications around the world: History, status and opportunity. *Journal of Wine Research, 29*(1), 1–25. https://doi.org/10.1080/09571264.2018.1433138
Mozell, M. R., & Thach, L. (2014). The impact of climate change on the global wine industry: Challenges and solutions. *Wine Economics and Policy, 3*(2), 81–89. https://doi.org/10.1016/j.wep.2014.08.001
Nidumolu, R., Prahalad, C. K., & Rangaswami, M. R. (2009). Why sustainability is now the key driver of innovation. *Harvard Business Review, 87*(9), 56–64.
OIV. (2004). *Development of sustainable viticulture*. Resolution. CST 1/2004. International Organisation of Vine and Wine, Paris, France.
OIV. (2008). *OIV guidelines for sustainable viticulture: Production, processing and packaging of products*. Resolution. CST 1/2008. International Organisation of Vine and Wine, Paris, France.
OIV. (2023). *State of the world vine and wine sector in 2022*. International Organisation of Vine and Wine. Retrieved from: https://www.oiv.int/sites/default/files/documents/2023_SWVWS_report_EN.pdf
OIV. (2016). *OIV guide for the implementation of principles of sustainable vitiviniculture*. International Organisation of Vine and Wine. Retrieved from: https://www.oiv.int/standards/oiv-guide-for-the-implementation-of-principles-of-sustainable-vitiviniculture-
Rinaldi, S., Bonamente, E., Scrucca, F., Merico, M. C., Asdrubali, F., & Cotana, F. (2016). Water and carbon footprint of wine: Methodology review and application to a case study. *Sustainability, 8*(7), 7. https://doi.org/10.3390/su8070621
Sabir, A. (2018). Sustainable viticulture practices on the face of climate change. *Agricultural Research and Technology, 17*(4), 556033. https://doi.org/10.19080/ARTOAJ.2018.17.556033
Salas, J., & Farreras, V. (2022). ¿Avanzamos hacia una vitivinicultura sostenible? Un estudio exploratorio de la Industria del vino de Argentina. *Estudios económicos, 39*(79), 127–167.
Schäufele, I., & Hamm, U. (2017). Consumers' perceptions, preferences and willingness-to-pay for wine with sustainability characteristics: A review. *Journal of Cleaner Production, 147*, 379–394. https://doi.org/10.1016/j.jclepro.2017.01.118
Schultze, S. R., & Sabbatini, P. (2019). Implications of a climate-changed atmosphere on cool-climate viticulture. *Journal of Applied Meteorology and Climatology, 58*(5), 1141–1153. https://doi.org/10.1175/JAMC-D-18-0183.1
Soceanu, A., Dobrinas, S., Sirbu, A., Manea, N., & Popescu, V. (2021). Economic aspects of waste recovery in the wine industry. A multidisciplinary approach. *Science of the Total Environment, 759*, 143543. https://doi.org/10.1016/j.scitotenv.2020.143543
Sustentavid. (2023). The Code – Sustentavid. Retrieved from: https://www.sustentavid.org/en/el-codigo/
United Nations. (1987). *Report of the World Commission on Environment and Development: Our common future*. Retrieved from: https://sustainabledevelopment.un.org/content/documents/5987our-common-future.pdf
Valenzuela, L., & Maturana, S. (2016). Designing a three-dimensional performance measurement system (SMD3D) for the wine industry: A Chilean example. *Agricultural Systems, 142*, 112–121. https://doi.org/10.1016/j.agsy.2015.11.011

Chapter 4
Tropical Viticulture in Brazil: São Francisco Valley as an Important Supplier of Table Grapes to the World Market

Patricia Coelho de Souza Leão and Jullyanna Nair de Carvalho

4.1 The Brazilian Vitiviniculture

Viticulture has great socio-economic relevance in Brazil, representing the main source of income in many producing regions, such as Serra Gaúcha, Rio Grande do Sul, and different municipalities in the state of São Paulo, where small family farms predominate. In the Submédio São Francisco Valley, the Northeast region of Brazil, the profile of this productive sector is characterized by small, medium, and large farms producing table grapes, juices, or wines, whose activities have contributed to confirming viticulture as an important generator of employment and income in the country (Mello & Machado, 2020; da Silva et al., 2019).

Brazil has the peculiarity of being the only country where the three types of viticulture (temperate, subtropical and tropical) are found, associated with different edaphoclimatic conditions and production systems (da Silva et al., 2019; Pereira, 2020). Therefore, in Brazil, viticulture presents characteristics that vary according to the producing region, with specificities in phenology and duration of the production cycle, harvest time, cultivars, management, and market.

Temperate climate viticulture is characterized by an annual cycle followed by a dormancy period induced by low winter temperatures.

Tropical and subtropical viticulture can be defined in five types according to the Multicriteria Climatic Classification (Tonietto & Carbonneau, 2004): tropical dry, tropical wet, tropical alternatively dry/wet, sub-tropical alternatively dry/wet, and sub-tropical dominantly wet. They include many countries and very particular

P. Coelho de Souza Leão (✉)
Embrapa Semiárido, Petrolina, PE, Brazil
e-mail: patricia.leao@embrapa.br

J. Nair de Carvalho
UNIBRAS, Juazeiro, BA, Brazil

G. Gutiérrez Gamboa, M. Fourment (eds.), *Latin American Viticulture Adaptation to Climate Change*, https://doi.org/10.1007/978-3-031-51325-1_4

tropical viticultures in Brazil, Peru, Venezuela, Colombia, Guatemala, India, Thailand, and others characterized as subtropical in the South of Brazil, Uruguay, Korea, Japan and others.

Subtropical viticulture is practiced in regions of mild and short winters with one productive cycle per year; however, with the use of different management practices, two vegetative cycles are performed, obtaining two harvests per year. Tropical viticulture is characterized by not having a defined winter without the minimum accumulation of cold hours necessary to break the physiological dormancy of the buds.

The implementation and evolution of new technologies, such as the development of cultivars, along with the improvement of viticultural practices, have contributed to the success and expansion of Brazilian tropical viticulture.

The traditional temperate climate viticulture is in the South and Southeast regions of Brazil. Tropical viticulture is concentrated in the Northeast region, while subtropical viticulture is in the Southeast, Midwest, and Northeast regions of the country, especially in high-altitude microclimates (EPAMIG et al., 2020).

The production is differentiated according to the growing region: in the South, the cultivation of American and hybrid grapes for the manufacture of juices and wines predominates, and in the other regions, the cultivation is of American (*Vitis labrusca* and hybrids) and European (*Vitis vinifera*) table grapes, both for the domestic market and for export (Gazolla et al., 2020). The trend is that Brazil will continue to specialize its production of seedless table grapes to meet the international market and that, in the South, the area cultivated with vinifera grapes will increase to the detriment of the cultivation of American and hybrid grapes (Gazolla et al., 2020).

The area harvested with grapevines in Brazil, in 2022, was 74,520 ha. With the increase of this area each year, the table grape and its derivatives gain prominence, especially in the exports of fresh grapes from the states of Pernambuco and Bahia in the Submédio São Francisco Valley, as well as wines, sparkling wines, and juices produced in Rio Grande do Sul, which also registered an increase in exports (Kist et al., 2022).

Brazil occupied the 15th position in the world ranking of grape production and the 21st position in the ranking of areas cultivated with grapevines. Brazilian grape production grew by 5.92%, while the planted area decreased by 9.87% between the years 2010 and 2020. Brazil produced 1.84% of the world's grapes and occupied 1.06% of the world's grape-growing area in 2020. Despite not being among the world's largest producers of grapes, it is worth noting that Brazil is the country that stood out with the average productivity of vineyards in this period reaching 23 ton/ha, an increase of 16.84%, maintaining its world leadership status (FAO, 2022).

4.2 Tropical Viticulture in the Brazilian Semiarid Region

Most viticultural regions in the world are located between latitudes 40° to 50° N and 30° to 40° S, also known as the temperate climate belt (Kok, 2014). However, the last five decades have seen the expansion of world viticulture in tropical and subtropical climate countries in South America, Asia, and Africa (Brazil, Peru, India, Thailand, and Madagascar).

The vine adapts differentially in every climatic condition by imposing a particular management system, resulting in varied yield and quality of grapes. Thus, the focus here is going to be on the dry tropical viticulture, highlighting growing regions like Piura in northern Peru (05°11′40″ S 80°37′58″ W), Zulia in Venezuela (10°57′51″ N, 71°44′8″ W), and Petrolina and Juazeiro in the Northeast of Brazil (9°23′39″ S, 40°30′35″ W). Viticulture in these regions is the nearest to the Equator line in the world and has high similarities between them. São Francisco Valley can be considered an example of this group as one of the pioneers and most technically advanced for growing table grapes.

The production system in tropical conditions differs significantly from traditional viticulture in temperate climates. Thus, tropical and subtropical regions require other methods for breaking bud dormancy, pruning, training techniques, bud fertility, and vigor controls to achieve successful harvests, and thus make the crop viable in non-traditional regions (Jogaiah et al., 2013).

The Submédio São Francisco Valley in the Northeast of Brazil is the main production hub for table grapes in the country, also standing out as an important supplier of grapes in the Southern Hemisphere for the global market. The beginning of viticulture in the Submédio São Francisco Valley occurred with the implementation of public irrigation projects in the early 1960s. However, it was from the 1980s onwards that it expanded rapidly, with public and private investments that contributed to the technological advancement and development of viticulture in the Brazilian semiarid region, which found competitive advantages in the internal and external scenario in the tropical semiarid climate, with high temperatures, high insolation and availability of good quality water for irrigation (Soares & de Leao, 2009).

The intra-annual climatic variability results in changes in plant physiology, with variable responses in floral differentiation and bud fertility, phenology, growth of branches and berries, and especially in the physiology of ripening and consequently in the production of phenolic compounds, anthocyanins that give grapes more or less color, sugar content, and acidity in the fruit. Therefore, the volume and especially the quality of the grapes can vary greatly between the harvests of the first and second half of the year.

It is worth mentioning that the development of new cultivars, different management technologies for tropical regions, certification of viticultural products as integrated production, geographical indications and organic production were fundamental for the progress of viticulture in the region, especially in the pole Petrolina/Juazeiro. Public institutions, such as Embrapa and partners provided the technical and scientific support for this development.

The area cultivated with grapes in the Submédio São Francisco Valley had a growth of 11% in the period between 2010 and 2021, becoming 13.51% of the area with grape plantations in Brazil (IBGE, 2022). The production of grapes in the region had a high growth (89.62%) between the period 2010 and 2021, going from 241 to 457 thousand tons, which represented 26.83% of the Brazilian production

(Fig. 4.1a) (IBGE, 2022). However, a smaller growth of only 11% was observed concerning the cultivated area, which demonstrates the use of technologies that have significantly increased the yield of the vineyards. The average yield in this region is higher than the national and world averages, with larger volumes produced each year, and in the period between 2010 and 2021, the growth was 53.4%. In 2021, the average yield of grapes was 39.8 t/ha, 73.64% and 209.11% higher than the averages for Brazil and the world, respectively (Fig. 4.1b). Petrolina, State of Pernambuco, was the municipality with the highest average yield (47.6 t/ha), followed by Juazeiro, State of Bahia, which produced about 32 t/ha (IBGE, 2022). From this information, it can be concluded that the world highlights for the increase

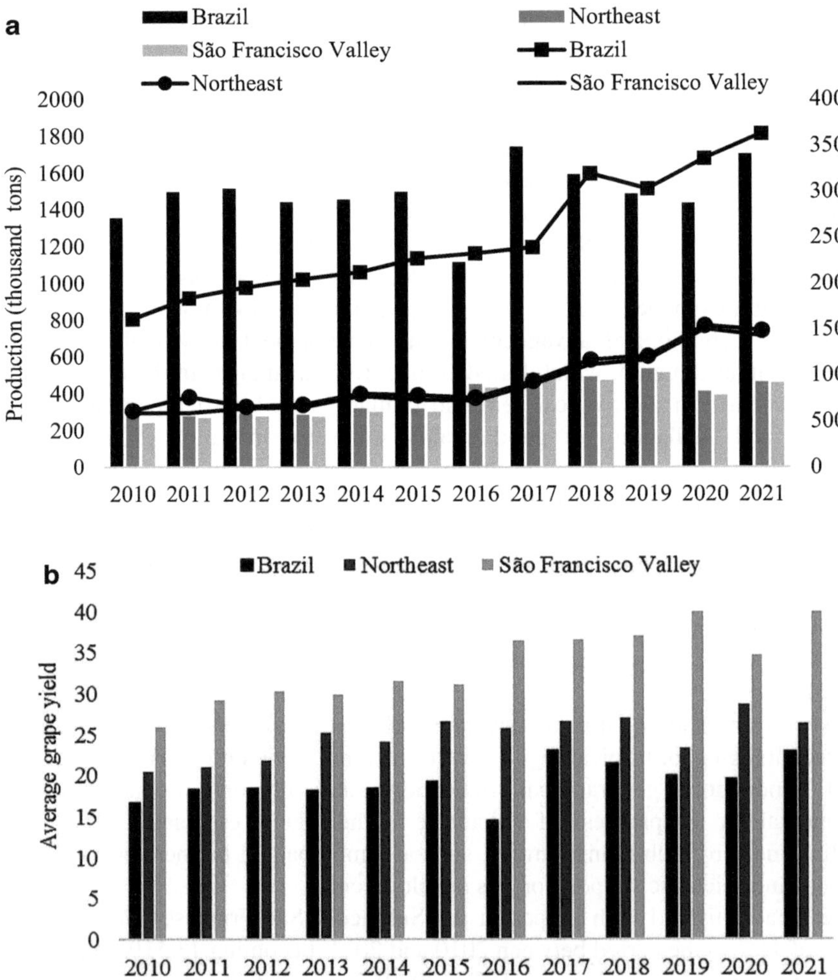

Fig. 4.1 The volume of grapes and value obtained for grape production (**a**) and average yield (**b**) of grapes in the São Francisco Valley, Northeast region, and Brazil. (Source: IBGE (2022))

in Brazilian yield is the high yield achieved in the Submédio São Francisco Valley, especially in the municipality of Petrolina.

The grape is one of the main fruits on the export agenda of the Submédio São Francisco Valley region. In the period from 2010 to 2021, the region increased by 23% the quantity of grapes exported. In 2022, 52,6 thousand tons were exported, collecting approximately 113,9 million dollars (MDCI, 2023), which corresponded to 98% of the volume and value of Brazilian grape exports. However, it is worth noting the importance of the domestic market in the dynamics of regional grape growing, for it absorbed about 381.8 thousand tons of grapes, which corresponded to 83.56% of production in the year 2021 (IBGE, 2022).

Finally, following the consumption trend of the world market for fresh fruit and in order to offer new cultivar alternatives to the domestic market, grape growers in the Submédio São Francisco Valley currently concentrate on the production of seedless grapes. This region is the main producer of seedless grapes in the country, supplying both domestic and foreign markets.

4.3 Climate, Soil, and Vegetation Characterization in the Brazilian Semiarid Region

The rainfall regime in the Brazilian semiarid region is marked by scarcity, spatial, and temporal irregularity and long periods of drought. Rainfall is concentrated in 4 months per year (December to April) with an annual average of less than 800 mm (Fig. 4.2). Evaporation rates are high ranging from 1200 to 3200 mm.year^{-1} due to the high solar radiation associated with the irregularity of the rainfall regime. The average annual temperatures in most of this region present values between 24 and 28 °C (Fig. 4.2). The relative humidity reaches lower values around 56% in the central part of the region and increases reaching 76–80% in the transition belt with the east coast (Moura et al., 2019).

In the semi-arid region, there is a great variation of lithology, original material, relief, and soil moisture that resulted in the presence of different classes of soils. In general, the soils are poor in organic matter, with low cation exchange capacity (CEC), low natural fertility, and highly variable textures according to the type of soil, from sandy or sandy-clayey in Latosols to clayey in Vertisols (Cunha et al., 2010).

The vegetational complex that corresponds to what is known as Caatinga has recently been categorized as a Phytogeographic Domain, and no longer as a Biome. The Domain can be characterized as xerophytic vegetation, low, with generally discontinuous canopy, and deciduous foliage in the dry season with great floristic and physiognomic variation along its range of occurrence. The most recent classification recognizes eight Ecoregions in the Caatinga, with distinct vegetational structures, understanding that the variation in vegetation structure is conditioned by relief, soil, and annual precipitation characteristics (Queiroz et al., 2017).

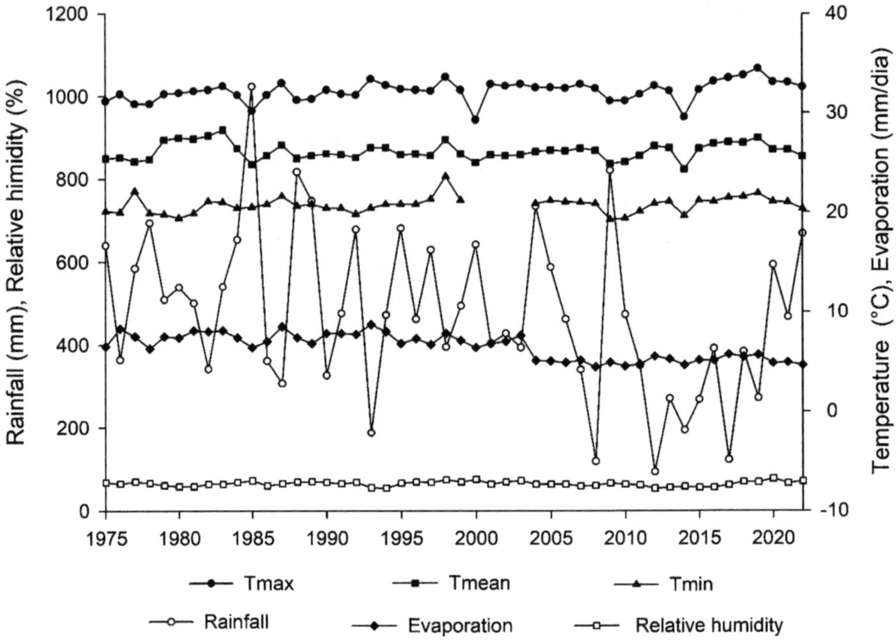

Fig. 4.2 Climate data (1975–2022) on rainfall (mm), mean, minimum, and maximum air temperature (°C), relative humidity (%) and evaporation (mm)

4.4 Table Grape Cultivars

Over this last decade, the table grape cultivars 'BRS Vitória' (Maia et al., 2012), 'BRS Isis' (Ritschel et al., 2013), 'BRS Nubia' (Maia et al., 2013), and more recently, 'BRS Melodia' (Maia et al., 2019) and 'BRS Tainá' (Leao et al., 2020) were released (Fig. 4.3). The new cultivars developed by Embrapa, especially 'BRS Vitória' have aroused great interest and rapidly expanded the cultivated areas, with an important economic and social impact on Brazilian tropical viticulture.

It is worth mentioning the importance of cultivars developed by different international breeding companies, among which the most important white grapes are Arra 15® (Grapa), Sugar Crisp® (Bloom Fresh) Autumm Crisp® (Sun Word) and Timpson® (Bloom Fresh), Sweet Globe® and Cotton Candy® (Bloom Fresh); among red grapes: Sweet Celebration®, Candy Snaps® (Bloom Fresh), Timco® (Bloom Fresh), Scarlotta Seedless® (Sun World) (Fig. 4.4) and black grapes: Sweet Saphire® (Bloom Fresh), Sable® and Midnight Beauty® (Sun World). Besides these most important ones, many others are being produced on a smaller scale or still being evaluated in test blocks by companies in commercial areas. The cultivars are available only to licensed growers and the contracts imply the payment of royalties on the volumes produced or commercialized and may vary according to the genetic company.

Fig. 4.3 Grape cultivars developed by Embrapa: 'BRS Vitória' (**a**), 'BRS Isis' (**b**), 'BRS Nubia' (**c**), 'BRS Clara' (**d**), 'BRS Melodia' (**e**) and 'BRS Tainá' (**f**). (Pictures: Patrícia Coelho de Souza Leão (**a–c**, **e**, **f**) and Embrapa (**d**))

Fig. 4.4 Cultivars from international private breeding programs: Arra 15® (**a**), Sugar Crisp® (**b**), Autumn Crisp® (**c**), Timpson® (**d**), Sweet Celebration® (**e**), Scarlotta Seedless® (**f**). (Pictures: Amanda Rodrigues (**a**), Bloom, Fresh (**b**, **e**), Andrea Pavesi (**c**, **f**) and Bloom Fresh (**d**))

The cultivars should present, as one of its main characteristics, the high fertility of buds, preferably in the basal buds, allowing short or medium pruning with about five buds per stick. In addition, they must present productive stability to allow the production of two harvests per year with average yields above 25 tons/ha in each harvest, avoiding the problem of alternating productivity between harvests, which is common in most traditional cultivars

4.5 Production Systems

The grapevine production system, especially in tropical conditions, is a complex system where pruning and irrigation are crucial factors to enable fruit production. However, commercial production with high productivity and quality adequate for the market requires the adoption of a set of management practices that include the choice of rootstock appropriate for each cultivar of grapevine, conduction system and planting density, rational management of water and nutrients through fertigation, phytosanitary control by adopting the precepts of the integrated management of pests and diseases, soil management practices and control of spontaneous weeds, and a set of practices performed on the canopy and clusters, which include different operations of green pruning, as well as the use of growth regulators with different objectives.

The conduction system used varies according to several factors, but mainly the type of grape and purpose of production (consumption in nature or processing of wines and juices). In the Submédio São Francisco Valley, the trellis is predominant for table grape production, while for winemaking, three trellising systems can be found in the region: trellis, espalier and lyre. The latter has as its main advantage in the increase in productivity compared to espalier.

Pruning practices include dry pruning, which is done on woody branches after the resting period, and green pruning operations that are performed on the grapevine canopy to eliminate shoots, branches, secondary shoots, tendrils, leaves, and inflorescences and are carried out during the vegetative growth cycle.

For the production of two harvests per year, mixed pruning is used, when the same plant simultaneously combines spurs (branches with two to three buds) close to the main branch and canes or shoots (branches with more than four buds), that is, the production units are composed of one or more spurs and, in general, two to three canes (Fig. 4.5a). Pruning can be performed at any time of the year. Still, it is recommended a minimum interval of 40 days between harvest and pruning of the next cycle to promote the complete maturation of the branches and storage of reserves before the production pruning of the next cycle.

Green pruning gathers a set of operations performed on the grapevine canopy to eliminate organs such as shoots, stems, branches, tendrils and inflorescences, and tip of the shoots (Fig. 4.5b), and are performed during the vegetative growth cycle. These practices can affect bud fertility, increased cluster and berry size, ripening evolution, and color development. They modify the microclimate inside the

Fig. 4.5 Production pruning and budding stage (**a**) and tip of the shoots (**b**). (Pictures: Patrícia Coelho de Souza Leão (**a**) and César Mashima (**b**))

vineyard, allowing more aeration and luminosity and lower relative humidity, increasing the efficiency of phytosanitary control of diseases and pests.

In table grapes, the appearance of the bunch and berries influence the acceptance and choice of the consumer. Therefore, the correct implementation of bunch management practices is essential since they directly affect characteristics such as weight, size, shape, compactness, ripeness, and color of the grapes. It is up to the grape grower to decide whether to perform them, depending on the economic aspects and requirements of the destination market of the grape, as well as the requirements of each cultivar.

The management practices that have a direct action on the improvement of the quality of the bunches are the following: thinning and detaching of bunches, decompaction of bunches and thinning of berries, ringing of the stem and/or branches, and application of growth regulators.

The protected cultivation or covering of the vineyard with plastics such as raffia and polyethylene or 100% polyethene has increased in recent years, the main objective being the protection of the bunches against damage caused by rainfall, such as abortion at flowering and fruit set, the incidence of disease, cracking of berries and rotting of ripe grapes. Despite the high investment in covering the vineyard, the return on investment is quick for grapes with high market value, especially for export.

The table grapes are harvested according to the characteristics of each cultivar, considering in general minimum values of soluble solids of 15°Brix, and a ratio (brix/acidity ratio) of minimum 20. Therefore, the monitoring of the content of soluble solids and titratable acidity is necessary as the grapes ripen. Before the harvest, it is important to clean the bunches to eliminate rotten berries, stained and with mechanical or physiological damage. The grapes are packed in the packing house and stored in cold storage or harvested and transported immediately to the domestic market.

4.6 Final Considerations

The diversification and improvement of management practices of grapevine cultivars observed in the last decade resulted in positive impacts in strengthening of the production chain of tropical grape growing in Brazil. The offer of new cultivars consolidated the consumption of seedless grapes in Brazil and reduced the volumes imported, especially from Chile. In the foreign market, one notices the trend in recent years of regular supply of different types of table grapes throughout the year, including the months of the first semester when supply is restricted in importing markets (Lima et al., 2019). In addition, there is growing interest in diversity, i.e., grapes with different shapes, colors and especially unique and exotic flavors. *Gourmet* type grapes are mainly aimed at niche markets with higher sales values and show a strong growth trend. However, cultivars that have white color, large berries, and firm, crunchy, neutral, and pleasant flavor continue to be the ones that find more space in foreign markets.

The availability of many cultivars has required adjustments in the production system. Aspects such as choice of rootstock, management system, protected cultivation, spacing, types of pruning, density of shoots and bunches, use of growth regulators, water and nutrient management, phytosanitary control, determination of the harvest point, and use of technologies to increase shelf life need to be defined considering the different responses and requirements of each one of the table grape cultivars.

In addition, technological challenges of tropical viticulture in the semi-arid region are also those common to other Brazilian fruit growing chains: seedling promotion and certification; development and use of inputs and biological/natural control agents; adding value to fruits and their derivatives; reduction of post-harvest losses; development of new packaging and forms of commercialization attractive to the consumer; varietal diversification; structuring of observatories, registers and databases for decision making by the actors of the productive chains; expansion of the use of precision fruit farming tools; mechanization in fruit farming and the use of support equipment for monitoring, based on the adoption of information and knowledge technologies; strengthening the intensive use of warning systems and damage mitigation technologies by reducing the impact of climate changes and losses by biotic and abiotic factors associated with climate; and developing and promoting geographical indications of fruits (Lima et al., 2019). Besides, global issues of world agriculture in this twenty-first century are related to mitigation of climate change impacts and sustainability.

Therefore, these demands highlight the importance of the organization of the different links of the production chain, the associations and cooperatives of small and medium producers, and the strengthening of partnerships between public and private institutions to ensure the sustainable growth of table grape production in the Brazilian semiarid region to meet the growing demands of high-quality grapes for national and international markets.

References

Cunha, T. J. F., Petrere, V. G., Silva, D. J., Mendes, A. M. S., de Melo, R. F., de Oliveira Neto, M. B., da Silva, M. S. L., & Álvarez, I. A. (2010). Principais solos do semiárido tropical brasileiro: caracterização, potencialidades, limitações, fertilidade e manejo. In I. B. Sá & P. C. G. da Silva (Eds.), *Semi árido Brasileiro: Pesquisa, Desenvolvimento e Inovação* (p. 402). Embrapa Semiárido.

da Silva, J. N., Ponciano, N. J., Melo Souza, C. L., de Souza, P. M., & Hespanhol Viana, L. (2019). Caracterização da viticultura tropical nas regiões norte e noroeste fluminense. *Revista Brasileira de Fruticultura, 41*(6), 1–8.

EPAMIG. (2020). *Informe Agropecuário. Vinhos de inverno* (Vol. 41, p. 312). Empresa de Pesquisa Agropecuária de Minas Gerais. Retrieved from: https://ainfo.cnptia.embrapa.br/digital/bitstream/item/219233/1/Art-9-IA312-25-nov-2020.pdf

FAO. (2022). The Food and Agriculture Organization of the United Nations. Retrieved from: http://www.fao.org/faostat/en/#data

Gazolla, R., Gründling, R. D. P., & Aragao, A. A. (2020). A produção e o comércio internacional de uva. *Revista Brasileira de Agrotecnologia, 10*(3), 68–74.

IBGE. (2022). Instituto Brasileiro de Geografia e Estatística. Levantamento Sistemático da Produção Agrícola. Retrieved from: https://sidra.ibge.gov.br/home/lspa/brasil

Jogaiah, S., Oulkar, D., Vijapure, A., Maske, S. R., Sharma, A. E., & Somkuwa, R. R. (2013). Influence of canopy management practices on fruit composition of wine grape cultivars grown in semi-arid tropical region of India. *African Journal of Agricultural Research, 8*, 3462–3472.

Kist, B. B., Carvalho, C., de Beling, R. R., & Machado, M. O. (2022). *Anuário Brasileiro de Horti & Fruti 2022* (pp. 1–96). Gazeta Santa Cruz.

Kok, D. (2014). A review on grape growing in tropical regions. *Turkish Journal of Agricultural and Natural Sciences*. Special Issue, 1236–1241.

Leao, P. C. S., Borges, R. M. E., Melo, N. F., de Barbosa, M. A. G., Lima, M. A. C., de Ferreira, R. C. F., & Biasoto, A. C. T. (2020). *'BRS Tainá': nova cultivar de uvas sem sementes de cor branca para o Submédio do Vale do São Francisco*. Embrapa Semiárido. 14 p. (Embrapa Semiárido. Circular Técnica, 122). Retrieved from: https://ainfo.cnptia.embrapa.br/digital/bitstream/item/215646/1/BRS-Taina-2020.pdf

Lima, M. A. C., de Guerra, C. C., Bianchini, F., & de Leao, P. C. (2019). *Pesquisa, desenvolvimento e inovação para a produção tropical de uvas para mesa, vinho e suco: situação atual e oportunidades* (p. 62). Embrapa Semiárido. (Embrapa Semiárido. Documentos, 293).

Maia, J. D. G., Ritschel, P. S., Camargo, U. A., de Souza, R. T., Fajardo, T. V. M., de Naves, R. L., & Girardi, C. L. (2012). *'BRS Vitória': Nova cultivar de uva de mesa sem sementes com sabor especial e tolerante ao míldio* (p. 12). Embrapa Uva e Vinho. (Embrapa Uva e Vinho. Comunicado Técnico, 126). Retrieved from: https://www.infoteca.cnptia.embrapa.br/infoteca/bitstream/doc/941724/1/cot126.pdf

Maia, J. D. G., Ritschel, P. S., Camargo, U. A., de Souza, R. T., Fajardo, T. V. M., de Naves, R., & Girardi, C. L. (2013). BRS Núbia Nova Cultivar de Uva de Mesa com Sementes e Coloração Preta Uniforme. Comunicado Técnico, 139, Embrapa Uva e Vinho, Bento Gonçalves, RS, p. 12.

Maia, J. D. G., Ritschel, P. S., Camargo, U. A., de Souza, R. T., Grohs, D. S., & Fajardo, T. V. M. (2019). *'BRS Melodia': Nova cultivar de uvas sem sementes, com sabor especial de mix de frutas vermelhas, recomendada para cultivo na Serra Gaúcha, em cobertura plástica* (p. 22). Embrapa Uva e Vinho. (Embrapa Uva e Vinho. Circular Técnica, 144). Retrieved from: https://ainfo.cnptia.embrapa.br/digital/bitstream/item/203207/1/Circ-Tec144.pdf

MDCI. (2023). Comex stat. Retrieved from: O Bioma aatinga. Retrieved from: www.acaatinga.org.br/wpcontent/uploads/Conhe%C3%A7a_e_Conserve_a_Caatinga_-_Volume_1__O_Bioma_Caatinga.pdf

Mello, L. M. R., & Machado, C. A. E. (2020). Viticultura brasileira: panorama 2019. Comunicado Técnico 214. Embrapa. Bento Gonçalves, ISSN: 1808–6802.

Moura, M. S. B., Espínola Sobrinho, J., Silva, T. G. F., & Souza, W. M. (2019). Aspectos meteorológicos do Semiárido brasileiro. In: XIMENES, Luciano Feijão; Silva, Maria Sonia Lopes da; BRITO, Luiza Teixeira de Lima (ed.). Tecnologias de convivência com o semiárido brasileiro. Fortaleza: Banco do Nordeste, 2019. p. 1116. (ISBN: 978-85-68360-28-6).

Pereira, G. E. (2020). The three different winegrowing zones in Brazil according to climate conditions and vine managements. In A. M. Jordao & R. V. Botelho (Eds.), *Vitis: Biology and species*. Nova Science Publishers. Chapter 6.

Queiroz, L. P. D., Cardoso, D., Fernandes, M. F., & Moro, M. F. (2017). Diversity and evolution of flowering plants of the Caatinga domain. In J. M. C. Silva, I. R. Leal, & M. Tabarelli (Eds.), *Caatinga* (pp. 23–63). Springer.

Ritschel, P. S., Maia, J. D. G., Camargo, U. A., de Souza, R., Fajardo, T. V. M., de Naves, R. L., & Girardi, C. L. (2013). *BRS Isis: nova cultivar de uva de mesa vermelha, sem sementes e tolerante ao míldio* (p. 20). Embrapa Uva e Vinho. (Embrapa Uva e Vinho. Comunicado Técnico, 143).

Soares, J. M., & de Leao, P. C. S. (2009). *A vitivinicultura no Semiárido brasileiro* (p. 756). Embrapa Informação Tecnológica; Embrapa Semiárido.

Tonietto, L., & Carbonneau, A. (2004). A multicriteria climatic classification system for grape growing regions wordwide. *Agricultural Forest and Meteorology, 124*(1–2), 81–97.

Chapter 5
Heavy Metal Stress Response in Plants and Their Adaptation

Gustavo Brunetto, Daniela Guimarães Simão, Luciane A. Tabaldi, Paulo A. A. Ferreira, Edicarla Trentin, Carina Marchezan, Tadeu Luis Tiecher, Eduardo Girotto, Lessandro De Conti, Cledimar Rogério Lourenzi, Kleber Resende Silva, Anderson C. R. Marques, Letícia Morsch, Allan Augusto Kokkonen, Stefano Cesco, and Tanja Mimmo

5.1 Introduction

Anthropogenic activities add significant amounts of metallic elements to the soil, especially heavy metals, which, when observed at levels above those considered adequate in soils and/or water springs, can cause toxic effects on plants, animals, microorganisms and human beings. In plants, the symptoms of toxicity initially may appear in the roots, through morphological and anatomical modifications (Guimarães et al., 2016; Somavilla et al., 2018; Morsch et al., 2022). For this reason, water absorption and nutrients may be impaired, which may cause changes to the nutritional concentration of shoot organs (Amari et al., 2017; De Conti et al., 2018, 2020; Trentin et al., 2022). Also, biochemical and physiological changes may

G. Brunetto (✉) · D. G. Simão · L. A. Tabaldi · P. A. A. Ferreira · E. Trentin · C. Marchezan
A. A. Kokkonen
Universidade Federal de Santa Maria (UFSM), Santa Maria, Rio Grande do Sul, Brazil

T. L. Tiecher · E. Girotto
Instituto Federal do Rio Grande do Sul (IFRS), Porto Alegre, Brazil

L. De Conti
Instituto Federal Farroupilha (IFFarroupilha), Santa Maria, Brazil

C. R. Lourenzi
Universidade Federal de Santa Catarina (UFSC), Florianópolis, Brazil

K. R. Silva · L. Morsch
Instituto Tecnológico Vale (ITV), Belém, Brazil

A. C. R. Marques
Universidade Regional Integrada do Alto Uruguai e das Missões (URI),
Frederico Westphalen, Brazil

S. Cesco · T. Mimmo
Free University of Bozen/Bolzano, Bolzano, Italy

G. Gutiérrez Gamboa, M. Fourment (eds.), *Latin American Viticulture Adaptation to Climate Change*, https://doi.org/10.1007/978-3-031-51325-1_5

occur in the shoot of plants, especially because of the excess of heavy metals in soils and, consequently, in the tissue (Tiecher et al., 2016a, b; Trentin et al., 2019; Hammerschmitt et al., 2020; Schwalbert et al., 2021). All this may compromise the plant development, the yield, and even depreciate its quality (Couto et al., 2018; Hammerschmitt et al., 2020; Schwalbert et al., 2021). The different plant species can possess strategies to avoid the absorption of heavy metals (Montiel-Rozas et al., 2016; De Conti et al., 2020; Trentin et al., 2022). However, different internal mechanisms to avoid transport between organs have also been described (Mwamba et al., 2016; Zhou et al., 2017; Xiao et al., 2020). Thus, some plant species can grow and produce properly even in soils with an excess of heavy metals in soils or tissue (Zhou et al., 2017; Xiao et al., 2020). This chapter aims to present sources of heavy metals in soils, the main modifications caused by heavy metals in roots, as well as physiological and biochemical changes in plants. Moreover, adaptation strategies are also outlined with particular emphasis on the subtropical/tropical soils and the plant species growing in these substrates.

5.2 Sources and Excess of Heavy Metals in Environments

The application of pesticides represents a cause of increased heavy metal content in soils. In subtropical regions, such as the South of Brazil, fruit crops receive frequent applications of fungicides that can cause an increase in the levels of heavy metals in the soil, especially those belonging to the chemical group of dithiocarbamates (e.g. mancozeb and propineb) and the inorganic Cu-based ones (e.g. Cu sulfate, Cu oxychloride and Cu hydroxide), because they contain Cu, Zn and Mn in their molecular formulas. These fungicides are used for the control of leaf diseases, especially downy mildew (*Plasmopara viticola*) in orchards (Brunetto et al., 2017). In a single cycle of grapevine culture, for example, the sum of fungicide applications can cause the input of 6.76 kg Cu/ha by year and 2.00 kg Zn/ha by year (Tiecher et al., 2022), which often causes excessive accumulation of these metals in vineyard and orchard soils (Brunetto et al., 2014a; Miotto et al., 2017; Korchagin et al., 2020; Cesco et al., 2021). The substitution of fungicides containing Cu by those containing Zn to a lesser extent also contributes to the increase in the contents of other heavy metals in soil, such as Zn (Brunetto et al., 2014a, 2017; Cambrollé et al., 2015; Tiecher et al., 2016a, b). Table 5.1 presents the total Cu and Zn contents in soils cultivated with fruit trees.

With the accumulation of heavy metals in soils cultivated with fruit trees, especially above the adsorption capacity of the soil, leaching phenomena are expected, causing contamination of surface and groundwater. A large potential for toxicity to fruit trees, grapevines, and cover crops used in orchards should also be considered (Brunetto et al., 2014a). Also, Cu and Zn accumulation in the vineyard and orchard soils can also be promoted by the continued application of organic by-products, such as animal waste. In the latter, the source of metals is essentially the animal diet. This effect has been observed in pasture and grain production areas with a history

Table 5.1 Total contents of Cu and Zn in soils cultivated with fruit trees in the southern region (Brazil)

References	Soil type	Age of area	Depth	Total Cu	Total Zn
		years	cm	mg/kg	
		Vitis vinifera			
Mirlean et al. (2007)	Oxisol	5	0–5	402.0	176.4
		40		1006.8	199.1
	Spodosol	61		1979.2	204.2
		100		2214.7	240.0
	Entisol	20		50.9	16.2
		45		561.4	51.6
Brunetto et al. (2014a)	Ultisol	14	0–20	19.7	15.8
		30		73.1	16.8
Brunetto et al. (2014b)	Inceptisol	4	0–20	244.4	151.2
		6		263.3	158.2
		10		290.8	179.9
Couto et al. (2015)	Alfisol	95	0–20	463.3	96.7
Brunetto et al. (2018a)	Ultisol	4	0–20	48.7	122.3
		15		52.6	65.3
Hummes et al. (2019)	Entisol	120+	0–20	1151.4	36.9
	Oxisol	10		66.7	11.6
Korchagin et al. (2020)	Entisol	122	0–20	1819.0	151.0
		Malus domestica			
Brunetto et al. (2018b)	Inceptisol	8	0–20	62.0	49.1
		18		72.5	107.1

of application of animal waste (liquid and solid) and organic compost (Girotto et al., 2010; Tiecher et al., 2013; Couto et al., 2018). The same can happen in vineyards and orchards, in conventional, integrated, or organic production systems, which have long histories of organic fertilizer applications, especially when the doses added to the soil are defined without technical criteria.

Heavy metal levels in soils are also influenced by vineyard age, fungicide application history, application frequency, soil type, organic matter content, and soil management (Brunetto et al., 2016; Bortoluzzi et al., 2019). In particular, the age of the vineyard is one of the main indicators, because of the high use of fungicides in viticulture. In young vineyards, cultivated in places with no history of cupric fungicide application, Cu contents in soils are generally low, as observed by Hummes et al. (2019). In a vineyard with five (5) years of cultivation, a Cu content extracted by Mehlich-1 of 12.4 mg/kg was observed. On the other hand, Casali et al. (2008) observed a Cu content extracted by EDTA of 583 mg/kg in a vineyard with 40 years of cultivation, while Korchagin et al. (2020), in a vineyard with 120 years of cultivation, observed a Cu content extracted by Mehlich-1 of 1300 mg/kg. The history of cultivation and, consequently, of fungicide applications is also related to the average rainfall of a specific region. Mirlean et al. (2007) observed Cu contents of 75 to 500 mg/kg in soils from vineyards located in regions where the average annual rainfall varies from 350 to 750 mm. While the Cu content observed in a vineyard located in the southern region of Brazil, with rainfall ranging from 1700 to 2000 mm, was 3200 mg/kg.

5.3 Excess of Heavy Metals and Morphological and Anatomical Changes in Roots

In viticulture, the accumulation of heavy metals in the soil, such as Cu and Zn, is frequently reported in numeral areas across the world (Beygi & Jalali, 2019; Brunetto et al., 2016, 2018a; Campillo-Cora et al., 2019; Miotto et al., 2017; Sonoda et al., 2019). The roots, the first plant organ to come into contact with the metal, exhibit early signs of toxicity. Several structural alterations in this crucial plant region have been observed in studies analyzing the bioaccumulation of these metals in grapevines and associated cover crops (Guimarães et al., 2016; Ambrosini et al., 2018; Castro et al., 2021; Morsch et al., 2022).

Considering the morphological aspects, roots of vines and cover crops that grow under high concentrations of Cu and Zn often show similar and easily noticeable symptoms, such as short, rigid and darkened roots with a larger diameter (Yang et al., 2011; Ambrosini et al., 2015; Guimarães et al., 2016; Tiecher et al., 2018; Morsch et al., 2022). Figure 5.1 show fragments of roots in a differentiated region, where lateral roots develop; it is noted that the roots are thicker and darker under influences of Cu (Fig. 5.1b) and Zn (Fig. 5.1c). Copper can also promote lateral root development in grapevines (Fig. 5.1b; Morsch et al., 2022) and in cover crops such as black oat (Guimarães et al., 2016). Many of these lateral roots, however, do not develop completely, resulting in swelling in both the apical (Kopittke et al., 2009; Morsch et al., 2022) and distal areas of the root apex (Fig. 5.1b). On the other hand, Zn seems to reduce the number of lateral roots in grapevines (Fig. 5.1c; Silva & Simão, 2022).

As previously noted, the reduction in root length is one of the most evident symptoms of Cu and Zn phytotoxicity. This effect can be explained anatomically by the decrease in cell division frequency associated with cell differentiation at the root apex. In this context, a decrease in mitotic activity of the root meristem of grapevines and cover crops has been reported as an impact of Cu and Zn accumulation (Juang et al., 2012; Ambrosini et al., 2015; Guimarães et al., 2016; Castro et al., 2021; Morsch et al., 2022). Copper may also be involved with anatomical changes in the root cap of grapevines (Ambrosini et al., 2015) and cover crops, especially, when cultivated in soil (Guimarães et al., 2016). This results in a reduction in the size of the root cap, which may also affect the protection of the root apex during its growth in the soil (Lynch et al., 2012), and the subsequent development of the root axis.

The anatomical changes of the root apex modify the arrangement of differentiated tissues in the distal regions of the root. For example, high contents of Cu

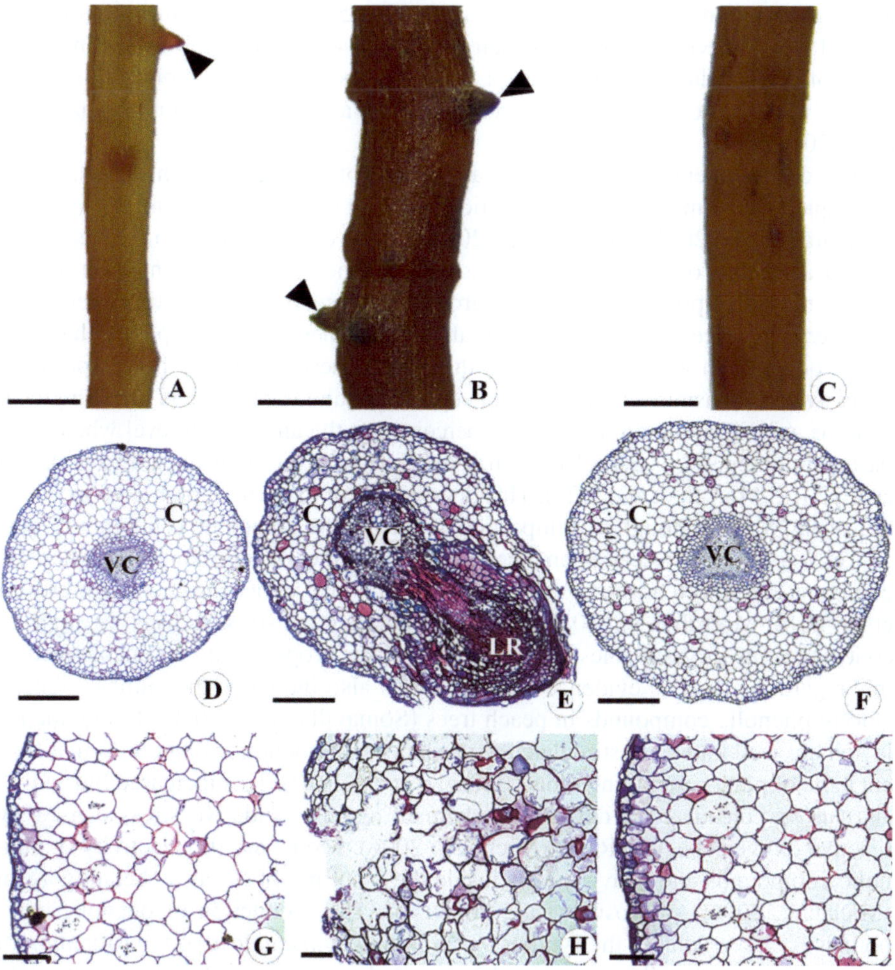

Fig. 5.1 Structural changes in grapevine roots by heavy metals (Roots of rootstock SO4 - *Vitis berlanderi* x *V. riparia*; (**a**, **d**, and **g**) Control treatment; (**b**, **e** and **h**) High concentrations of Cu; (**c**, **f** and **i**) High concentrations of Zn). (**a–c**) Morphological aspects in distal portions. (**d–f**) General anatomical aspects, in cross sections. (**g–i**) Details of epidermal and cortical regions, in cross sections. (Arrowheads indicate the development of lateral roots; C, cortex; LR, lateral root; VC, vascular cylinder). Bars: **a–c** = 1 mm; **d–f** = 200 µm; **g–i** = 50 µm

increased the areas of the cortex and vascular cylinder in grapevine roots, resulting in an increase in the organ's diameter (Ambrosini et al., 2015, 2018). The disorganization of cortical cell layers in the apical region promotes this increase in thickness in the distal portions. In some cases, the development of larger cells in the cortex can also be observed (Fig. 5.1d, e). Regarding Zn, studies have shown that a high content of this metal also increases the root diameter of grapevines (Castro et al., 2021). In some varieties, the greater diameter of the roots seems to occur due to the greater number of cortical cell layers (Fig. 5.1f), although more studies are needed for grapevines as well as their associated cover crops. From a functional point of view, a larger diameter root could compensate for a shorter axis, since shorter roots represent a smaller area for water and nutrient uptake (Ambrosini et al., 2018).

The development of lateral roots, as a result of high Cu contents, occurs as a consequence of a modified meristematic activity in the vascular cylinder (Fig. 5.1e; Guimarães et al., 2016; Morsch et al., 2022), which can contribute to the absorptive role of a shorter root axis in a compensatory manner. Early differentiation of root tissues may also promote secondary growth in roots, a trait recently observed in some grapevine genotypes (Morsch et al., 2022). These authors reported that both main root growth and lateral root production in these plants are highly prone to excessive Cu. Plasmolysis in grapevine epidermal and cortical cells under high Cu doses is another symptom that can be perceived at the anatomical level when comparing treatments (Fig. 5.1g, h) and may interfere with root uptake pathways (Juang et al., 2012; Morsch et al., 2022). This does not seem to occur at high Zn concentrations (Fig. 5.1i), although it is important to expand the grapevine genotypes tested.

Yet, in grapevine research, an increase in the number of cells containing phenolic compounds is reported when roots develop under high Cu concentrations (Ambrosini et al., 2015; Morsch et al., 2022). Studies evaluating the stress induced by high Zn concentrations in grapevines and associated cover crops should be encouraged in this regard, as there is evidence that this metal also increases the number of cells storing phenolic compounds in peach trees (Somavilla et al., 2018). Heavy metals in soil are well known to stimulate the production of phenolic compounds in different parts of plants, including roots (Bouazizi et al., 2010). This increased biosynthesis can be considered a defense mechanism against oxidative stress caused by reactive oxygen species (ROS) (Michalak, 2006). Moreover, the reactivity of phenolic compounds and heavy metals inside the plant might act as a chelating agent (Michalak, 2006; Ambrosini et al., 2016a, b). The production of phenolic compounds is also crucial in the synthesis of lignin, which can lead to the formation of more resistant cell walls, preventing heavy metal entrance and distribution throughout the plant (Michalak, 2006).

All of these results, demonstrating the phytotoxic effects of heavy metals on the root system, indicate that anatomical changes are caused by cellular events that operate on mitosis and differentiation processes, modulating the structure of each root, as well as the root system as a whole. It is worth mentioning that an overview of some of the phytotoxic effects related to high concentrations of Cu and Zn in the roots, mainly in grapevines, has been presented here due to the importance of these metals in agricultural soils and vineyards. Thus, we highlight the grapevines as appropriate models for evaluating structural responses to heavy metals. The structural changes in roots might compromise the transport of water and nutrients to the other parts of the plant, affecting the shoot's equilibrate development and, as a result, the crop's yield potential and reproduction capacity. Therefore, structural alterations in the root system caused by heavy metals can help in understanding plant development under such stress conditions, as well as serve as markers of metal contamination levels in a specific soil.

Nevertheless, strategies for reducing the impact of heavy metals on cultivated plants, such as evaluating the tolerance of genotypes/cultivars, should be considered. Heavy metal effects on grapevine genotypes might vary; some are more tolerant than others (Castro et al., 2021; Morsch et al., 2022). The Magnolia genotype (*Vitis rotundifolia*), for example, was the most tolerant in the latter investigation, which evaluated three grapevine genotypes in response to excess Cu; the roots of this genotype did not show severe structural alterations (Morsch et al., 2022). The impact of P in minimizing Cu phytotoxicity was also evaluated in the same study. Phosphorus can mitigate Cu toxicity, showing that the reduced size of the root cap, early tissue differentiation, and development of lateral roots close to the root meristem were minimized (Morsch et al., 2022). Beneficial effects of P supply on root anatomy have also been observed in cover crops when cultivated with excess Cu (Guimarães et al., 2016). Furthermore, it has already been observed that liming also minimizes the morphological and anatomical alterations caused by excess Cu in grapevine roots, which happens because of the increase in pH, which decreases Cu availability, but also because of the increase of Ca and Mg in the soil (Ambrosini et al., 2015). This indicates that other elements associated with heavy metals can mitigate phytotoxic effects, and are also strategies to extend the use of vineyard soils.

5.4 Aspects of the Excess Heavy Metals in Mineral Nutrition

The uptake of chemical elements by the plant root system occurs through three main mechanisms: (i) the apoplastic route in which water and ions move through a continuous system of cell walls, without crossing any cell membranes as it crosses the root cortex until the element is blocked at Caspary striation; (ii) the symplastic route in which water and ions, once inside a root cell, flow between cells through plasmodesmata that interconnect neighboring cells; and (iii) the transmembrane route in which water and ions move across the plasma membranes of neighboring cells, with a short stay in the cell wall space. The movement of any chemical element across the cell membrane is accomplished through ion transporters and channels, which regulate the elements' entry into the cytoplasm of cells (Marschner, 2012).

In this context, excessive heavy metal uptake can compromise the selective permeability of membranes, causing imbalances in the uptake of other elements (Tiecher et al., 2018; De Conti et al., 2019; Marastoni et al., 2019). The negative interferences of phytotoxicity on ion homeostasis are a direct effect of excess heavy metals in the root growth medium. However, the loss of efficiency of the photosynthetic process and the allocation of energy to survival mechanisms in stressed plants can promote secondary effects on nutrition by limiting the availability of energy to other physiological processes, such as nutrient uptake and translocation, which occurs predominantly in an active way (Marschner, 2012).

The impact of heavy metal toxicity on plant nutrition is directly dependent on the type of contaminating element considered, the level of contamination, and the plant/cultivar species considered. In fact, plant species often differ in their tolerance mechanisms activated to mitigate damage (Zhang et al., 2014). In general, uptake of the micronutrients Cu^{2+}, Fe^{2+}, Mn^{2+} and Zn^{2+} tend to be impacted to a greater magnitude by heavy metal toxicity, due to competition in the root apoplast (Rabêlo & Borgo, 2016).

P is the macronutrient that frequently has its uptake affected in plants subjected to heavy metal toxicity, especially Cu and Zn. This is probably related to the physiological and morphological changes in the root system, especially the reduction of growth and root volume, which is a symptom of toxicity frequently reported in plants grown under Cu (Ambrosini et al., 2015; Guimaraes et al., 2016; De Conti et al., 2018, 2021) and Zn (Tiecher et al., 2017, 2018; Somavilla et al., 2018) toxicity conditions. Moreover, it has been recently demonstrated that direct Cu action on the uptake mechanisms (e.g. transporters) can be at the base of the limited root acquisition of P in Cu toxicity (Feil et al., 2020). A role played by an alteration of root plasma membrane permeability induced by Cu toxicity cannot also be excluded. Since the main mechanism of P supply is diffusion, which occurs in ml proportion in soil, and the nutrient has low mobility in soil, reducing the volume of soil explored by the root system can significantly impact the root uptake. In Fig. 5.2a, it is possible to observe the reduction in the P levels in the tissues of the part area of several plant species when grown under conditions with high levels of Cu in the soil. *Paspalum plicatulum* was the only species that did not show a reduction in P levels as a function of increasing Cu levels in the soil, indicating the influence of some

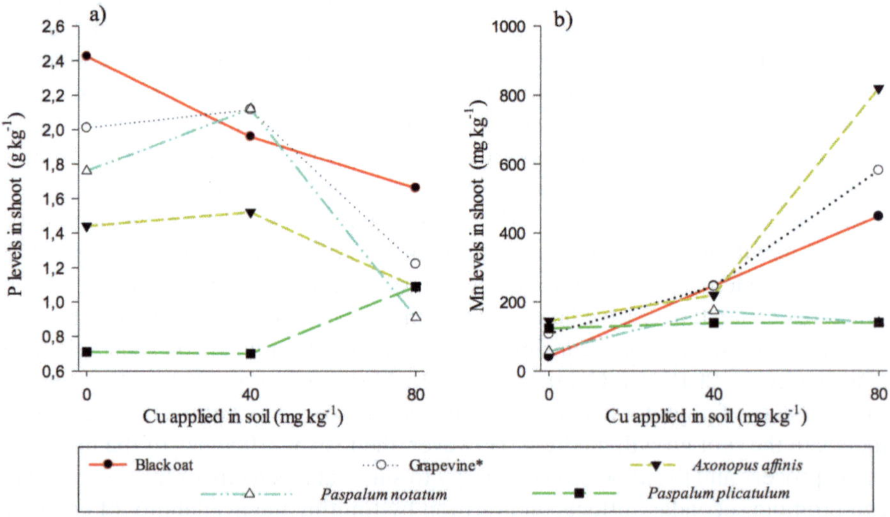

Fig. 5.2 The concentration of P (**a**) and Mn (**b**) in the shoot of different plant species grown in soil with increasing doses of Cu. *Grapevines are grown in monocropping. (Source: adapted from De Conti et al. (2018, 2019, 2021))

mechanism of tolerance. This species has been identified with a great capacity for employment in the phytoremediation of Cu-contaminated soils in southern Brazil, in a study conducted by De Conti et al. (2019, 2021), given its ability to tolerate high levels of Cu contamination.

The additional application of P in Cu-contaminated soils has shown positive effects on the growth of some plant species, highlighting the losses in P nutrition in plants grown in Cu-contaminated soils. The additional application of P in Cu-contaminated soils is pointed out as a remediation strategy in environments contaminated by this heavy metal (Baldi et al., 2018; Ferreira et al., 2018). However, the short-term exhaustion of the non-renewable sources to produce phosphate fertilizers raises the question of the environmental sustainability level of this practice. Moreover, similar to grapevines (*Vitis* spp.), many plant species used as a tolerance mechanism to excess heavy metals the accumulation of the contaminating metal at the root level, limiting, thus, the metal translocation to shoot (Juang et al., 2012; Cambrollé et al., 2015; Tiecher et al., 2017, 2018). This phenomenon occurs predominantly in the apoplast of the roots, i.e., in recalcitrant tissues and without metabolic activity. Allied with this, the compartmentalization of cationic elements in organelles of low metabolic activity of the root system may occur, in part, through complexation with phosphate anions, compromising the distribution of the nutrient over long distances between plant organs, and may induce a nutritional imbalance in organs of the shoot (Dresler et al., 2014; Baldi et al., 2018).

Contamination of the soil by heavy metals may promote an increase in the availability of other cationic elements by substituting the exchange sites of the soil colloids (Sposito, 1989). Elements with similar ionic radius and the same valence may compete for primary absorption sites or nutrient transport zones in the roots, causing excessive absorption and/or translocation of other elements. This influence can be observed in some plant species in Fig. 5.2b, such as black oat (*Avena strigosa*), grapevine, and *Axonopus affinis,* where the increase in Cu levels in the soil promoted an increase in the Mn levels in the shoot. In soils characterized by multielemental contamination, the increase in shoot translocation of some of the contaminants is often reported in black oat plants grown in soil contaminated by Cu and Zn (Tiecher et al., 2016a, b). On the other hand, depending on the type of contaminant and the plant species considered, a decrease in Mn contents may occur, as observed in forage grasses grown in soils contaminated by Cd (Zhang et al., 2014), as well as by Cu (Kopittke et al., 2009) and Ni (Kopittke et al., 2010).

Soil contamination with heavy metals also promotes damage to soil fauna, altering the abundance, richness, and diversity of many organisms, which can negatively affect important processes such as mineralization of organic constituents and biological N fixation, causing detrimental indirect nutritional effects to plants (Lecerf et al., 2021; Hammami et al., 2022). Biological N fixation is responsible for providing much of the N required by plants. However, in soils contaminated with heavy metals, the symbiotic association can be severely compromised, as verified by Hammami et al. (2022) in beans (*Phaseoulus vulgaris*) grown in heavy metal-contaminated soils.

In some cases, there may be an increase in the concentration of nutrients in the biomass of plants submitted to stress by an excess of heavy metals, when compared to the control treatment. This is mainly related to the lower biomass production of these plants, resulting in a nutrient concentration effect in the tissues. Thus, the nutritional effects promoted by heavy metal toxicity may contribute, directly or indirectly, to the reduction of plant growth and/or productivity.

5.5 Excess of Heavy Metals and Changes in Physiological and Biochemical Variables

Numerous species of cultivated and wild plants grown in environments characterized by an excess of metals can present several alterations in various morphological, physiological, and biochemical processes, such as changes in nutrient concentration, reduction in the concentration of photosynthetic pigments, root morphology and production, activation of the enzymatic and non-enzymatic antioxidant system, leaf area, and leaf dry matter. Thus, as will be highlighted, metal toxicity causes numerous alterations simultaneously.

5.5.1 Excessive Content of Metals in the Plant on Photosynthetic Characteristics

Excessive concentrations of metals in the plant act negatively on the photochemical and biochemical phases of photosynthesis, with reflections on carbon assimilation and production of photo-assimilates (glucose, fructose and sucrose). Moreover, the redistribution of photosynthates from the leaves to the roots is also then affected. Excess of metals in plants can reduce photosynthetic activity in leaves because of photosystems inhibition (Broadley et al., 2007). This phenomenon is ascribed to the limitation in the formation and/or alteration of the structure of the main (chlorophyll *a*) and accessory (chlorophyll *b* and carotenoids) photosynthetic pigments, increasing fluorescence energy losses. As a consequence, plants exhibit lower ATP and NADPH formation, and lower stomatal conductance, growth, CO_2 assimilation rate, and dry matter production (Ambrosini et al., 2018; Ferreira et al., 2018).

This is mainly due to (i) changes in the composition of chloroplast membranes; (ii) the formation of complexes between the metal and the pigment, such as the replacement of Mg in the chlorophyll molecule by Cu or Zn; and (iii) lipid peroxidation of chloroplast membranes through the formation of reactive oxygen species (ROS) (Gill & Tuteja, 2010),

In the photosynthesis process, after the absorption of light energy by the photosynthetic pigments of the antenna complex, there are three ways of dissipating this energy: (i) photochemical dissipation, where the light energy is used in the photochemical processes of photosynthesis through resonance; (ii) non-photochemical dissipation, which is the production of heat in the form of infrared radiation; and (iii) fluorescence, which is the emission of energy in the visible region. However, under stress conditions caused, for example, by the accumulation of heavy metals in plant tissues, there will be a lower proportion of the energy being allocated to photosynthesis and, proportionally, greater dissipation in the forms of heat and fluorescence of chlorophyll *a* (Yruela, 2009; Kabata-Pendias, 2011).

The main photosynthetic parameters that can be determined using fluorometers assess the functional performance of photosystem II (PSII), which can be indicative of changes caused by metal toxicity, such as initial fluorescence (F_0), maximum fluorescence (F_m), PSII maximum quantum yield (F/F_{vm}), non-photochemical quenching (NPQ), and electron transport rate (ETR), which are negatively influenced by the excess of metals in the plant tissues (Ambrosini et al., 2018; Trentin et al., 2019, Tiecher et al., 2017). In this regard, it is widely reported that higher fluorescence energy losses are associated with lower photosynthetic pigment concentrations and reduced maximum quantum yield of PSII. As a consequence, plants exhibit lower ATP and NADPH formation, carbon assimilation, growth, and dry matter production.

Associated with this phenomenon, plants with toxicity symptoms due to the excess of metals show higher values of non-photochemical quenching (NPQ) as a way to protect the leaves from luminosity-induced damage, such as increased ROS formation (Tiecher et al., 2017). Usually, this occurs due to reduced F/F_{vm} values, indicating that plants are dissipating energy in the form of heat, probably due to the damage to the photosynthetic apparatus caused by the destruction of chloroplasts.

Other photosynthetic parameters that change when plants are under metal-toxic conditions are net photosynthetic rate (*A*), intercellular CO_2 concentration (Ci), transpiration rate (*E*), stomatal conductance (Gs) and water use efficiency (WUE). Under metal stress conditions, in general, there is a reduction in the activity of enzymes involved in carbon fixation, such as ribulose-1,5-biphosphate carboxylase oxygenase (Rubisco), reducing *A*. As the rate of CO_2 assimilation decreases, Ci values tend to increase. Consequently, the transpiration rate increases, reducing Gs and reducing WUE.

5.5.2 Changes in Growth Caused by the Excess of Metals

Metal toxicity in general is also related to reduced leaf size, plant height growth and biomass production, and visual symptoms of chlorosis, brown spots and necrotic areas (Cambrollé et al., 2011; Yang et al., 2011). Plant biomass production is severely affected under conditions of an excess of these heavy metals, reducing both root and shoot dry mass. The inhibition of cell elongation and division, which is caused by the excess of these heavy metals, also accounts for the significant reduction in biomass production. On the other hand, some plant species possess tolerance mechanisms, which include a reduction of metal translocation to the aerial part because the accumulation of the element normally occurs in larger quantities in the root apoplast of the plants.

Under heavy metal stress conditions, there is an increase in abscisic acid (ABA) concentration in plants. Some results indicate that there is a strong correlation between high levels of ABA and a reduction in plant stress. ABA is a phytohormone that plays an important role in heavy metal tolerance, as it is directly involved in the expression of genes encoding enzymes and proteins involved in stress defense (Yang et al., 2011). In addition, with the increase in ABA concentration, the signaling of other phytohormones, such as jasmonic acid, may occur. This plant hormone is also key in plant defense and is normally synthesized in response to biotic and abiotic stresses, such as elemental toxicity. Jasmonic acid is believed to mediate heavy metal-induced gene expression.

It is a known fact that increasing ABA content in the plant promotes stomatal closure, reducing the process of leaf transpiration. There is some evidence that this process results in the restriction of metal transport from the root to the shoot (Bücker-Neto et al., 2017). Nevertheless, the interaction of phytohormones in the processes of plant responses to heavy metal toxicity is not fully understood, but evidence shows that ABA and jasmonic acid are involved in heavy metal detoxification (Wang et al., 2021; Li et al., 2022).

5.5.3 Enzymatic and Non-enzymatic Antioxidant System Under Metal Stress

High levels of metals in plant tissues can stimulate the formation of reactive oxygen species (ROS), such as superoxide anion ($O_2^{\bullet-}$), hydrogen peroxide (H_2O_2), and the hydroxyl radical ($OH^{\bullet}$), which are agents that can cause major damages to plant cell membranes.

This condition can create oxidative stress in the plant with reductions in root and shoot growth. However, the plant can use mechanisms to combat this eventual stress and prevent the negative effects caused by ROS by developing a complex antioxidant defense system, including enzymatic antioxidants, where the enzymes superoxide dismutase (SOD), catalase (CAT) and ascorbate peroxidase (APX) can be highlighted (Fatima & Ahmad, 2005; Mittler, 2002). SOD is crucial for the removal of $O_2^{-\bullet}$ in the compartments where the radicals are formed. The breakdown of $O_2^{-\bullet}$ is always accompanied by the production of H_2O_2 (Fig. 5.3), which acts as both an oxidant and a reductant (Schoonen et al., 2010). H_2O_2 is less damaging and reactive than $O_2^{-\bullet}$when accumulated in plant tissues and can be eliminated by catalases and peroxidases (Mittler, 2002; Smirnoff & Arnaud, 2019).

The Cu tolerance mechanisms in ground cover plants grown in vineyards, among other factors, for example, consist of increasing the activity of the antioxidant enzymes SOD and POD to contrast the excessive production of ROS (Silva et al., 2021). Therefore, these enzymes may be physiological parameters that plants use to

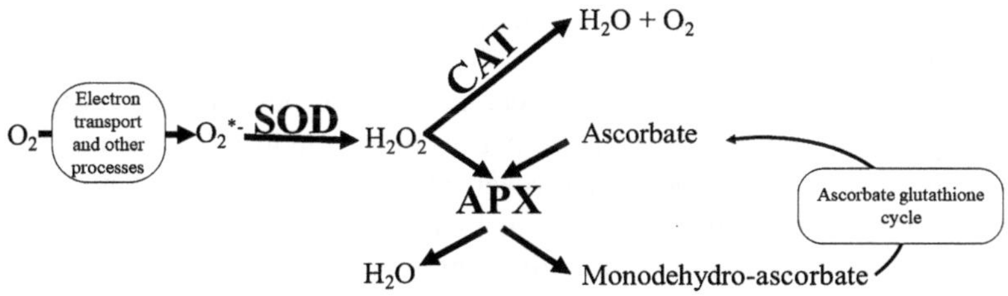

Fig. 5.3 Representation of the defense mechanisms using the antioxidant enzymes SOD, CAT and APX against ROS

cope with oxidative stress resulting from heavy metal toxicity. On the other hand, the increase in heavy metal availability may promote a reduction in the activity of antioxidant enzymes, causing damage to cellular components. SOD activity was lower in peach leaves growing in soils supplemented with Cu (Hammerschmitt et al., 2020), maximizing the toxic effects of this metal on plants.

In addition to the enzymatic antioxidant defense system, the non-enzymatic antioxidant ones are also fundamental to cells. These latter (among others, for example, ascorbic acid (AsA), glutathione (GSH), α-tocopherol, and carotenoids. AsA and GSH) are found in high concentrations in chloroplasts and other cellular compartments. AsA is associated with the removal of H_2O_2 via ascorbate peroxidase, as well as reacting with superoxide radicals and hydroxyl radicals. It is also involved in the regeneration of another non-enzymatic antioxidant, α-tocopherol.

The non-protein thiol groups, among them GSH, are known to play a central role in metal response mechanisms in plants. GSH is a sulfur-containing tripeptide and has been considered to be a very important antioxidant involved in cellular defense against toxic agents, e.g., under conditions of metal toxicity. In response to the stresses, plants increase the activity of GSH biosynthetic enzymes and, consequently, GSH concentrations. In addition, GSH is a precursor in the synthesis of phytochelatins and maintains the cellular redox state. A high level of thiol groups may enable metabolites to function in the detoxification of ROS and free radicals.

5.6 Adaptation Strategies of Species to Excess Heavy Metals

Regarding the defense strategies used by the plants to tolerate the excess of metals, there are a number of factors to consider, such as (i) the mechanisms of absorption, transport, and accumulation of the elements in the plant tissues, (ii) the primary mechanisms of toxicity at the molecular, cellular, and subcellular level, (iii) secondary mechanisms of interference with the functional processes of the plants, (iv) the homeostatic response mechanisms that, in some cases, lead to tolerance mechanisms against the heavy metal (Barceló & Poschenrieder, 1990). These mechanisms protect the cells by avoiding the acquisition, translocation and accumulation of excessive quantities of the free ions, thus increasing the tolerance of the plants to phytotoxicity.

Some plants can accumulate heavy metals inside or outside their tissues due to their ability to adapt to the chemical properties of the environment, acting like passive receptors for these elements. However, plants can also exert control over the uptake and translocation of toxic elements by means of several specific physiological reactions. With this, plants exhibit different tolerance mechanisms in response to excess heavy metals, including different strategies such as: (i) selective exclusion of metals; (ii) change in the uptake capacity, where the hindrance in the uptake may occur as a result of altered membrane permeability; (iii) increased exudation of chelating substances limiting the phytodisposability of the metal to the plant; (iv) restriction of metal transport; (v) retention of the metal at the root level and, or, at the sieve elements of the xylem; (vi) immobilization in the cell wall; (vii) associations with mycorrhizae; (viii) biochemical mechanisms—changes in the forms of compartmentalization of heavy metal, immobilization of metal in the vacuole, changes in cellular metabolism and intercellular production of binding compounds with the formation of toxic metal sequestering and inactivating compounds and; (ix) tolerance of the enzymatic system to metals (Shaw, 1989). Regardless of the mechanisms involved, resilient plants, which are endemic to polluted soils, have been found to activate significant ecophysiological adaptation, manifesting resistance to soil contamination with heavy metals (Raskin et al., 1994).

5.6.1 Mechanisms to Prevent Heavy Metal Uptake and Translocation

The mechanisms that prevent/reduce heavy metal uptake by plants are related to the increase in dissolved organic compounds (DOC) concentration and pH values, through the exudation of low molecular weight organic compounds by roots in the rhizosphere region (Chaignon et al., 2009; De Conti et al., 2018); alteration in plasmatic membrane permeability; binding capacity between the element and the cell wall and the storage of excess metal in the cell vacuole (Kabata-Pendias, 2011; Montiel-Rozas et al., 2016).

One of the plant strategies to prevent the uptake of toxic elements involves exclusion mechanisms, in which plants prevent the entry of elements into the cytosol through the exudation of low molecular weight organic compounds, which protect the internal enzymatic activity from the destructive effects of heavy metals (Montiel-Rozas et al., 2016). Among the main chelating substances exuded by roots are low molecular weight organic acids, such as phenolic compounds, organic acids (such as malate, citrate and oxalate), amino acids and sugars, whose exudation is increased under conditions of nutritional disturbances due to toxicity or deficiency of chemical elements (Meier et al., 2012; Montiel-Rozas et al., 2016; Zafari et al., 2016).

In contaminated soils, where the mobility of dominant metals is probably the least determinant factor for the rate of uptake, organic molecules having metal-chelating properties can play an important role in reducing the root acquisition of these elements (Shaw, 1989). Moreover, the qualitative and quantitative pattern of the root exudation process is strongly specific to each plant species (Dresler et al., 2014). Concerning the chemical properties of these root exudates, low molecular weight organic acids are characterized by their high ability to form metal cation complexes (Montiel-Rozas et al., 2016; Zafari et al., 2016). The complexing ability of these ligands is attributed to the presence of specific functional groups (mainly carboxyl (-COOH) and hydroxyl (OH)), which promote interactions between heavy metals and complexing agents. The increase in the proportion of complexed chemical species, to the detriment of the free form in the soil solution, reflects in the limitation of its bioavailability (Dresler et al., 2014; Brunetto et al., 2016).

In a study conducted by De Conti et al. (2018), from the ion speciation of the solution of soil contaminated with Cu (40 and 80 mg/kg) grown with black oats for 53 days, there was observed an increase in the proportion of Cu chemical species complexed and reduction of Cu^{+2} in the rhizosphere, when compared to non-rhizospheric soil. The study of these organic compounds that make up the rhizosphere of numerous plant species and the connection between these compounds and the reduction in the uptake of heavy metals by plants is of great interest and may represent a step forward in phytoavailability.

The excess of heavy metals acquired by plants can be immobilized in the cell wall or stored in the vacuole, avoiding translocation to the aerial part of the plants. The cell wall is composed of polysaccharides, such as cellulose, hemicellulose, and pectin, which contain many active groups (such as hydroxyl, amine, and carboxyl) (Zhou et al., 2017). These active groups can participate in a number of reactions, such as ion exchange, crystallization, precipitation, adsorption, and complexation, thus altering the state of heavy metals within the plant and promoting a limitation in the amounts of free ions in root tissues, with a consequent reduction in their translocation to aerial part (Wang et al., 2015; Leclercq-Dransart et al., 2019). In addition to these substances, pectins and histidines have also been reported to participate in the immobilization of heavy metals in the cell wall, reducing toxicity to plants (Leita et al., 1996). The sequestration of heavy metals in the vacuole is one way to remove excess free metal ions from the cytosol favoring cellular metabolism (Assunção et al., 2003; Sheoran et al., 2010; Thiesen et al., 2023). Vacuoles are cellular organelles with low metabolic activity (Denton, 2007). Thus, the compartmentation of these complex metal ions in vacuoles is a step of the tolerance mechanism of hyperaccumulating plants.

The retention of heavy metals to negatively charged groups in the cell wall and vacuole of the roots has a negative impact on the concentration levels of the metal present in the shoot, as described by Thiesen et al. (2023), evaluating the subcellular distribution of Mn (300–900 μM) in grasses (*P. notatum, P. plicatulum, A. strigosa, L. multiflorum*) cohabiting vineyards in the Campanha Gaúcha region; in *B. napus* plants grown in nutrient solution with increasing concentrations of Cu and Cd (0, 50 and 200 μM) (Mwamba et al., 2016); in *B. napus* cultivars grown with a high concentration of Cu and Cr (200 μM), (Li et al., 2019); in castor oil plant (*Ricinus communis* L.) grown in soil with 380 mg Zn/kg and doses of Cd (0; 0.5; 5 and 25 mg/kg) (He et al., 2020); and in wheat (*Triticum aestivum* L.) grown in nutrient solution with doses of Zn (0, 1, and 10 μM) and N (0.5; 7.5; and 15 mM) (Nie et al., 2017).

Metal accumulation is regulated by physiological, biochemical, and genetic processes in the plant (Baker, 1987). The accumulation of metals in all plant organs does not have the same meaning. Usually, the root is the priority organ for entry and accumulation (Barceló & Poschenrieder, 1990). For example, in tomato plants, the highest concentrations of Cu, Ni, Cr, Mn, and Pb were reported in order of root > leaf > stem > fruit (Topal et al., 2022).

5.6.2 Metabolic Mechanisms

Plants can exhibit different tolerance mechanisms in response to excess heavy metals, including the production of intracellular compounds that can form bonds with the metals, such as the formation of peptides rich in thiol groups (phytochelatins and metallothioneins), chelation by organic acids and amino acids, and compartmentation of metals into subcellular structures (Santos et al., 2006; Emamverdian et al., 2015; Hasan et al., 2017).

The chelating organic compounds synthesized by the plants can contribute to metal detoxification by reducing the concentration of free metal in the cytosol, thus limiting its reactivity and solubility (Ding et al., 2015). In plants, the main classes of known heavy metal chelators include phytochelatins (PCs), metallothioneins (MTs), organic acids and amino acids (Santos et al., 2006; Emamverdian et al., 2015). Phytochelatins can be a detoxifying agent for some heavy metals, thanks to their ability to bind these elements and, thus, avoid their reaction with the free sulfur groups of vital proteins and enzymes (Grant et al., 1998). Phytochelatins are formed by three amino acids: glutamate (Glu), cysteine (Cis) and glycine (Gli), with Glu and Cis linked via a γ-carboxylamide (Santos et al., 2006). According to Inouhe (2005), phytochelatins have as a general structure (γ-glutamyl-cysteine) n-glycine ($n = 2–11$), as well as variants with repeated γ-glutamyl-cyteinyl units, which are formed in plants and yeast. Phytochelatins are capable of binding to various metals by means of sulfhydryl and carboxyl residues. Phytochelatins are found complexed with Cd to form low and high molecular weight complexes (LMW and HMW, respectively). Moreover, the LMW complexes are formed in the cytosol and subsequently transported to the vacuole when Cd^{2+} and S^{2-} are incorporated to produce HMW complex, which represents the main form of Cd storage (Santos et al., 2006).

Plants, when exposed to toxic levels of Cu for long periods, tend to synthesize compounds rich in thiol groups, which could complex the metal preventing its toxic action (Ding et al., 1994). It has been shown that an excess of Cu induces the accumulation of phytochelatin that, then, can remove Cu from its free fraction via the metal complexation (Lwalaba et al., 2020). Phytochelatins may possess prominent functions in Cu and Cd tolerance. Different studies have shown that the concentration of phytochelatins increases in response to the uptake of these elements, both in non-tolerant and tolerant plants, but intolerant plants, the synthesis of phytochelatins seems to be faster and in greater quantities (Barceló & Poschenrieder, 1990; Salt & Rauser, 1995).

It is also well demonstrated that in plant tissues, there are several transporters involved in metal uptake, translocation and vacuolar compartmentation. At the plasma membrane, there are the heavy metal transport ATPases (HMAs), the ZIP family (ZRT/IRT-like proteins), the cation diffusion facilitator (CDF) family, and Nramps (natural resistance-associated macrophage protein). Interestingly, some transporters of the HMA family are located in the vacuole membrane to transport Cd and Zn into the vacuoles of root cells, and thus prevent their transport to the shoot. This is the case in *Arabidopsis thaliana* AtHMA3 and rice OsHMA3 which can compartmentalize in the vacuoles of some metals or only Cd, respectively.

Like HMA transporters, the ZIP (iron-regulated transporter-like proteins) ones located at the plasma membrane play a crucial role in regulating metal homeostasis in plant cells. They transport metals from the extracellular space or from organelles to the cytoplasm (Ajeesh Krishna et al., 2020). Recently, Liu et al. (2019) reported that OsZIP1 prevents the excessive accumulation of Zn, Cu, and Cd in plants. However, heavy metal transport occurs not only by HMA and ZIP transporters but also thanks to Nramps (natural resistance-associated macrophage protein) and yellow stripe-like transporters (YSL). In *A. thaliana*, AtNramp3 and AtNramp4 mobilize vacuolar Fe to enhance early plant development under Fe deficiency (Lanquar et al., 2005). In addition, the tonoplast AtNramp3 regulates metal accumulation and alters the expression of the primary root transporter for Fe uptake (IRT1) and by the action of ferric reductase (FRO2), mobilizing vacuolar stores of metal to the cytosol (Thomine et al., 2000, 2003). Regarding YSL transporters, the Ni/Fe transporter TcYSL3 from the hyperaccumulator plant *T. caerulescens* and the peanut Cu transporter AhYSL3.1 regulate metal homeostasis in the vascular tissues of hyperaccumulator plants through the transport of nicotianamine (NA)-metal chelates (Gendre et al., 2007; Dai et al., 2018).

Increased concentrations of heavy metals in cells result in the production of ROS, which causes oxidative stress, peroxidation of membrane lipids and result in reduced membrane selectivity and integrity, and consequently, damage to cellular constituents, such as the photosynthetic apparatus, proteins and nucleic acids (Mittler et al., 2004; Yruela, 2005). Thus, plants can activate an antioxidant defense system to neutralize ROS, which can be enzymatic or non-enzymatic. The main antioxidant enzymes are catalases (CAT), superoxide dismutases (SODs), ascorbate peroxidases (APX), dehydroascorbate reductases (DHAR), guaiacol peroxidases (GPOD), glutathione reductases (GR), glutathione peroxidases (GPX), glutathione-S-transferases (GST), and monodehydroascorbate reductases (MDHAR) (Mittler et al., 2004; Gill & Tuteja, 2010; Sharma et al., 2012). In addition, plants can synthesize non-enzymatic antioxidants, such as ascorbate (AA), glutathione (GSH), carotenoids, alkaloids, tocopherols, proline, and phenolic compounds (such as flavonoids, tannins, and lignin), to aid the free radical detoxification (Gill & Tuteja, 2010; Sharma et al., 2012). The balance between antioxidant defense system synthesis and ROS formation is important to maintain the balance in cells and to avoid oxidative stress in plants due to excess heavy metals. The synthesis of antioxidant enzymes in response to the Cu and Zn excess in grapevine (*Vitis vinifera*) plants to control ROS formation was observed in studies by Tiecher et al. (2017, 2018) and also observed in oat plants (*Avena strigosa* Schreb.) (Tiecher et al., 2016a, b). However, at high concentrations of Cu and Zn, the synthesis of antioxidant enzymes is not sufficient to balance and control the formation of ROS, consequently causing damage to cells and phytotoxicity symptoms.

5.7 Conclusion and Future Perspectives

Agricultural practices can be one of the causes of the increased levels of heavy metals in soils, and, when these contents are excessive, significant damage is observed at the root level, with changes in morphology and anatomy. This phenomenon limits the volume of soil explored by the roots, the acquisition of water and nutrients, and causes physiological and biochemical damage, which impairs crop development and yield capacity. However, some plant species activate specific strategies to shrink the acquisition as well as the allocation of the metals to the shoot. Considering the levels of heavy metals nowadays found in cultivated soils, a limitation in the application at the field scale of the Cu-bearing products (pesticides or waste) is desirable for more sustainable soil management. In addition, a better knowledge of the Cu toxicity phenomena in plants as well as the metal thresholds in soil inducing the toxicity, appears to be crucial to decide when and how to intervene. In this respect, there are several approaches already proposed (ameliorants by using, for example, limestone, organic compost, etc.). However, the definition of the types to use, the doses to apply, and the frequency of these applications to the soil appear to be fundamental for an effective solution to the problem. Moreover, the breeding program aimed at obtaining crop cultivars or rootstocks more tolerant to this nutritional disorder certainly appears indispensable.

References

Ajeesh Krishna, T. P., Maharajan, T., Victor Roch, G., Ignacimuthu, S., & Antony Ceasar, S. (2020). Structure, function, regulation and phylogenetic relationship of ZIP family transporters of plants. *Frontiers in Plant Science, 11*, 662.

Amari, T., Ghnaya, T., & Abelly, C. (2017). Nickel cadmium and lead phytotoxicity and potential of halophytic plants in heavy metal extraction. *South African Journal of Botany, 111*, 99–110.

Ambrosini, V. G., Rosa, D. J., Prado, J. P. C., Borghezan, M., Melo, G. W. B., Soares, C. R. F. S., Comin, J. J., Simão, D. G., & Brunetto, G. (2015). Reduction of copper phytotoxicity by liming: A study of the root anatomy of young vines (*Vitis labrusca* L.). *Plant Physiology and Biochemistry, 96*, 270–280.

Ambrosini, V. G., Rosa, D. J., Basso, A., Borghezan, M., Pescador, R., Miotto, A., Melo, G. W. B., Soares, C. R. F. S., Comin, J. J., & Brunetto, G. (2016a). Liming as an ameliorator of copper toxicity in Black oat (*Avena strigosa* Schreb.). *Journal of Plant Nutrition, 40*, 404–416.

Ambrosini, V. G., Soriani, H. H., Rosa, D. J., Tiecher, T. L., Girotto, E., Simão, G. D., Melo, G. W. B., Zalamena, J., & Brunetto, G. (2016b). Impacto do excesso de cobre e zinco no solo sobre videiras e plantas de cobertura. In G. W. B. Melo, J. Zalamena, G. Brunetto, & C. A. Ceretta (Eds.), *Calagem, adubação e contaminação em solos cultivados com videiras* (pp. 91–110). Embrapa uva e Vinho.

Ambrosini, V. G., Rosa, D. J., de Melo, G. W. B., Zalamena, J., Cella, C., Simão, D. G., Silva, L. S., Santos, H. P., Toselli, M., Tiecher, T. L., & Brunetto, G. (2018). High copper content in vineyard soils promotes modifications in photosynthetic parameters and morphological changes in the root system of 'Red Niagara' plantlets. *Plant Physiology and Biochemistry, 128*, 89–98.

Assunção, A. G., Schat, H., & Aarts, M. G. (2003). *Thlaspi caerulescens*, an attractive model species to study heavy metal hyperaccumulation in plants. *New Phytologist, 159(2)*, 351–360.

Baker, A. J. (1987). Metal tolerance. *New Phytologist, 106*, 93–111.

Baldi, E., Miotto, A., Ceretta, C. A., Quartieri, M., Sorrenti, G., Brunetto, G., & Toselli, M. (2018). Soil-applied phosphorous is an effective tool to mitigate the toxicity of copper excess on grapevine grown in rhizobox. *Scientia. Horticulturae, 227*, 102–111.

Barceló, J. U. A. N., & Poschenrieder, C. (1990). Plant water relations as affected by heavy metal stress: a review. *Journal of Plant Nutrition, 13(1)*, 1–37.

Beygi, M., & Jalali, M. (2019). Assessment of trace elements (Cd, Cu, Ni, Zn) fractionation and bioavailability in vineyard soils from the Hamedan, Iran. *Geoderma, 337*, 1009–1020.

Bortoluzzi, E. C., Korchagin, J., Moterle, D. F., Rheinheimer, D. S., & Caner, L. (2019). Accumulation and precipitation of Cu and Zn in a centenarian vineyard. *Soil Science Society of America Journal, 83*, 492–502.

Bouazizi, H., Jouili, H., Geitmann, A., & El Ferjani, E. (2010). Structural changes of cell wall and lignifying enzymes modulations in bean roots in response to copper stress. *Biological Trace Element Research, 136*, 232–240.

Broadley, M. R., White, P. J., Hammond, J. P., Zelko, I., & Lux, A. (2007). Zinc in plants. *New Phytologist, 173*, 677–702.

Brunetto, G., Miotto, A., Ceretta, C. A., Schmitt, D. E., Heinzen, J., Moraes, M. P., Canton, L., Tiecher, T. L., Comin, J. J., & Girotto, E. (2014a). Mobility of copper and zinc fractions in fungicide amended vineyard sandy soils. *Archives of Agronomy and Soil Science, 60*, 609–624.

Brunetto, G., Schmitt, D. E., Comin, J. J., Miotto, A., Moraes, M. P., & Heinzen, J. (2014b). Frações de cobre e zinco em solos de vinhedos no Meio Oeste de Santa Catarina. *Revista Brasileira de Engenharia Agrícola e Ambiental, 18*, 805–810.

Brunetto, G., Melo, G. W. B., Terzano, R., Del Buono, D., Astolfi, S., Tomasi, N., Pii, Y., Mimmo, T., & Cesco, S. (2016). Copper accumulation in vineyard soils: Rhizosphere processes and agronomic practices to limit its toxicity. *Chemosphere, 262*, 293–307.

Brunetto, G., Ferreira, P. A. A., Melo, G. W., Ceretta, C. A., & Toselli, M. (2017). Heavy metals in vineyards and orchard soils. *Revista Brasileira de Fruticultura, 39*, 263–274.

Brunetto, G., Comin, J. J., Miotto, A., Moraes, M. P. D., Sete, P. B., Schmitt, D. E., Gatiboni, L. C., Melo, G. W. B., & Morais, G. P. (2018a). Copper and zinc accumulation, fractionation and migration in vineyard soils from Santa Catarina State, Brazil. *Bragantia, 77*, 141–151.

Brunetto, G., Benedet, L., Ambrosini, V. G., Comin, J. J., Melo, G. W. B., Santos, M. A. S., Lourenzi, C. R., Loss, A., Belli Filho, P., Schmitt, D. E., & Couto, R. R. (2018b). Copper and zinc fractions in the profile of an Inceptisol cultivated with apple in southern Brazil. *Bragantia, 77*, 333–347.

Bücker-Neto, L., Paiva, A. L. S., Machado, R. D., Arenhart, R. A., & Margis-Pinheiro, M. (2017). Interactions between plant hormones and heavy metals responses. *Genetics and Molecular Biology, 40*, 373–386.

Cambrollé, J., Mancilla-Leytón, J. M., Muñoz-Vallés, S., Luque, T., & Figueroa, M. E. (2011). Zinc tolerance and accumulation in the salt-marshshrub *Halimione portulacoides*. *Chemosphere, 86*, 867–874.

Cambrollé, J., García, J. L., Figueroa, M. E., & Cantos, M. (2015). Evaluating wild grapevine tolerance to copper toxicity. *Chemosphere, 120*, 171–178.

Campillo-Cora, C., Fernández-Calviño, D., Pérez-Rodríguez, P., Fernández-Sanjurjo, M. J., Núñez-Delgado, A., Álvarez-Rodríguez, E., et al. (2019). Copper and zinc in rhizospheric soil of wild plants growing in long-term acid vineyard soils. Insights on availability and metal remediation. *Science of the Total Environment, 672*, 389–399.

Casali, C. A., Moterle, D. F., Rheinheimer, D. S., Brunetto, G., Corcini, A. L. M., Kaminski, J., & Melo, G. W. B. (2008). Formas e dessorção de cobre em solos cultivados com videira na Serra Gaúcha do Rio Grande do Sul. *Revista Brasileira de Ciência do Solo, 32*, 1479–1487.

Castro, C., Carvalho, A., Pavia, I., Bacelar, E., & Lima-Brito, J. (2021). Grapevine varieties with differential tolerance to Zinc analysed by morpho-histological and cytogenetic approaches. *Scientia Horticulturae, 288*, 110386.

Cesco, S., Pii, Y., Borruso, L., Orzes, G., Lugli, P., Mazzetto, F., ... & Mimmo, T. (2021). A smart and sustainable future for viticulture is rooted in soil: How to face cu toxicity. *Applied Sciences, 11(3)*, 907.

Chaignon, V., Quesnoit, M., & Hinsinger, P. (2009). Copper availability and bioavailability are controlled by rhizosphere pH in rape grown in an acidic Cu-contaminated soil. *Environmental Pollution, 157(12)*, 3363–3369.

Couto, R. R., Benedet, L., Comin, J. J., Belli Filho, P., Martins, S. R., Gatiboni, L. C., Radetski, M., De Valois, C. M., Ambrosini, V. G., & Brunetto, G. (2015). Accumulation of copper and zinc fractions in vineyard soil in the mid-western region of Santa Catarina, Brazil. *Environmental Earth Sciences, 73*, 6379–6386.

Couto, R. R., Faversani, J., Ceretta, C. A., Ferreira, P. A. A., Marchezan, C., Facco, D. B., Garlet, L. P., Silva, J. S., Comin, J. J., Bizzi, C. A., Flores, E. M. M., & Brunetto, G. (2018). Health risk assessment and soil and plant heavy metal and bromine contents in field plots after ten years of organic and mineral fertilization. *Ecotoxicology and Environmental Safety, 153*, 142–150.

Dai, J., Wang, N., Xiong, H., Qiu, W., Nakanishi, H., Kobayashi, T., ... & Zuo, Y. (2018). The yellow stripe-like (YSL) gene functions in internal copper transport in peanut. *Genes, 9(12)*, 635.

De Conti, L., Ceretta, C. A., Tiecher, T. L., Silva, L. O. S., Tassinari, A., Somavilla, L. M., Mimmo, T., Cesco, S., & Brunetto, G. (2018). Growth and chemical changes in the rhizosphere of black oat (*Avena strigosa*) grown in soils contaminated with copper. *Ecotoxicology and Environmental Safety, 163*, 19–27.

De Conti, L., Ceretta, C. A., Melo, G. W., Tiecher, T. L., Silva, L. O. S., Garlet, L. P., Mimmo, T., Cesco, S., & Brunetto, G. (2019). Intercropping of young grapevines with native grasses for phytoremediation of Cu-contaminated soils. *Chemosphere, 216*, 147–156.

De Conti, L. et al. (2020). Iron fertilization to enhance tolerance mechanisms to copper toxicity of ryegrass plants used as cover crop in vineyards. *Chemosphere, 243*, 125298.

De Conti, L., Marques, A. C. R., Ceretta, C. A., Tarouco, C. P., Nicoloso, F. T., Ferreira, P. A. A., Tiecher, T. L., Tassinari, A., Bicalho da Silva, I. C., & Brunetto, G. (2021). Tolerance and phytoremediation potential of grass species native to South American grasslands to copper-contaminated soils. *International Journal of Phytoremediation, 23*(7), 726–735.

Denton, B. (2007). Advances in phytoremediation of heavy metals using plant growth promoting bacteria and fungi. MMG 445 *Basic Biotechnol., 3*, 1–5.

Ding, X., Jiang, J., Wang, Y., Wang, W., & Ru, B. (1994). Bioconcentration of cadmium in water hyacinth (*Eichhornia crassipes*) in relation to thiol group content. *Environmental Pollution, 84*(1), 93–96.

Ding, X., Hua, Y., Chen, Y., Zhang, C., & Kong, X. (2015). Heavy metal complexation of thiol-containing peptides from soy glycinin hydrolysates. *International Journal of Molecular Sciences, 16(4)*, 8040–8058.

Dresler, S., Hanaka, A., Bednarek, W., & Maksymiec, W. (2014). Accumulation of lowmolecular-weight organic acids in roots and leaf segments of *Zea mays* plants treated with cádmium and copper. *Acta Physiologiae Plantarum, 36*, 1565–1575.

Emamverdian, A., Ding, Y., Mokhberdoran, F., & Xie, Y. (2015). Heavy metal stress and some mechanisms of plant defense response. *Scientific World Journal, 2015*, 756120.

Fatima, R. A., & Ahmad, M. (2005). Certain antioxidant enzymes of *Allium cepa* as biomarkers for the detection of toxic heavy metals in wastewater. *Science of the Total Environment, 346*, 256–273.

Feil, S. B., Pii, Y., Valentinuzzi, F., Tiziani, R., Mimmo, T., & Cesco, S. (2020). Copper toxicity affects phosphorus uptake mechanisms at molecular and physiological levels in Cucumis sativus plants. *Plant Physiology and Biochemistry, 157*, 138–147.

Ferreira, P. A. A., Tiecher, T., Tiecher, T. L., de Melo Rangel, W., Soares, C. R. F. S., Deuner, S., Tarouco, C. P., Giachini, A. J., Nicoloso, F. T., Brunetto, G., Coronas, M. V., & Ceretta, C. A. (2018). Effects of *Rhizophagus clarus* and P availability in the tolerance and physiological response of *Mucuna cinereum* to copper. *Plant Physiology and Biochemistry, 122*, 46–56.

Gendre, D., Czernic, P., Conéjéro, G., Pianelli, K., Briat, J. F., Lebrun, M., & Mari, S. (2007). TcYSL3, a member of the YSL gene family from the hyper-accumulator *Thlaspi caerulescens*, encodes a nicotianamine-Ni/Fe transporter. *The Plant Journal, 49(1)*, 1–15.

Gill, S. S., & Tuteja, N. (2010). Reactive oxygen species and antioxidant machinery in abiotic stress tolerance in crop plants. *Plant Physiology and Biochemistry, 48*, 909–930.

Girotto, E., Ceretta, C. A., Brunetto, G., Rheinheimer, D. S., Silva, L. S., Lorensini, F., Lourenzi, C. R., Vieira, R. C. B., & Schmatz, R. (2010). Acúmulo e formas de cobre e zinco no solo após aplicações sucessivas de dejeto líquido de suínos. *Revista Brasileira de Ciência do Solo, 34*, 955–965.

Guimarães, P. R., Ambrosini, V. G., Miotto, A., Ceretta, C. A., Simão, D. G., & Brunetto, G. (2016). Black oat (*Avena strigosa* Schreb.) growth and root anatomical changes in sandy soil with different copper and phosphorus concentrations. *Water, Air, & Soil Pollution, 227*, 192.

Grant, C. A., Buckley, W. T., Bailey, L. D., & Selles, F. (1998). Cadmium accumulation in crops. *Canadian Journal of Plant Science, 78*(1), 1–17.

Hammami, H., Parsa, M., Bayat, H., & Aminifard, M. H. (2022). The behavior of heavy metals in relation to their influence on the common bean (*Phaseolus vulgaris*) symbiosis. *Environmental and Experimental Botany, 193*, 104670.

Hammerschmitt, R. K., Tiecher, T. L., Facco, D. B., Silva, L. O. S., Schwalbert, R., Drescher, G. L., Trentin, E., Somavilla, L. M., Kulmann, M. S. S., Silva, I. C. B., Tarouco, C. P., Nicoloso, F. T., Tiecher, T., Mayer, N. A., Krug, A. V., & Brunetto, G. (2020). Copper and zinc distribution and toxicity in 'Jade'/'Genovesa' young peach tree. *Scientia Horticulturae, 259*, 108763.

Hasan, M. K., Cheng, Y., Kanwar, M. K., Chu, X.-Y., Ahammed, G. J., & Qi, Z.-Y. (2017). Responses of plant proteins to heavy metal stress—A review. *Frontiers in Plant Science, 8*, 1492.

He, C., Zhao, Y., Wang, F., Oh, K., Zhao, Z., Wu, C., Zhang, X., Chen, X., & Liu, X. (2020). Phytoremediation of soil heavy metals (Cd and Zn) by castor seedlings: Tolerance, accumulation and subcellular distribution. *Chemosphere, 252*, 126471.

Hummes, A. P., Bortoluzzi, E. C., Tonini, V., Silva, L. P., & Petry, C. (2019). Transfer of copper and zinc from soil to grapevine-derived products in young and centenarian vineyards. *Water, Air & Soil Pollution, 230*, 150–161.

Inouhe, M. (2005). Phytochelatins. *Brazilian Journal of Plant Physiology, 17*, 65–78.

Juang, K. W., Lee, Y. I., Lai, H. Y., Wang, C. H., & Chen, B. C. (2012). Copper accumulation, translocation, and toxic effects in grapevine cuttings. *Environmental Science and Pollution Research International, 19*, 1315–1322.

Kabata-Pendias, A. (2011). *Trace elements in soils and plants* (4ª ed., p. 534). CRC Press.

Kopittke, P. M., Asher, C. J., Blamey, F. P. C., & Menzies, N. W. (2009). Toxic effects of Cu^{2+} on growth, nutrition, root morphology, and distribution of Cu in roots of Sabi grass. *Science of the Total Environment, 407*, 4616–4621.

Kopittke, P. M., Blamey, F. P. C., Kopittke, R. A., Asher, C. J., & Menzies, N. W. (2010). Tolerance of seven perennial grasses to high nickel in sand culture. *Environmental Chemistry, 7(3)*, 279–286.

Korchagin, J., Moterle, D. F., Escosteguy, P. A. V., & Bortoluzzi, E. C. (2020). Distribution of copper and zinc fractions in a Regosol profile under centenary vineyard. *Environmental Earth Sciences, 79*, 1–13.

Lanquar, V., Lelièvre, F., Bolte, S., Hamès, C., Alcon, C., Neumann, D., ... & Thomine, S. (2005). Mobilization of vacuolar iron by AtNRAMP3 and AtNRAMP4 is essential for seed germination on low iron. *The EMBO Journal, 24(23),* 4041–4051.

Lecerf, A., Cébron, A., Gilbert, F., Danger, M., Roussel, H., & Maunoury-Danger, F. (2021). Using plant litter decomposition as an indicator of ecosystem response to soil contamination. *Ecological Indicators, 125*, 107554.

Leclercq-Dransart, J., Demuynck, S., Waterlot, C., Bidar, G., Sahmer, K., Pernin, C., Deram, A., Lepretre, A., & Douay, F. (2019). Distribution of metals and cell wall compounds in leaf parts of three tree species suitable for the phytomanagement of heavy metal-contaminated soils. *Water, Air & Soil Pollution, 230*, 237.

Leita, L., De Nobili, M., Cesco, S., & Mondini, C. (1996). Analysis of intercellular cadmium forms in roots and leaves of bush bean. *Journal of Plant Nutrition, 19(3-4),* 527–533.

Li, L., Long, M., Islam, F., Farooq, M. A., Wang, J., Mwamba, T. M., Shou, J., & Zhou, W. (2019). Synergistic effects of chromium and copper on photosynthetic inhibition, subcellular distribution, and related gene expression in *Brassica napus* cultivars. *Environmental Science and Pollution Research, 26*, 11827–11845.

Li, Y., Zhang, S., Bao, Q., Chu, Y., Sun, H., & Huang, Y. (2022). Jasmonic acid alleviates cadmium toxicity through regulating the antioxidant response and enhancing the chelation of cadmium in rice (Oryza sativa L.). *Environmental Pollution, 304*, 119178.

Liu, X. S., Feng, S. J., Zhang, B. Q., Wang, M. Q., Cao, H. W., Rono, J. K., ... & Yang, Z. M. (2019). OsZIP1 functions as a metal efflux transporter limiting excess zinc, copper and cadmium accumulation in rice. *BMC Plant Biology, 19*, 1–16.

Lwalaba, J. L. W., Louis, L. T., Zvobgo, G., Richmond, M. E. A., Fu, L., Naz, S., ... & Zhang, G. (2020). Physiological and molecular mechanisms of cobalt and copper interaction in causing phyto-toxicity to two barley genotypes differing in Co tolerance. *Ecotoxicology and Environmental Safety, 187*, 109866.

Lynch, J., Marschner, P., & Rengel, Z. (2012). Effect of internal and external factors on root growth and development. In P. Marschner (Ed.), *Marschner's mineral nutrition of higher plants* (pp. 331–346). Academic Press.

Marastoni, L., Sandri, M., Pii, Y., Valentinuzzi, F., Brunetto, G., Cesco, S., & Mimmo, T. (2019). Synergism and antagonisms between nutrients induced by copper toxicity in grapevine rootstocks: Monocropping vs. intercropping. *Chemosphere, 214*, 563–578.

Marschner, H. (2012). *Marschner's mineral nutrition of higher plants* (3ª ed.). Academic Press. 651p.

Meier, S., Alvear, M., Borie, F., Aguilera, P., Ginocchio, R., & Cornejo, P. (2012). Influence of copper on root exudate patterns in some metallophytes and agricultural plants. *Ecotoxicology and Environmental Safety, 75*, 8–15.

Michalak, A. (2006). Phenolic compounds and their antioxidant activity in plants growing under heavy metal stress. *Polish Journal of Environmental Studies, 15*, 523–530.

Miotto, A., Ceretta, C. A., Girotto, E., Trentin, G., Kaminski, J., De Conti, L., Toselli, M., Baldi, E., & Brunetto, G. (2017). Copper accumulation and availability in sandy, acid, vineyard soils. *Communications in Soil Science and Plant Analysis, 48*, 1–7.

Mirlean, N., Roisenberg, A., & Chies, J. O. (2007). Metal contamination of vineyard soils in wet subtropics (southern Brazil). *Environmental Pollution, 149*, 10–17.

Mittler, R. (2002). Oxidative stress, antioxidants and stress tolerance. *Trends in Plant Science, 7*(9), 405–410.

Mittler, R., Vanderauwera, S., Gollery, M., & Van Breusegem, F. (2004). Reactive oxygen gene network of plants. *Trends Plant Science, 9*(10), 490–498.

Montiel-Rozas, M. M., Madejon, E., & Madejón, P. (2016). Effect of heavy metals and organic matter on root exudates (low molecular weight organic acids) of herbaceous species: An assessment in sand and soil conditions under diferent levels of contamination. *Environmental Pollution, 216*, 273–281.

Morsch, L., Somavilla, L. M., Trentin, E., Silva, K. R., Oliveira, J. M. S., Brunetto, G., & Simão, D. G. (2022). Root system structure as a criterion for the selection of grapevine genotypes

that are tolerant to excess copper and the ability of phosphorus to mitigate toxicity. *Plant Physiology and Biochemistry, 171*, 147–156.
Mwamba, T. M., Li, L., Gill, R. A., Islam, F., Nawaz, A., Ali, B., Farooqa, M. A., Lwalabaa, J. L., & Zhou, W. (2016). Differential subcellular distribution and chemical forms of cadmium and copper in Brassica napus. *Ecotoxicology and Environmental Safety, 134*, 239–249.
Nie, Z., Zhao, P., Wang, J., Li, J., & Liu, H. (2017). Absorption kinetics and subcellular fractionation of zinc in winter wheat in response to nitrogen supply. *Frontiers in Plant Science, 18*(8), 1435–1445.
Rabêlo, F. H. S., & Borgo, L. (2016). Changes caused by heavy metals in micronutrient content and antioxidant system of forage grasses used for phytoremediation: an overview. *Ciência Rural, 46,* 1368–1375.
Raskin, I., Kumar, P. N., Dushenkov, S., & Salt, D. E. (1994). Bioconcentration of heavy metals by plants. *Current Opinion in Biotechnology, 5(3),* 285–290.
Salt, D. E., & Rauser, W. E. (1995). MgATP-dependent transport of phytochelatins across the tonoplast of oat roots. *Plant Physiology, 107(4)*, 1293–1301.
Santos, F. S., Hernández-Allica, J., Becerril, J. M., Amaral-Sobrinho, N., Mazur, N., & Garbisu, C. (2006). Chelate-induced phytoextraction of metal polluted soils with *Brachiaria decumbens. Chemosphere, 65(1),* 43–50.
Schoonen, M. A., Harrington, A. D., Laffers, R., & Strongin, D. R. (2010). Role of hydrogen peroxide and hydroxyl radical in pyrite oxidation by molecular oxygen. *Geochimica et Cosmochimica Acta, 74(17)*, 4971–4987.
Schwalbert, R., Stefanello, L. O., Schwalbert, R. A., Tarouco, C. P., Drescher, G. L., Trentin, E., Nicoloso, F. T., Tassinari, A., Silva, I. B., Brunetto, G., & Nicoloso, F. T. (2021). Soil tillage affects soybean growth and promotes heavy metal accumulation in sedes. *Ecotoxicology and Environmental Safety, 216*, 112191.
Sharma, P., Jha, A. B., Dubey, R. S., & Pessarakli, M. (2012). Reactive oxygen species, oxidative damage, and antioxidative defense mechanism in plants under stressful conditions. *Journal of Botany, 2012*, 217037.
Shaw, J. (1989). *Heavy metal tolerance in plants: evolutionary aspects*. CRC press.
Sheoran, V., Sheoran, A. S., & Poonia, P. (2010). Role of hyperaccumulators in phytoextraction of metals from contaminated mining sites: a review. *Critical Reviews in Environmental Science and Technology, 41(2)*, 168–214.
Silva, K. R., & Simão, D. G. (2022). Alterações estruturais em videiras e plantas de cobertura do solo que coabitam vinhedos e pomares contaminados com cobre e zinco, com ênfase no sistema radicular. In G. Brunetto, E. Trentin, G. W. B. Melo, & E. Girotto (Eds.), *Contaminação em solos de pomares e vinhedos: causas, efeitos e estratégias de manejo* (pp. 243–263). Sociedade Brasileira de Ciência do Solo – Núcleo Regonal Sul.
Silva, I. C. B., Somavilla, A., Marques, V., Tarouco, C. P., Schwalbert, R., Trentin, E., Quadros, F. L. F., Ferreira, P. A. A., & Brunetto, G. (2021). Potential phytoremediation of pampa biome native and invasive grass species cohabiting vineyards contaminated with Cu in Southern Brazil. *Environmental Science and Pollution Research, 1*, 1–23.
Smirnoff, N., & Arnaud, D. (2019). Hydrogen peroxide metabolism and functions in plants. *New Phytologist, 221(3)*, 1197–1214.
Somavilla, L. M., Simão, D. G., Tiecher, T. L., Hammerschimitt, R. K., Oliveira, J. M. S., Mayer, N. A., Pavanello, E. P., Trentin, E., Belles, S. W., & Brunetto, G. (2018). Structural changes in roots of peach rootstock cultivars grown in soil with high zinc content. *Scientia Horticulturae, 237*, 1–10.
Sonoda, K., Hashimoto, Y., Wang, S. L., & Ban, T. (2019). Copper and zinc in vineyard and orchard soils at millimeter vertical resolution. *Science of the Total Environment, 689*, 958–962.
Sposito, G. (1989). *The chemistry of soils*. Oxford University Press. 277p.
Thiesen, L. A., Brunetto, G., Trentin, E., da Silva, A. A. K., Tabaldi, L. A., Schwalbert, R., ... & Nicoloso, F. T. (2023). Subcellular distribution and physiological responses of native and exotic grasses from the Pampa biome subjected to excess manganese. *Chemosphere, 310,* 136801.

Thomine, S., Wang, R., Ward, J. M., Crawford, N. M., & Schroeder, J. I. (2000). Cadmium and iron transport by members of a plant metal transporter family in Arabidopsis with homology to Nramp genes. *Proceedings of the National Academy of Sciences, 97(9),* 4991–4996.

Thomine, S., Lelièvre, F., Debarbieux, E., Schroeder, J. I., & Barbier-Brygoo, H. (2003). AtNRAMP3, a multispecific vacuolar metal transporter involved in plant responses to iron deficiency. *The Plant Journal, 34(5),* 685–695.

Tiecher, T., Brunetto, G., Ceretta, C. A., Comin, J. J., Girotto, E., Miotto, A., Moraes, M. P., Benedett, L., Ferreira, P. A., Lourenzi, C. R., & Couto, R. R. (2013). Forms and accumulation of copper and zinc in a sandy Typic Hapludalf soil after long-term application of pig slurry and deep litter. *Revista Brasileira de Ciência do Solo, 37,* 812–824.

Tiecher, T. L., Tiecher, T., Ceretta, C. A., Ferreira, P. A. A., Nicoloso, F. T., Soriani, H. H., Tassinari, A., Paranhos, J. T., De Conti, L., & Brunetto, G. (2016a). Plant Physiology and Biochemistry Physiological and nutritional status of black oat (*Avena strigosa* Schreb.) grown in soil with interaction of high doses of copper and zinc. *Plant Physiology and Biochemistry, 106,* 253–263.

Tiecher, T. L., Ceretta, C. A., Ferreira, P. A. A., Lourenzi, C. R., Tiecher, T., Girotto, E., Nicoloso, F. T., Soriani, H. H., De Conti, L., Mimmo, T., Cesco, S., & Brunetto, G. (2016b). The potential of *Zea mays* L. in remediating copper and zinc contaminated soils for grapevine production. *Geoderma, 262,* 52–61.

Tiecher, T. L., Tiecher, T., Ceretta, C. A., Ferreira, P. A. A., Nicoloso, F. T., Soriani, H. H., De Conti, L., Kulmann, M. S. S., Schneider, R. O., & Brunetto, G. (2017). Tolerance and translocation of heavy metals in young grapevine (*Vitis vinifera*) grown in sandy acidic soil with interaction of high doses of copper and zinc. *Scientia Horticulturae, 222,* 203–212.

Tiecher, T. L., Soriani, H. H., Tiecher, T., Ceretta, C. A., Nicoloso, F. T., Tarouco, C. P., Clasen, B. E., De Conti, L., Tassinari, A., Melo, G. W. B., & Brunetto, G. (2018). The interaction of high copper and zinc doses in acid soil changes the physiological state and development of the root system in young grapevines (*Vitis vinifera*). *Ecotoxicology and Environmental Safety, 148,* 985–994.

Tiecher, T. L., Girotto, E., Ceretta, C. A., Tassinari, A., Almeida, H., Rupp, L. C. D., Rafatti, L., Moterle, D. F., & Almança, M. A. K. (2022). Fontes de entrada de nitrogênio, fósforo, cobre e zinco em sistemas de produção agrícola. Em: Brunetto, G., Trentin, E., de Melo, G. W. B., & Girotto, E. (Orgs.), *Contaminação em solos de pomares e vinhedos: Causas, efeitos e estratégias de manejo, 1,* 49–81. Sociedade Brasileira de Ciência do Solo - Núcleo Regional Sul.

Topal, E. I. A., Topal, M., & Öbek, E. (2022). Assessment of heavy metal accumulations and health risk potentials in tomatoes grown in the discharge area of a municipal wastewater treatment plant. *International Journal of Environmental Health Research, 32(2),* 393–405.

Trentin, E., Facco, D. B., Hammerschmitt, R. K., Ferreira, P. A. A., Morsch, L., Belles, S. W., Ricachenevsky, F. K., Nicoloso, F. T., Ceretta, C. A., Tiecher, T. L., Tarouco, C. P., Berghetti, A. L. P., Toselli, M., & Brunetto, G. (2019). Potential of vermicompost and limestone in reducing copper toxicity in young grapevines grown in Cu-contaminated vineyard soil. *Chemosphere, 226,* 421–430.

Trentin, E., Cesco, S., Pii, Y., Valentinuzzi, F., Celletti, S., Feil, S. B., Zuluaga, M. Y. A., Ferreira, P. A. A., Ricachenevsky, F. K., Stefanello, L. O., De Conti, L., Brunetto, G., & Mimmo, T. (2022). Plant species and pH dependent responses to copper toxicity. *Environmental and Experimental Botany, 196,* 104791.

Wang, J., Su, L., Yang, J., Yuan, J., Yin, A., Qiu, Q., Zhang, K., & Yang, Z. (2015). Comparisons of cadmium subcellular distribution and chemical forms between low-Cd and high-Cd accumulation genotypes of watercress (*Nasturtium officinale* L. R. Br.). *Plant and Soil, 396,* 325–337.

Wang, Y., Mostafa, S., Zeng, W., & Jin, B. (2021). Function and mechanism of jasmonic acid in plant responses to abiotic and biotic stresses. *International Journal of Molecular Sciences, 22*(16), 8568.

Xiao, Z., Pan, G., Li, X., Kuang, X., Wang, W., & Liu, W. (2020). Effects of exogenous manganese on its plant growth, subcellular distribution, chemical forms, physiological and biochemical traits in Cleome viscosa L. *Ecotoxicology and Environmental Safety, 198,* 110696.

Yang, Y., Sun, C., Yao, Y., Zhang, Y., & Achal, V. (2011). Growth and physiological responses of grape (*Vitis vinifera* "Combier") to excess zinc. *Acta Physiologiae Plantarum, 33*, 1483–1491.
Yruela, I. (2005). Copper in plants. *Brazilian Journal of Plant Physiology, 17*, 145–156.
Yruela, I. (2009). Copper in plants: Acquisition, transport and interactions. *Functional Plant Biology, 36*, 409–430.
Zafari, S., Sharifi, M., Chashmi, N. A., & Mur, L. A. (2016). Modulation of Pb-induced stress in *Prosopis* shoots through an interconnected network of signaling molecules, phenolic compounds and amino acids. *Plant Physiology and Biochemistry, 99*, 11–20.
Zhang, X., Gao, B., & Xia, H. (2014). Effect of cadmium on growth, photosynthesis, mineral nutrition and metal accumulation of bana grass and vetiver grass. *Ecotoxicology and Environmental Safety, 106*, 102–108.
Zhou, J., Wan, H., He, J., Lyu, D., & Li, H. (2017). Integration of cadmium accumulation, subcellular distribution, and physiological responses to understand cadmium tolerance in apple rootstocks. *Frontiers in Plant Science, 8*, 966.

Chapter 6
The Cradle of Chilean Wine Industry? The Vitiviniculture of the Pica Oasis

Gastón Gutiérrez Gamboa, Victoria Contreras Cortez, Sergio Jara, Philippo Pszczólkowski, Irina Díaz-Gálvez, and Nicolás Verdugo-Vásquez

6.1 Introduction

The origin of the Chilean vitiviniculture is a controversial subject. Some professionals recognize that Francisco de Carabantes brought the first cuttings from Perú to Concepción giving rise to the national vitiviniculture (Hernández & Moreno, 2011). This hypothesis posed by the French naturalist Claude Gay was corrected by the national historiography (Lacoste et al., 2010). Chilean vitiviniculture originates in the sixteenth century, when Listán Prieto cuttings arrived from Spain and were planted in the Southern valleys of the Viceroyalty of Peru (Lacoste et al., 2010; Milla-Tapia et al., 2013), probably in the Codpa and Azapa basins that currently are in the Arica y Parinacota Region, Chile. Historic records have mentioned that the

G. Gutiérrez Gamboa (✉)
Instituto de Investigaciones Agropecuarias, INIA Carillanca, Vilcún, Chile

Escuela de Agronomía, Facultad de Ciencias, Ingeniería y Tecnología, Universidad Mayor, Temuco, Chile
e-mail: gaston.gutierrez@inia.cl

V. C. Cortez · S. Jara
Cooperativa Lagar de Los Oasis, Pica, Región de Tarapacá, Chile

P. Pszczólkowski
Pontificia Universidad Católica de Chile, Facultad de Agronomía e Ingeniería Forestal, Departamento de Fruticultura y Enología, Santiago, Chile

I. Díaz-Gálvez (✉)
Instituto de Investigaciones Agropecuarias, INIA Raihuen, San Javier, Chile
e-mail: idiaz@inia.cl

N. Verdugo-Vásquez (✉)
Instituto de Investigaciones Agropecuarias, INIA Intihuasi, La Serena, Chile
e-mail: nicolas.verdugo@inia.cl

G. Gutiérrez Gamboa, M. Fourment (eds.), *Latin American Viticulture Adaptation to Climate Change*, https://doi.org/10.1007/978-3-031-51325-1_6

vine was probably introduced and established in the oasis of Matilla and Pica toward the end of the sixteenth century (Bermúdez Miral, 1987) by the Spanish conquerors. The wine was one of the most important agricultural products in the following centuries for Tarapacá (Billinghurst, 1893; Bermúdez Miral, 1987). The Pica and Matilla oasis and the Quisma valley are administratively situated in Tarapacá Region, Pica commune, Chile, and are defined as the Pica Oasis.

Northern Chile is rarely cited for their wine production despite their great historic importance on the subject dating back more than 400 years (Castro et al., 2015; Gutiérrez-Gamboa et al., 2023). History of wine production in this zone is not taught in the universities that teach viticulture and enology, probably because of the French paradigm phenomena (Pszczólkowski & Lacoste, 2016), in which the Chilean elites direct their looks toward European products to the detriment of the regional products (Núñez, 2016). Notably, toward the seventeenth century, the agricultural product that put Tarapacá in contact with the great centers of the Viceroyalty of Peru, was the wine (Bermúdez Miral, 1987; Rice, 1997). However, the mining industry was developed strongly in Tarapacá, rebuilding the economy of Pica, Matilla, and Quisma from agriculture to mining in the following decades (Bermúdez Miral, 1987). Despite this, wine production strongly continued in Tarapacá from the nineteenth century, during which the main vineyards were in Pica, Matilla, and Quisma (Basadre, 1884; Bermúdez Miral, 1987). In addition, the local population mentions that the wines obtained from a vineyard in Matilla and produced by Medina Hermanos were awarded gold medals in a world wine exhibition in Seville (Spain) in 1903, and in Paris (France) in 1907 (Daponte Araya, 2000).

Scarce knowledge about the wine heritage of northern Chile can be explained by the social, political, and economic consequences some of them derived from the War of the Pacific (Castro, 2013). The last harvest in Tarapacá took place in 1937, mainly due to the following events: (i) the water expropriation of Chintaguay stream that derived the water from Matilla and Quisma to Iquique in the 1920s; (ii) the anti-alcohol campaign promoted by Arturo Alessandri Palma (president of Chile in the 1920s) that resulted in the application of a series of taxes to wine production in Tarapacá, which was promoted by some big winegrowers from the Central valleys; (iii) the Chileanization of Tarapacá, in which Chilean civil and military forces promoted the acculturation of Peruvian population by attacking, intimidating, expelling and murdering Peruvians or their descendants who decided to stay in these lands after the War of the Pacific (Núñez, 1985; Bermúdez Miral, 1987; González Miranda, 1995); many of them cultivated vines in the territory.

Pica viticulture was developed in the heart of the Atacama Desert at an altitude close to 1225 m above sea level under water scarcity. Subtropical conditions characterize the viticulture of northern Chile, providing a series of physiological problems in vine production that comprise markedly acrotony, heterogenous maturity, and early decay of grapevines (Gutiérrez-Gamboa et al., 2023). Soils of these viticultural zones are sandy, saline, with low organic matter and alkaline. Grapevines are irrigated from the oasis water obtained from filtration galleries, a technology brought by the Spanish conquerors. These galleries consist of an almost horizontal tunnel dug underground until it reaches a water-bearing zone (Lictevout et al., 2020). The orchards also are irrigated using water obtained from deep well drillings.

Tarapacá Region holds a valuable diversity of grapevine genetic material adapted to extreme desert conditions, including unknown vine genotypes, minority, and Criolla varieties (Milla-Tapia et al., 2013; Franck et al., 2020). Some genetic materials are more adapted to others to the desert pedoclimatic conditions (Bavestrello-Riquelme et al., 2012; Franck et al., 2020), which considerably affect the breeding of plant material. Small wine producers from Pica, Quisma, and Matilla that belong to the "Cooperativa Lagar de los Oasis" are propagating grapevine plant material from a survey of genetic materials tested earlier with the "Instituto de Investigaciones Agropecuarias" (INIA Chile). Some of the members of this cooperative are descendants of the first winegrowers of Tarapacá, who seek to recover and preserve the winemaking history from their ancestral legacy (Billinghurst, 1893; Bermúdez Miral, 1987).

Chilean viticultural valleys located in the Atacama Desert are not recognized by the actual Decree Law that establishes the Chilean viticultural zoning (BCN, 2020), probably due to the scarcity of the surface planted with vines (SAG, 2021). This chapter aims to provide descriptive information about the vitiviniculture of the Pica and Matilla oasis and Quisma valley, touching upon some aspects of agronomy and enological issues that must be considered by the wine professionals and consumers for the vitivinicultural recovery and development of this territory.

6.2 Groundwater Used for Vine Irrigation in Pica Vitiviniculture

The vines cultivated in Pica Oasis are irrigated with water obtained from filtration galleries (Lictevout et al., 2020) and water obtained from the deep well drillings. The filtration galleries or "*socavones*" consist of an ancient water management system developed in ancient times by Persians (Mostafaeipour, 2010), and brought to Pica by the Spanish conquerors. This system is used to provide a reliable supply of water in arid and semi-arid climates, where precipitations are scarce (Mostafaeipour, 2010). The system consists of an underground and almost horizontal tunnel with vertical shaft wells, which tap and drain groundwater to the earth surface (Lictevout et al., 2020). The water flows downward by gravity in the gallery composition as is shown in Fig. 6.1, which allows it to supply water for domestic purposes and irrigate downslope lands (Lictevout et al., 2020).

Table 6.1 shows water chemical parameters of a sample collected in the Pica Oasis. Based on this, electrical conductivity (EC) was higher than the adequate levels to irrigate, whereas boron (B) content is higher than the damage levels for plant growth. Lictevout et al. (2020) reported that EC varied from 0.4 to 2.2 dS/m in

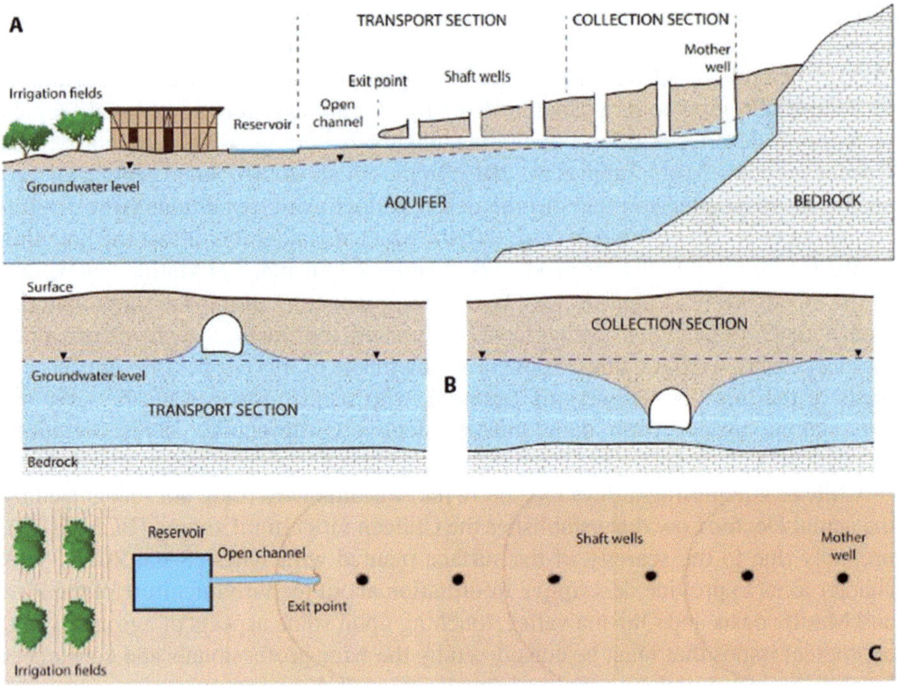

Fig. 6.1 A filtration gallery. (Figure obtained from those published by Lictevout et al. (2020), which was modified from English (1968) and Naghibi et al. (2015)). (**a**) Longitudinal section; (**b**) cross section; (**c**) aerial view

water samples of Pica, which inversely coincided with water temperature. These authors also reported that pH levels of groundwater from Pica ranged from 7.2 to 9.1, which was considerably higher than to those exposed in Table 6.1. In another study, Herrera Apablaza et al. (2018) reported that Pica groundwaters are adequate for irrigation in terms of EC, sodium adsorption ratio (SAR) and residual sodium carbonate (RSC), but not by sodium (Na), B, lithium (Li), potential salinity (PS) and soluble sodium percentage (SSP). These authors also reported that dissolved ions are accumulated in the surface aquifer, which was attributed to the sulphated dissolution by alluvium in the area, bad practices in irrigation-fertilization and the aridity condition of the zone. Finstad et al. (2016) reported that salt distribution varies predictably with depth and soil age, and the most soluble compounds are concentrated nearest to the land surface. Based on the author's findings, mineral sulfate tends to decrease with decreasing soil depth, following a pattern indicative of Rayleigh-like fractionation as solute-rich waters migrate toward the land surface (Finstad et al., 2016).

Table 6.1 Soil and water physicochemical parameters of samples obtained from a vineyard cultivated in Pica and their sufficient parameters

			Sufficient level according to texture	
	Unit	Values	Sandy loam to sandy silt loam[a]	Silt loam to clay loam
Soil chemical parameters				
pH	–	7.7	5.8–7.0	5.8–6.0
Electrical conductivity	dS/m	2.1	<1.5	<1.5
Organic matter	%	0.36	>1.5	>1.5
Inorganic nitrogen	mg/kg	15	10–20	15–25
Extractable phosphorus	mg/kg	4	>5	>8
Exchangeable cations				
Cation-exchange capacity	cmol (+)/kg	7.4	8.0–15.0	15.0–30.0
Calcium base saturation	% CEC	60	60–65	55–65
Magnesium base saturation	% CEC	20.1	12–15	10–15
Potassium base saturation	% CEC	2	2–3	3–4
Water chemical parameters			**Sufficient level**	**Damaging values**
Electrical conductivity	dS/m	1.78	<0.75	>3.0
pH	–	6.6	–	–
Temperature	°C	18.7	–	–
Boron	mg/L	2.66	<0.50	>2.0

According to the data published by Hirzel (2014)
[a]Soil texture of the Pica sample contains 88% of sand, 3% of silt and 9% of clay

6.3 Soil Physicochemical Conditions of Pica Vitiviniculture

Table 6.1 shows the soil physicochemical parameters of a vineyard planted in the Pica Oasis. Soils of these viticultural zones are sandy (>80% of sand), saline (2.1 dS/m), with low organic matter (<0.4%) and alkaline (pH > 7) (Table 6.1).

Salinity is the process of salt accumulation in soils (Corwin, 2021). Salinity occurs in zones where the water evaporation from soils exceeds precipitation and dissolved salts in the soil solution tend to concentrate at the soil surface (Corwin, 2021; Keller, 2020), Notably, the hyperarid Atacama Desert is characterized by having local basins with surficial salt crusts, where shallow groundwater drives the major soil processes (Finstad et al., 2016). Soil and water electrical conductivity contents in Pica samples were higher than the adequate levels (Table 6.1). The threshold above which salinity starts to affect *Vitis vinifera* growth and yield formation is approximately 2 dS/m, and above 16 dS/m the vines cannot survive (Zhang et al., 2002; Keller, 2020). Since dissolved ions decrease the osmotic water potential, electrical conductivity is also a measure of water potential (Keller, 2020). Naturalized genotypes selected in northern Chile have shown a great tolerance to

Fig. 6.2 Boron toxicity (in **a**, **b**) and magnesium deficiency (in **c**) in European grapevine varieties growing in the Pica Oasis

drought and water deficit (Bavestrello-Riquelme et al., 2012; Milla-Tapia et al., 2013; Franck et al., 2020). "R32" genotype (naturalized) exhibited higher water productivity, trunk circumferences, yield, and pruning weights when grafted with Syrah compared to other rootstocks (Villalobos-Soublett et al., 2022).

Soil physicochemical parameters shown in Table 6.1 usually lead to deficiencies on magnesium (Mg) and toxicity by boron (B), sodium (Na^+) and chlorine (Cl^-) in grapevines (Keller, 2020; Gutiérrez-Gamboa et al., 2023). B is an essential element for plants since it allows to sustain cell wall structure and growth (Keller, 2020). B transport is a process mediated not only by passive diffusion but also by specialized transporters, whose activity is regulated in response to B soil conditions (Miwa & Fujiwara, 2010). B excess is not excluded by the vine roots, which increase toxicity under high B levels in the soil (Keller, 2020). B toxicity considerably decreases plant growth, mainly of shoots due to its effects on internode shortening and causes chlorosis beginning at the leaf tip and margins of the mature leaves (Sarafi et al., 2017). Based on an empirical point of view, the Criolla varieties established in Pica are less sensitive to boron toxicity than the European grapevine varieties (Fig. 6.2). Mg is a structural component of chlorophyll, and it is involved in the protein production (Keller, 2020). Mg deficiency is common in vines growing in sandy and calcareous soils since Ca^{2+} and K^+ availability slow down Mg uptake due to competition among these cations for root uptake (Huber & Jones, 2013). Mg deficiency symptoms appear as chlorotic discoloration of the interveinal areas of mature leaves (Fig. 6.2) since Mg is more mobile in the phloem than calcium (Ca) (Keller, 2020).

6.4 Climatic Conditions in Pica

The climatic information available for the Pica Oasis was obtained from meteorological stations located in different productive sites (Arenas Charlín et al., 2021), which allowed to calculate different meteorological and bioclimatic indices based on to the data reported by Gutiérrez-Gamboa et al. (2018). Currently, there is no historical information for a period of 30 years or more which would allow to define trends on meteorological and bioclimatic variables (Verdugo-Vásquez et al., 2023). The oldest climatic records found in Pica correspond to 2011, and from twenty years or more, it is possible to have enough information to estimate trends and evaluate the change or not in viticultural suitability of the Pica Oasis.

The meteorologic and bioclimatic indices, commonly used for the classification of viticultural zones, were calculated using information data available until 2022 and are shown in Table 6.2. Table 6.2 showed a high annual thermal amplitude, which is a typical phenomenon occurring in desert areas. In addition, Pica Oasis presented a low accumulation of annual rainfall, much lower than the annual evapotranspiration. Thus, agricultural development depends mainly on water resources available.

Based on the calculated bioclimatic indices, the Growing Season Temperature (GST) was 20.2 °C, which makes the Pica Oasis a "hot" climate for wine production (Jones & Davis, 2000). Several grapevine varieties for grape wine and table grape production could be suitable for cultivation under a hot climate (Jones & Davis, 2000). The accumulation of Growing Degree Days (GDD) was 2122 heat units, which allows to classify Pica Oasis as an "IV Region" (Amerine & Winkler, 1944). Viticultural zones classified in this region are suitable to produce wines by using late ripening grapevine varieties, such as Cabernet Sauvignon, Cabernet Franc, Sangiovese, Zinfandel, among others (Amerine & Winkler, 1944). Currently, no information is available on the classification of different bioclimatic indices for the suitability of Criolla varieties that are adapted to pedoclimatic conditions of Pica Oasis. Cold Night Index (CI) was 13.1 °C, which classifies the Pica Oasis in the CI+1 classification as "cold nights," which is considered favorable for wine production. CI considers the average minimum temperature during the final part of the maturation period (Tonietto & Carbonneau, 2004). The aim of this index is to improve the assessment of the quality potential for a particular site in relation to secondary metabolite accumulation such as phenolic and volatile compounds (Tonietto & Carbonneau, 2004).

Table 6.2 Climatic indices obtained from climatic information available in Pica

Climatic index	Period[a]	Value (unit)
Maximum temperature	Annual mean	28.7 (°C)
Minimum temperature	Annual mean	11.0 (°C)
Thermal amplitude	Annual mean	17.7 (°C)
Precipitation	Annual accumulation	5,1 (mm)
Growing season temperature	01 October to 30 April	20.2 (°C)
Growing degree days (at 10 °C)	01 October to 30 April	2.122 (heat units)
Cool night index	01 March to 31 March	13.1 (°C)
Evapotranspiration	Annual accumulation	2308 (mm)
Chilling hours (at 7 °C)	01 May to 31 July	67.7 (hours)

[a]Data correspond to the average obtained during the years 2011–2022

The bioclimatic indices shown above allow to define the Pica Oasis as the warmest productive area of Chile, which would expand the diversity of climates of the national production (Verdugo-Vásquez et al., 2023). Therefore, the development and production of grapevine material adapted to the "warm" climatic conditions of the Pica Oasis should be considered by the funders' programs. On the other hand, there is a low risk of frost in this area (there is only a record of 1 day with frost in the period 2011 to 2022 reaching a minimum temperature of −1.1 °C) and the accumulation of chilling hours is low, which usually results in problems in budburst and maturity heterogeneity (Gutiérrez-Gamboa et al., 2023).

6.5 The Vitiviniculture of Pica

The surface of planted vines registered in Tarapacá Region is 3.95 ha, which are found in Pozo Almonte (3.93 ha) and Pica (0.12 ha) locations (SAG, 2021). The planted vineyards in Pica have been mostly developed using Criolla varieties, such as Blanca Ovoide, Canela, and Moscatel Rosada, including Listán Prieto and some unknown vine genotypes, which are trellised in a vertical shoot position system or an overhead trellis system using artisanal wooden structures (Fig. 6.3a). Some famous grapevine varieties were also established in Pica, such as Cabernet Sauvignon, Carmenère, Merlot and Syrah (SAG, 2021), which are not empirically adapted to pedoclimatic conditions of the zone due to their scarce tolerance to boron toxicity and drought (Fig. 6.2). Moreover, it is possible to find individual vines in the gardens of citizens and in the green areas of Pica in association with ancient mango trees, which are trellised in a bush system or growing freely, respectively. The grape production in Pica is developed in association with different fruit species, such as mango (*Mangifera indica*), passion fruit (*Passiflora edulis*), guava (*Psidium guajava*), orange (*Citrus × sinensis*) and lemon (*Citrus × limon*) (Fig. 6.3c). Other

Fig. 6.3 Grape production in Matilla (**a, b**) and Pica (**c**) developed in association with vegetables (**a, b**) and fruits (**c**)

horticultural species, such as watermelons (*Citrullus lanatus*) and melons (*Cucumis melo*) are planted in association with vines (Fig. 6.3a). Notably, this production presents a common agronomic management that coincided with those exposed by Gutiérrez-Gamboa et al. (2023) in vine growing in the Codpa valley. Some other particularities characterize the vine production in Pica (Fig. 6.3). Some viticulturists protect the clusters by covering them or the whole vines using Raschel nets against bird damage (Fig. 6.3a, c). However, cluster covering produces dehydration and diseases in berries, negatively affecting vine production. Another biotic factor that also affects the Pica viticulture is the powdery mildew (*Uncinula necator*) that should be controlled in the vineyard. In addition, the vines are manually cultivated by small viticulturists, giving rise to a heterogeneous production at the vineyard level due to the low varietal purity at planting.

Pica commune has historical vestiges of a great development of the vitiviniculture since it is possible to find the presence of ancient wine presses, clay vessels, and stills closed down by the Government of Chile during the Northern Chileanization process (Fig. 6.4). Ancient wine presses were built by carving an area of the bedrock to create a flat surface surrounded by short walls (Walsh & Zorn, 1998). The flattened surface was used for treading and the wall kept the grape juice within the press (Walsh & Zorn, 1998). The grape juice was introduced into clay vessels that were buried close to half their dimension. This infrastructure was probably built in the colonial period since drainage studies reveal that investment in built infrastructure was heaviest in the sites close to the cores of the principal states in colonial Peru (Weaver, 2020). These buildings account for extensive agro-industrial complexes with wine presses, storage, fermentation structures, and distillery apparatus (Weaver et al., 2019).

"Lagar de Matilla" is a historical monument and site museum located in the Matilla Oasis within the commune of Pica, Tarapacá Region, Chile. "Lagar de Matilla" belongs to the set of national monuments of Chile since 1977 by virtue of the 746 Supreme Decree of October 5, 1977, which was placed in the category "Historical Monument." This heritage property represents one of the main

Fig. 6.4 Historical vestiges of vitiviniculture in Pica. (**a**) Presence of old wine presses and clay vessels in the Historical Monument Lagar de Matilla; (**b**) clay vessels found in a front garden in Pica; (**c**) still closed by the Government of Chile during the Northern Chileanization process

testimonies of Matilla wine production during the colonial era. In this site, most of the clay vessels have the year of elaboration written on them (Fig. 6.4a), which dates to the eighteenth century, specifically from 1756 until 1767 (Daponte Araya, 2000). The ancient wine presses are currently out of use and of a total of fifteen that exist in Pica and Matilla (Fig. 6.5), only six of them are in good condition to be studied (Daponte Araya, 2000). The wine presses were constructed mainly by using adobe and stuccoes from anhydrite mortar, an abundant raw material that characterizes most of the buildings in Pica (Daponte Araya, 2000). In addition, a trunk of a large tree of 9.8 m stands out among the components of the "Lagar de Matilla" wine press that worked as a grape pressing to wine elaboration (Daponte Araya, 2000). Notably, the "Lagar de Matilla" is exposed to the community in the center town of Matilla and is maintained by the residents of the sector.

One of the main precedents that demonstrates the importance of the wine industry in the Pica Oasis was the recent discovery of a buried wine bottle featuring Peruvian iconography on its label (Fig. 6.6a). In early June 2022, a resident of the town of Matilla handed over to the members of the "Cooperariva Lagar de los Oasis" a wine bottle he found buried in the midst of the Matilla desert. This glass bottle, which still contains liquid inside, presumably wine, is corked and also bears a label with "*Vino de Pica*" (Wine of Pica) written on it along with Peruvian flags and symbols (Fig. 6.6b). Interestingly, the Peruvian flag depicted on the label design does not correspond to the current national emblem but rather corresponds to the flag used in Peru during the presidency of José Bernardo de Tagle, who served as president between 1822 and 1824.

The members of the "Cooperativa Lagar de los Oasis" have an interesting experience in winemaking for which they used Criolla and European grapevine varieties, including unknown vine genotypes (Fig. 6.7). European grapevine varieties are not

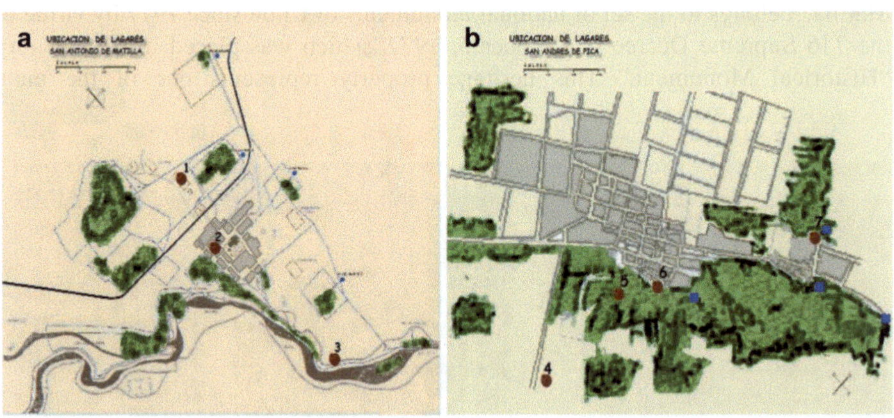

Fig. 6.5 Ancient wine presses found in Matilla (**a**) and Pica (**b**). Numbers and circles correspond to wine presses: (1) Lagar de la Familia Contreras, (2) Lagar de Medina Hermanos, (3) Lagar de la Botijería, (4) Lagar de Jesús María, (5) Lagar de La Comunidad, (6) Lagar de la Familia Zavala, (7) Lagar del Sector Miraflores. Blue, green, grey, and black symbols correspond to water from gallery filtrations, crops, buildings, and streets, respectively

Fig. 6.6 Bottle (**a**) of wine discovered and unearthed from the soils of Matilla, including its old (**b**) and restored (**c**) label design

Fig. 6.7 Criolla wines produced in Pica Oasis. Bottle of Blanca Ovoide sweet wine in the hands of the viticulturist Jorge Moya (**a**). Glass of wine containing the Canela and Listán Prieto blend in the hands of Rómulo Contreras viticulturist (**b**). Glass of wine containing Blanca Ovoide sweet wine (**c**). Glass of wine containing wine from an unknown vine genotype

well adapted to desert conditions, which considerably limits wine production. Notably, the wines produced in Pica using Criolla grapevine varieties have unique enological characteristics that usually results in an interesting salty profile. Blanca Ovoide grapes were vinified as sweet wine (Fig. 6.7a), since culturally it is more appreciated by the local consumers. This wine features a deep golden color. The nose is fruity, with notes of peaches, orange blossom, honey, cinnamon, nutmeg, and vanilla. In the mouth, the relaxing sweetness and salinity of the wine are well balanced. Grapes from Canela and Listán Prieto were used to produce a light-red blend (Fig. 6.7b). This wine features a deep and brilliant ruby color. In the nose, the exuberance of raspberry, plum, and currant emerges. Subtle notes of roses and vanilla are also present. In the mouth, the wine is fresh, with great intensity of acid and salty tastes. An orange wine was made using Moscatel Rosada grapes (Fig. 6.7c). This wine features an intense salmon color. In the nose, citrus notes such as lemon, orange and grapefruit stand out. Subtle peach notes are also noticeable. The body is intense and presents a fresh acidity that fills the mouth. Grapes from an abundant unknown genotype planted in Pica were vinified as white wine (Fig. 6.7d). This wine features a pale-yellow color and is characterized by intensely fruity aromas, highlighting notes of citrus, peaches, pineapple, and green apple. Subtle notes of orange blossom, jasmine, thyme, and geranium are also perceived. In the mouth, the wine is fruity, with salty tastes standing out.

6.6 Technical Conclusions

Subtropical conditions characterize the viticulture of Pica, providing a series of physiological problems in vine production that comprises a development of markedly acrotony, berry heterogenous maturity, and an early decay of vines. Pica viticulture is also affected by biotic factors, mainly by bird damage, powdery mildew, and bunch rot diseases. Soils of Pica are sandy, saline, with low organic matter and alkaline, whereas the water used for irrigation contains high levels of boron and electrical conductivity, resulting in boron toxicity and magnesium deficiency in vines, mostly in the European grapevine varieties.

6.7 Socioeconomic Considerations

Pica vitiviniculture holds a valuable history that dates back from the sixteenth century, which is an unknown subject both nationally and internationally. Toward the seventeenth century, wine was the most important agricultural product that put Tarapacá in contact with the great centers of the Viceroyalty of Peru. Nevertheless, social, political, and economic consequences derived from the water expropriation of Chintaguay stream, the increase of taxes to viticulturists, and the Chileanization of Tarapacá led to the disappearance of this activity, giving place to the last harvest in 1937 in Tarapacá. Despite this, some of the descendants of the first winegrowers of Tarapacá are seeking to preserve their wine history following their ancestral legacy.

This chapter provides valuable information about Pica vitiviniculture, adding preliminary data for the contribution to the diversity and heritage culture of the Chilean wine industry. However, it should be noted that there are several difficulties that put Pica viticulture at risk of disappearing: (i) there is no validated technical knowledge for the viticultural and oenological production of Criolla varieties and unknown vine genotypes in the zone, which requires urgent basal funding; (ii) the training received from technocratic experts and winemakers from the Chilean central valley has lacked empathy in relation to the thoughts and historic procedures of small producers; (iii) Pica's distance from urban centers is far, so the freight necessary to obtain materials and supplies is extremely expensive and is sometimes not available; (iv) the high temperatures that are recorded throughout the season make it difficult to store the wines; and (v) the producers are elderly and their children migrate to urban centers for greater economic well-being.

Therefore, it is imperative to start a technical intervention to support vitiviniculture in Pica Oasis but also to develop a political intervention that goes to the aid of this activity that is growing but still at high risk of disappearing.

Acknowledgments The authors are grateful for the support of the "Cooperativa Lagar de los Oasis" members, especially to Felipe Loayza, Andrés Loayza, Rómulo Contreras, Jorge Moya, Italo Manniello and Rafael Angulo (R.I.P). The authors also would like to thank the students of the agricultural section of "Liceo Bicentenario Padre Alberto Hurtado Cruchaga de Pica," and especially its director, Carolina Guerrero. In addition, the authors want to thank CORFO, CONADI, INDAP, and Fundación Educacional Collahuasi for their commitment to rescuing the wine heritage of the Pica Oasis. Finally, the authors want to thank all the residents of the Pica Oasis, especially Jorge Alache, Verne Angel, Ximena Loayza, Calogero Santoro, Lautaro Núñez, Luis Briones (R.I.P), and Enelidolfo O'Ryan, for their contribution to this research.

References

Amerine, M. A., & Winkler, A. J. (1944). Composition and quality of musts and wines of California grapes. *Hilgardia, 15*, 493–675.

Arenas Charlín, J., Rojas Arias, M., & Ugrinovic Guagama, V. (2021). *Análisis agroclimático de liceo "Padre Alberto Hurtado Cruchaga", comuna de Pica, provincia del Tamarugal, región de Tarapacá*. Universidad Arturo Prat.

Basadre, M. (1884). *Riquezas peruanas: colección de artículos descriptivos escritos para "La Tribuna"* (p. 1884). Imprenta de La Tribuna.

Bavestrello-Riquelme, C., Cavieres, L., Gallardo, J., Ibacache, A., Franck, N., & Zurita-Silva, A. (2012). Evaluation of drought stress tolerance in four naturalized grapevine genotypes (*Vitis vinifera*) from northern Chile. *Idesia, 30*, 83–92.

BCN. (2020). *Decreto 464. Establece zonificación vitícola y fija normas para su utilización*. Ministerio de Agricultura. Biblioteca del Congreso Nacional.

Bermúdez Miral, O. (1987). *El oasis de Pica y sus nexos regionales*. Ediciones Universidad de Tarapacá.

Billinghurst, G. E. (1893). La irrigación en Tarapacá. Patrimonio Cultural Común, Imprenta y Librería Ercilla.

Castro, L. (2013). Chilenization, taxes and the rental of agricultural lands in the Andean Area of Tarapacá (Northern Chile, 1883–1932). *História Unisinos, 17*(2), 146–157.

Castro, A., Mujica, F., & Argandoña, F. (2015). Between Pintatani and Codpa. Landscape and typical products in peasants stories, 1847–2013. *Revista Iberoamericana de Viticultura, Agroindustria y Ruralidad, 2*(6), 70–86.

Corwin, D. L. (2021). Climate change impacts on soil salinity in agricultural areas. *European Journal of Soil Science, 72*(2), 842–862.

Daponte Araya, J. F. (2000). La fiesta de la vendimia en los oasis de Pica y Matilla. Sincretismo de tres culturas. Degree Thesis. Universidad Católica de Valparaíso. Facultad de Filosofía y Educación. Instituto de Música.

English, P. W. (1968). The origin and spread of Qanats in the old world. *Proceedings of the American Philosophical Society, 112*(3), 170–181.

Finstad, K., Pfeiffer, M., McNicol, G., Barnes, J., Demergasso, C., Chong, G., & Amundson, R. (2016). Rates and geochemical processes of soil and salt crust formation in Salars of the Atacama Desert, Chile. *Geoderma, 284*, 57–72.

Franck, N., Zamorano, D., Wallberg, B., Hardy, C., Ahumada, M., Rivera, N., et al. (2020). Contrasting grapevines grafted into naturalized rootstock suggest scion-driven transcriptomic changes in response to water deficit. *Scientia Horticulturae, 262*, 109031.

González Miranda, S. (1995). El poder del símbolo en la Chilenización de Tarapacá. Violencia y nacionalismo entre 1907 y 1950. *Revista de Ciencias Sociales, 4*(5), 42–56.

Gutiérrez-Gamboa, G., Carrasco-Quiroz, M., Martínez-Gil, A. M., Pérez-Álvarez, E. P., Garde-Cerdán, T., & Moreno-Simunovic, Y. (2018). Grape and wine amino acid composition from Carignan noir grapevines growing under rainfed conditions in the Maule Valley, Chile: Effects of location and rootstock. *Food Research International, 105*, 344–352.

Gutiérrez-Gamboa, G., Contreras Cortez, V., Jara, S., Verdugo-Vásquez, N., & Pszczólkowski, P. (2023). Andean viticulture: Edaphoclimatic characterization of vineyards located in the Codpa valley. *Revista Iberoamericana de Viticultura, Agroindustria y Ruralidad, 10*(28), 26–40.

Hernández, A., & Moreno, Y. (2011). The origins of Chilean wine. Curicó, Maule, Itata, Biobío. Origo Ediciones.

Herrera Apablaza, V., Gutiérrez Roa, N., Córdova Molina, S., Luque Marín, J., Idelfonso Carpanchay, M., Flores Riveras, A., & Romero., L. (2018). Calidad del agua subterránea para el riego en el Oasis de Pica, norte de Chile. *Idesia, 36*(2), 181–191.

Hirzel, J. (2014). *Diagnóstico nutricional y principios de fertilización en frutales y vides* (2nd ed.). Trama Impresores. Colección Libros INIA – Instituto de Investigaciones Agropecuarias.

Huber, D. M., & Jones, J. B. (2013). The role of magnesium in plant disease. *Plant and Soil, 368*, 73–85.

Jones, G. V., & Davis, R. E. (2000). Climate influences on grapevine phenology, grape composition, and wine production and quality for Bordeaux, France. *American Journal of Enology and Viticulture, 51*, 249–261.

Keller, M. (2020). *The science of grapevines* (3rd ed.). Academic Press. Elsevier.

Lacoste, P., Yuri, J. A., Aranda, M., Castro, A., Quinteros, K., Solar, M., Soto, N., Gaete, J., & Rivas, J. (2010). Grape varieties in Chile and Argentina (1550–1850). *Genealogy of Torrontés. Mundo Agrario, 10*(20), 1–36.

Lictevout, E., Abellanosa, C., Maass, C., Pérez, N., Yáñez, G., & Véronique, L. (2020). Exploration, mapping and characterization of filtration galleries of the Pica Oasis, northern Chile: A contribution to the knowledge of the Pica aquifer. *Andean Geology, 47*(3), 529–558.

Milla-Tapia, A., Gómez, S., Moncada, X., León, P., Ibacache, A., Rosas, M., et al. (2013). Naturalised grapevines collected from arid regions in Northern Chile exhibit a high level of genetic diversity. *Australian Journal of Grape and Wine Research, 19*, 299–310.

Miwa, K., & Fujiwara, T. (2010). Boron transport in plants: Co-ordinated regulation of transporters. *Annals of Botany, 105*(7), 1103–1108.

Mostafaeipour, A. (2010). Historical background, productivity and technical issues of qanats. *Water History, 2*, 61–80.

Naghibi, S. A., Pourghasemi, H. R., Pourtaghi, Z. S., & Rezaei, A. (2015). Groundwater qanat potential mapping using frequency ratio and Shannon's entropy models in the Moghan watershed, Iran. *Earth Science Informatics, 8*(1), 171–186.

Núñez, L. (1985). Recuérdalo, aquí estaba el lagar: la expropiación de las aguas del valle de Quisma (I Región). *Revista Chungará, 14*, 157–167.

Núñez, E. (2016). The European Paradigm in the wine and spirits industry in Chile: Oporto, Jerez and Cognac made in Mitjans Company. *Revista Iberoamericana de Viticultura, Agroindustria y Ruralidad, 9*(3), 130–146.

Pszczólkowski, P., & Lacoste, P. (2016). Native varieties, an opportunity for Chilean pisco. *Revista de la Facultad de Ciencias Agrarias. Universidad Nacional de Cuyo, 48*(1), 239–251.

Rice, P. M. (1997). Wine and Brandy production in Colonial Peru: A historical and archaeological investigation. *Journal of Interdisciplinary History, 27*(3), 455–479.

SAG. (2021). Catastro vitícola nacional 2021. División de Protección Agrícola-Forestal y Semillas Subdepartamento de Viñas, Vinos y Bebidas Alcohólicas. Servicio Agrícola y Ganadero.

Sarafi, E., Siomos, A., Tsouvaltzis, P., Chatzissavvidis, C., & Therios, I. (2017). Boron toxicity effects on grafted and non-grafted pepper (*Capsicum annuum*) plants. *Journal of Soil Science and Plant Nutrition, 17*(2), 441–460.

Tonietto, J., & Carbonneau, A. (2004). A multicriteria climatic classification system for grapegrowing regions worldwide. *Agricultural and Forest Meteorology, 124*, 81–97.

Verdugo-Vásquez, N., Orrego, R., Gutiérrez-Gamboa, G., Reyes, M., Zurita Silva, A., Balbontín, C., Gaete, N., & Salazar-Parra, C. (2023). Climate trends and variability in the Chilean viticultural production zones during 1985–2015. *OENO one, 57*(1), 345–362.

Villalobos-Soublett, E., Verdugo-Vásquez, N., Díaz, I., & Zurita-Silva, A. (2022). Adapting grapevine productivity and fitness to water deficit by means of naturalized rootstocks. *Frontiers in Plant Science, 13*, 870438.

Walsh, C., & Zorn, J. R. (1998). New insights from old wine presses. *Palestine Exploration Quarterly, 130*(2), 154–161.

Weaver, B. J. M. (2020). An archaeology of the aesthetic: Slavery and politics at the Jesuit vineyards of Nasca. *Cambridge Archaeological Journal, 31*(1), 1–18.

Weaver, B. J. M., Muñoz, L. A., & Durand, K. (2019). Supplies, status, and slavery: Contested aesthetics of provisioning at the Jesuit haciendas of Nasca. *International Journal of Historical and Archaeology, 23*, 1011–1038.

Zhang, X., Walker, R. R., Stevens, R. M., & Prior, L. D. (2002). Yield-salinity relationships of different grapevine (*Vitis vinifera* L.) scion-rootstock combinations. *Australian Journal of Grape and Wine Research, 8*(3), 150–156.

Chapter 7
Recovering the Asoleado: A Heritage of the Rainfed of Maule Valley

Marisol Reyes Muñoz and José Lladser Urzúa

7.1 Introduction

Since the times of Chilean Independence, when viticulture was incipient, the *asoleado*—a Spanish word for something that has been exposed to the sun—was emerging as a wine of great interest. Back in the nineteenth century, it was gaining a space of recognition in the national market, becoming highly valued among the consumers (Lacoste et al., 2016). For its producers, small rainfed winemakers, the "asoleado" was a good alternative to obtain a valuable product from the País variety (Listan Prieto) of grape, which was in those years the most planted in the rainfed lands (Rojas, 1897). However, towards the end of the nineteenth century, several factors aligned to remove this wine from the market and even from popular memory.

The transformations that modified the landscape and wine production of Chile seem to have been alien to the rainfed and traditional producers of País grapes (Del Pozo, 2014), who continued to cultivate their centuries-old vines formed in a head or gobelet, generally using little technology and producing wines that are sold mainly in bulk. Although this traditionality is sometimes surveyed and valued in patrimonial terms (Rojas, 2021), there is no correlation between this and its impact on the national market, with the País variety receiving the lowest prices, as grapes and even when presented as wine, according to the Oficina de Estudios y Políticas Agrarias (ODEPA, 2023).

To this unfavorable commercial situation, we should add the climate change, with extreme temperatures that affect the cycle of the vine and low rainfall, which increasingly converge in the difficulty of generating profitability in traditional rainfed viticulture. In this scenario, the production of "asoleado" wine, which seeks to

M. Reyes Muñoz (✉) · J. Lladser Urzúa
Instituto de Investigaciones Agropecuarias (INIA), Villa Alegre, Chile
e-mail: mreyes@inia.cl

G. Gutiérrez Gamboa, M. Fourment (eds.), *Latin American Viticulture Adaptation to Climate Change*, https://doi.org/10.1007/978-3-031-51325-1_7

dehydrate the grapes to concentrate the sugars and achieve a wine of the generous type, could offer an alternative to enhance and make profitable the production of the País grape, as it surely did in the past. However, to propose plans and actions that allow this revaluation, it is necessary to review the history and know the current state of the production of "asoleado" in its area of origin.

This chapter makes a historical review of the production of "asoleado" and its importance for rainfed economics, and seeks to introduce the sociocultural and production conditions behind this type of wine through an exploratory analysis that allows to visualize the current state and the possibilities offered by this product as an alternative to small winemakers, who face an increasingly adverse climatic scenario in the rainfed fields.

7.2 Materials and Methodology

Different bibliographic sources were reviewed to allow us to know the history of the "asoleado," its relationship with the area of origin and the environment in which it developed, also exploring the changes in the climatic conditions of its production area.

In parallel, a primary information survey was carried out through the elaboration and application of a semi-structured survey, which allowed us to characterize the current state of the production of "asoleado" and its producers. The survey was conducted in the communes of San Javier and Cauquenes, in the Maule region. Although the bibliography shows Cauquenes as the main place of origin of the "asoleado," the commune of San Javier was part of the Province of Maule, which included both communes until 1873 (Gundian, 1884, cited by Rojas & Espina, 2015).

A non-probabilistic technique was used to determine the sample of the target group to be evaluated, because there weren't enough records with information of "asoleado" producers in the region. This sampling was carried out for convenience and considered 18 respondents, who were located through the web, from the INIA Cauquenes enology laboratory data and mostly by information provided by the producers themselves. The quantitative information obtained was analyzed with descriptive statistics with measures of central tendency, and the qualitative information through a coding of the open questions. Then, we proceeded to give names to the general patterns found on common responses, and then list these patterns and assign them a numerical value.

In addition to this, a chemical analysis of alcohol content, pH, and reducing substances was conducted on a sample of 12 wines from some of the surveyed producers. The purpose was to establish the basic oenological parameters to characterize a fortified wine. This was followed by a tasting session, for which a panel of three wine professionals was assembled. The panel consisted of individuals involved in enological journalism, wine sales in specialized stores, and sommelier activities.

7.3 Results and Discussion

7.3.1 Historical Overview

According to current Chilean legislation, "asoleado" is a sweet wine of the generous type, produced and bottled in the rainfed area between the Mataquito River in the north and the Biobío River in the south, from vines planted in the same area (SAG, 2009). The elaboration of this wine goes back to the times of national independence, when the proclamation of Bernardo O'Higgins as Supreme Director of the State was celebrated, in whose banquet he included the most select delicacies of the time and the toasts were made with chacolí from Santiago and "asoleado" from Concepción, in addition to other peninsular wines (Pérez, 1886).

The name "asoleado" refers to the fact that, for its elaboration, the grapes are partially dehydrated in the sun, and then vinified. According to O'Higgins (Mujica et al., 2019), only 4 days were enough since the grapes were cut and laid in the sun and then vinified. However, Gay (1855) pointed out that the process of sunning—"asolear"— took between 15 and 20 days, and that the quality of the product obtained was better when the sunning of grapes was extended to 25 days. There was not much care during the process; this was done outdoors without much concern for rain, dew, or herbs that could emerge during the sunning process (Gundian, 1884, cited by Rojas & Espina, 2015).

The País variety, one of the oldest in Chile, has been the most used in the production of "asoleado" (Rojas, 1897; Couyoumdjian, 2006). In the first half of the nineteenth century, it was classified as very vigorous and rustic, well adapted to various types of soil and resistant to drought (Rojas, 1897). Their wines, however, were described as poor in aroma and without any remarkable quality, except when they were intended for the elaboration of "asoleado." It could even be compared to port when the grapes came from rainfed hills (Rojas, 1897). Other varieties of ancient presence in the country, such as Moscatel de Alejandría or Torrontés, were also indicated as suitable for producing liqueur or generous types of wines (Rojas, 1897).

Cauquenes and Concepción are historically recognized as the places of origin of this wine (Gundian, 1884, cited by Rojas & Espina, 2015; Lacoste et al., 2016). Special emphasis was placed on the virtues of Cauquenes as an "asoleado" producing area towards the end of the nineteenth century, when it was seen with an enormous potential to produce Bordeaux, (jerez) sherry, port and champagne. (Gundian, 1884, cited by Rojas & Espina, 2015; Rojas, 1897). In that area, far from large urban centers, with few and bad transport routes, the producers of País grapes, small and with few economic resources, found in the "asoleado" an alternative that allowed them to reach the most populated places with greater purchasing power, where they could achieve higher prices for a product of higher quality and valuation (Gundian, 1884, cited by Rojas & Espina, 2015; Rojas, 1897).

Between 1850 and 1880, "asoleados" wines occupied a privileged place among the capital's elite and other important cities such as Valparaíso, Chillán or La Serena. Its arrival from Cauquenes was widely publicized and its price far exceeded the prices of dry wine (Lacoste et al., 2016). This golden age of the "asoleado" would come to an end after the Pacific War, when a greater economic bonanza allowed the import and consumption of European products, leaving aside the "asoleados" for European products such as champagne or sherry-jerez (Couyoumdjian, 2006). Additionally, national elites, like other Latin American elites, strove to imitate European consumption habits and detach themselves from local customs (Lacoste et al., 2016). Falsification had also done its part, taking advantage of the prestige gained by the producers of Cauquenes and Concepción; many producers from other regions of the country saturated the market and contributed to the weakening of a product built with sacrifice and effort (Lacoste, et al. 2016).

In 1938, through the Law of Alcohols, the denomination of origin of fortified wines of Cauquenes was granted to the fortified and liqueurs wines, which were produced between the Maule and Itata rivers and which were elaborated by experimental stations, agricultural schools owned by the State, or by associations of producers and winegrowers, supervised in their technical aspect by the Ministry of Agriculture (Reyes & Lavín, 2022). However, this Law made no specific mention of the word "asoleado." Probably for this reason, between 1930 and 1950, several companies made use of the concept in their labels to promote their products, even when they did not come from the area that had given prestige to this wine, did not have the quality of it (Lacoste et al., 2016) and even used varieties that were not even present in the rainfed lands of Cauquenes (Pinochet, 1941).

Even while the fortified and liqueur wines of Cauquenes were protected around 1938 (Ministerio de Agricultura, 1938), it was not until 1979 that the Alcohol Law explicitly protected the "asoleado" (Ministerio de Agricultura, 1979), reserving that denomination for the fortified wines produced and bottled in the rainfed area between the Mataquito River to the north, as indicated by the previous law, only this time extended to the Biobío River in the south.

In Cauquenes, the production of "asoleado" was part of the viticultural identity, and efforts were made to maintain and improve the quality of making wines. In 1939 the "Cooperativa Vitivinícola de Cauquenes" was created, and a production of 200,000 L of "asoleado" wines were declared back in 1943 (Lomas de Cauquenes, 2023). On another side, in the Cauquenes Wine Experimental Station, various tests were carried out in elaboration techniques that would optimize its quality and guarantee its stability over time (Pinochet, 1941). These works bore fruit, as evidenced by obtaining first place in the National Wine Exhibition of Santiago in 1932 and then again in 1972, both with the "asoleado" from the Experimental Station. Also, it gained recognition by winning the silver medal in the World Competition of Budapest with the white "asoleado" of 1962 harvest (Reyes & Lavín, 2022). For decades, the Experimental Station continued with the production of "asoleado," where the wine was aged for at least 4 years and their work contributed to maintaining a favorable image of the product until the 90s, when the commercial production of this and other wines were completed (Reyes & Lavín, 2022).

In the second half of the nineteenth century, Chilean viticulture would be divided into two from the Maule River to the north, it is possible to find a wine region with a predominance of French vines, cutting-edge technologies and large investments, while between the Maule and Biobío rivers, rainfed vines would continue to prevail, using traditional methodologies based on the País vine (Briones, 2008). To these days, the catalogue of varieties present in the Maule and Biobío regions has been expanded and state-of-the-art technology has been incorporated into wine production. However, the País variety led in the head continues to be cultivated in the rainfed of the communes of San Javier and Cauquenes, which concentrate more than 40% of the national surface of this variety (ODEPA, 2021).

As in the rest of the country, climate change has manifested in the rainfed. Looking at the maximum temperatures recorded from 1964 to 2021, we can see an increase that approaches 3 °C. In the same trend, days with temperatures above 25 and 30 °C have increased by 20 and 33 more days, respectively (Reyes & Salazar, 2023). Rainfall has fallen to such levels that, in the last 20 years, only five times have rains been recorded reaching 650 mm, which is what was considered as normal rainfall in the rainfed agroclimatic of Cauquenes. This situation presents a problem for the production and profitability of rainfed country grape producers, whose grapes and wines are the worst paid in the national market, according to ODEPA (2023) in its Wine Bulletin.

7.3.2 Current State of "Asoleado" Production and Its Producers

The search considered the location of 24 individuals who were identified as "asoleado" producers. Out of these, five were no longer producing, and one was out of the country, so they were not surveyed. The respondents were mainly from the commune of Cauquenes ($n = 10$), San Javier ($n = 6$), and Empedrado ($n = 2$). More than half of the producers own properties ranging from 0.6 to 6.0 ha, although we also found producers with larger areas. Among these, we came across two properties, one of 140 ha and another of 840 ha. Regarding the vineyards from which the grapes for "asoleado" are obtained, 50% of them are smaller than 1.5 ha. In general, most of the small producers of País grapes in the communes of San Javier and Cauquenes have vineyard areas smaller than 5 ha (Reyes, 2020); however, this situation was expected. The majority of the respondents (44%) produce between 1000 and 1240 L per season, followed by 400–850 L (22%) (Table 7.1). There was one producer who made 20,000 L; however, they purchased grapes externally, so the area dedicated to "asoleado" was not considered in the evaluation. In general,

Table 7.1 Frequency of liters produced, liters produced, and grape requirement for estimated sun-drying area

Frequency (N)	Liters produced	Grape requirement (kg)
3	100–300	250–750
4	400–850	1.000–2.125
8	1.000–1.240	2.500–3.100
2	2.000–3.500	5.000–8.750
1	20.000	Buy to others

producers calculate the number of grapes to sun-dry in order to produce the desired volume. This calculation considers a net yield ranging from 30% to 40% of the initial harvested weight. Based on the reported liters and common production values for the type of vineyards owned by the producers, it was estimated that the majority (67%) do not use more than 17% of the vineyard area for "asoleado" and, in no case, exceeds 33%.

More than half of the producers are over 60 years old, with an age range between 33 and 86 years. Forty-four percent of the producers indicated that they have completed high school, while 33% of them completed primary education or part of it. Only two individuals have technical studies, and one has a higher education degree. Lastly, one person mentioned having no formal education.

To understand what "asoleado" means to the producers, they were openly asked about the significance of this wine for them and the importance of producing it. Since the responses were open-ended, seven common patterns were identified based on the frequency of their mention (Table 7.2). These seven patterns were grouped into three main motivations: (a) "Commercial" motivations, which are purely related to selling the product and its market positioning. (b) "Psychological/emotional/familial" motivations, which refer to the emotional memories of family experiences and personal enjoyment of the production activity. (c) "Preservation of culinary heritage" motivations, which relate to the idea of passing on knowledge about the production process and the socio-cultural importance of the product to current and future generations.

Based on the relative importance of motivations for producing "asoleado," five profiles of producers were determined (Table 7.3). In general, the commercial aspect takes precedence, followed by the interest in preserving heritage, and thirdly, a combination of commercial interest with psychological/emotional/familial motivation (Table 7.3). The commercial motivation, driven by the higher price and demand for "asoleado" compared to dry wine, has been a fact since the nineteenth century. While producers also express motivation for heritage preservation, for several of the surveyed individuals, the survival of their vineyard depends on the sale of "asoleado."

Trends in motivational factors were analyzed based on the age of the producers, considering a broad generational division at 50 years old. Although there were only

Table 7.2 Motivations and common patterns detected in "asoleado" production

Common patterns	Frequency (*N*)	Motivation
Preservation of heritage tradition	6	Preservation of culinary heritage
Easy sales and/or demand	8	Commercial
Good price	3	
Gender-specific market niche	2	
To share with family and/or friends	3	Psychological/emotional/familial
Opportunity to improve condition of País grapes	2	
For the pleasure of making them	2	

Table 7.3 Number of producers by the grouping of motivations in "asoleado" production

Type of motivation	Frequency (*N*)
Commercial	6
Preservation of culinary heritage	5
Commercial and psychological/emotional/familial	4
Psychological/emotional/familial	2
Commercial and preservation of culinary heritage	1
Total	*18*

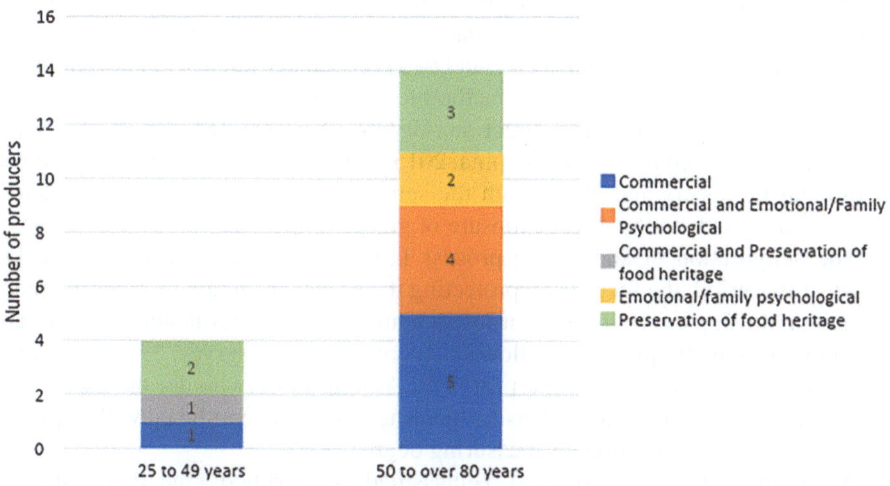

Fig. 7.1 Producers grouped by age ranges regarding their motivations for "asoleado" production

four individuals below the age of 50, it can be observed that both the commercial aspect and the preservation of culinary heritage are present in both age groups (Fig. 7.1). Notably, in the older age range, commercial motivation and psychological/emotional/familial motivation stand out. This can be explained by the fact that asoleado has been present throughout their lives, evoking memories and experiences that evoke warm aspects of home and family life.

It was asked whether the production of "asoleado" was a family tradition, to which 67% responded "no", and 33% responded "yes". Regarding how many years they have been producing "asoleado," most respondents (39%) have a trajectory of 5–10 years, 22% have been producing for less than 5 years, and the same percentage has been in the industry for over 20 years. Finally, 17% indicated having 11–20 years of producer experience. This indicates that most producers have a medium to short trajectory in producing this type of wine. However, "asoleado" has been present in their daily lives, aligning with the previously mentioned psychological/emotional/familial motivation.

Regarding how they learned the production technique, 44% learned it on their own initiative, and 22% learned it through their family. Learning from colleagues and/or friends, through consultancy and other means account for 11%.

The variety used for making "asoleado" is primarily "País," as indicated by all the surveyed producers. Fourteen producers exclusively use this variety, while four include another white variety, with Moscatel de Alejandría or Italia being used in two cases. Semillón, Torontel, and Sauvignon Blanc are also utilized.

The harvest timing is mainly determined by taste, with soluble solids (°Brix) being the second indicator, and personal perception being the third (Fig. 7.2a). Regarding the method of sun-drying or dehydrating the grapes to concentrate the sugar (Fig. 7.2b), most of the producers directly sun-dry the grapes, either by hanging them in tents on the ground (50%), directly on the ground (11%), on a roof (11%), or by placing the cut clusters back on the vines (6%). Only two producers indicated drying them indoors (11%), and two practice late harvest (11%). Although the legislation does not specify the method to be used for sugar concentration, traditional accounts mention the direct sun-drying of grapes (Mujica et al., 2019; Gundian, 1884, cited by Rojas & Espina, 2015), which is also defended by the more traditional producers and aligns with the semantic logic of the word "asoleado," which corresponds to the direct exposure of grapes to the sun.

The dehydration or sun-drying process lasts from 2 weeks to over 4 weeks. During this period, the focus is on protecting the grapes from rain by covering them at night and using covers to prevent insect damage. During fermentation, there are generally no specific protocols followed. According to the survey, 39% of the producers do not follow any specific protocols. Among those who do, the most common parameters monitored are density and/or temperature, combined with applying preservatives, yeast addition, or measuring degrees Brix.

According to the law, "asoleado" is classified as a fortified wine, and as such, it must have a minimum alcohol content of 14% achieved through natural fermentation. However, based on chemical analysis conducted on 11 samples, the average

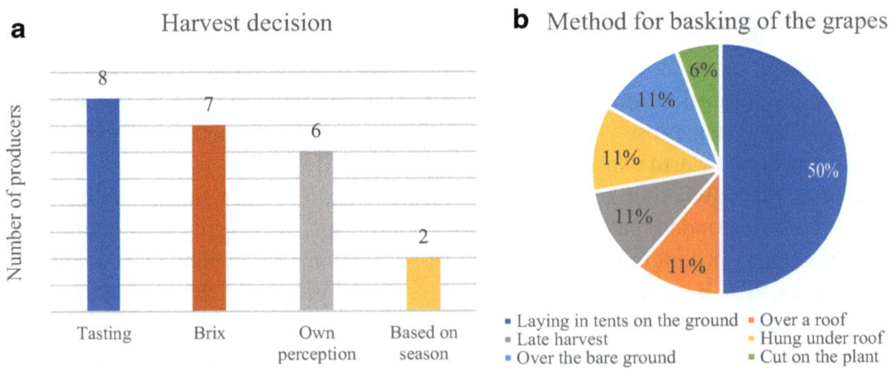

Fig. 7.2 Indices for determining the harvest timing (**a**) and methods used for sun-drying or dehydrating the grapes (**b**)

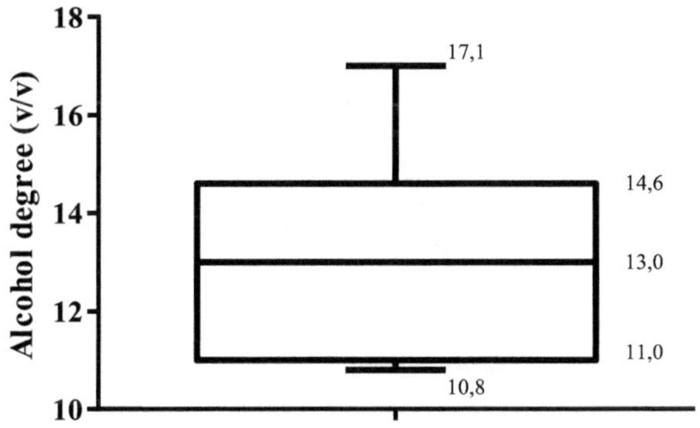

Fig. 7.3 Dispersion measurements of "asoleado" wine samples according to their alcohol content

alcohol content was 13.1% (Fig. 7.3). Only three of the samples exceeded the minimum required by law, and three were very close to the required alcohol level. It was also observed that wines made from grapes dried directly under the sun tended to have higher alcohol levels and pH than those dried using other dehydration techniques. The addition of alcohol, known as "encabezado," which would classify the wine as a liqueur wine rather than a fortified wine, was mentioned by only two producers, and only one of them considered it a regular practice. This practice was recommended to ensure consistent quality and stability over time (Pinochet, 1941).

Most producers (55.6%), especially smaller ones, store their sun-dried wines in plastic containers. Those who produce on a larger scale use stainless steel tanks. Only three producers age their wines in barrels as an aging technique, and this process lasts for 3–5 years. The use of clay amphorae is also mentioned among the production techniques used.

The wine is mainly sold in the same year it is produced, and this is precisely one of the reasons for producing sun-dried wine, as mentioned by the producers. They think it allows for quicker sales and better prices than dry wine. It was only mentioned that the wine that couldn't be sold in the same season is kept for the following year.

Although the law does not explicitly require aging for sun-dried wine, at the Cauquenes Experimental Station, guaranteed aging of 4 years was considered a characteristic of its quality and highly valued by consumers (Pinochet, 1941). It was thought that aging significantly improved the organoleptic quality of sun-dried wines.

To analyze the perceptions of quality, a cross-referencing of information was conducted. Firstly, producers were asked about the main quality attributes of their sun-dried wine, and secondly, a panel of wine professionals conducted a tasting of 14 wines. This sample was divided into three groups: aged samples, 2022 production samples, and 2023 production samples. Most of the producers identified the characteristic of sweetness as the most outstanding attribute of their wines. In the second place, color was mentioned, followed by flavor, closely followed by alcohol content (Fig. 7.4). The aroma of "sun-dried" was mentioned as a distinct aroma, described as the particular smell that the wine acquires when the grapes are

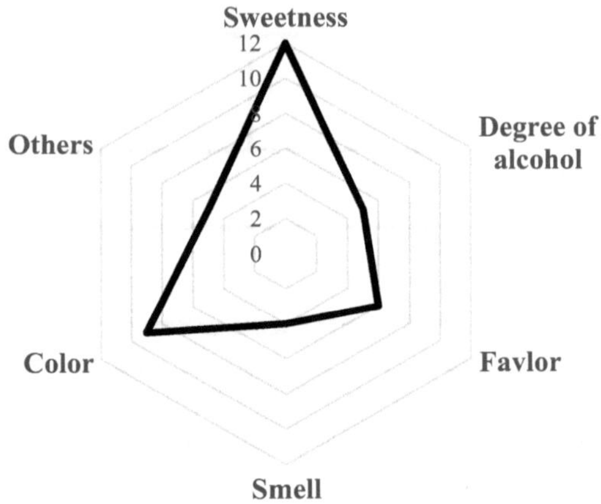

Fig. 7.4 Frequency of quality attributes highlighted by producers regarding their sun-dried wines

Table 7.4 Frequency of percentage range for "asoleado" wine sales in different commercial channels and percentage of sales by format

Percentage range of sales	Property or home (*N*)	Market or similar (*N*)	Restaurants (*N*)	Others (*N*)
[1–25%]	3	2	2	–
[26–50%]	3	3	1	3
[51–75%]	2	–	–	–
[76–100%]	8	3	–	1
Total of mentions	*16*	*8*	*3*	*4*

sun-dried. In the "other" category, the natural condition of the wine was emphasized, referring to wines that are not corrected or preserved during production.

Due to the variability in sales formats and the number of producers using each format (Table 7.4), the analysis of this and their prices were conducted separately, calculating measures of central tendency for each format. Most of the producers sell in bulk, with the most common price being $3000 per liter (Fig. 7.5a), and most of those who use this channel sell almost exclusively from their homes. The second most used format is the 750 mL bottle (Fig. 7.5b), with a median price of $5000 and a price range between $2500 and $10,000. The 375 mL bottle is used by three producers and records the highest sales prices, with a maximum of $24,000 and a minimum of $5000 (Fig. 7.5c). Notably, in this case, those who age their wine fall into this category, which directly influences the higher price. The least frequent format, with two mentions, is the 500 mL bottle (Fig. 7.5d).

A set of questions was conducted to understand the perspective of continuity and future perspectives from the point of view of those who are currently involved in the production of "asoleado." As producers immersed in agri-food systems that are

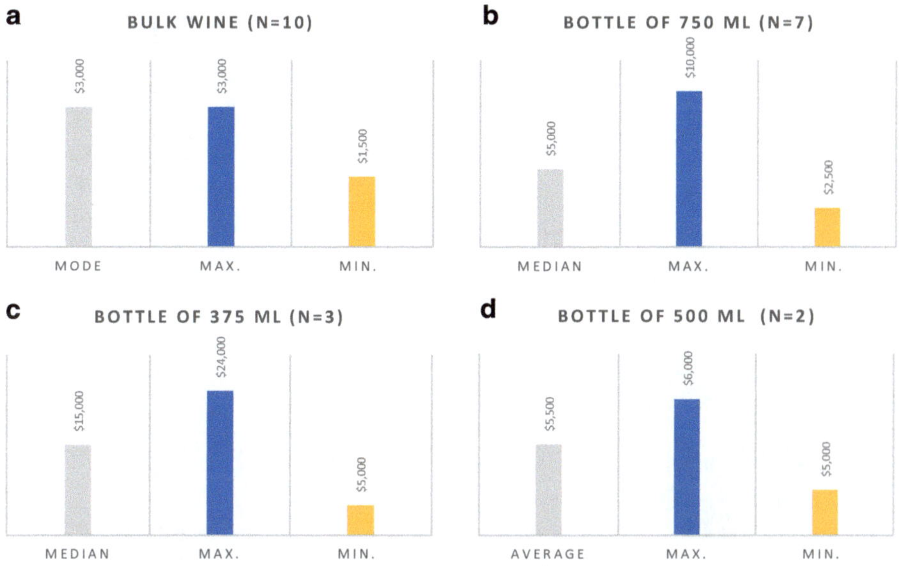

Fig. 7.5 Central tendency statistics for each sales format of "asoleado". (**a**) Represents the statistics on bulk wine sales. (**b**) Represents the statistic on the bottle of 750 ml sales. (**c**) Represents statistic on the 375 ml bottle sales and (**d**) represents the statistics of the 500 ml bottle sales

constantly subjected to climate change and facing a business with fragile stability and profitability, this aspect becomes highly relevant.

Half of the producers consider that the price they receive for their product is adequate, however, 44% consider it to be very low in relation to the work involved (Fig. 7.6a). Older producers are more likely to consider the price low, but it is still more attractive than that of dry wine because "asoleado" has higher demand. There is some uncertainty about the future price, as only a third of the producers think it may increase (Fig. 7.6b). Most of them intend to continue producing "asoleado" (Fig. 7.6c), however, over half of them do not have anyone in their family circle to continue with the production (Fig. 7.6d), especially those older producers whose children already have their own occupations and none have opted for vineyard work and its products.

Another important aspect is the perception that producers have of "asoleado" as a heritage food, indicating that the product has been transmitted transgenerationally and provides identity to rural families in a locality and region, with cultural and identity importance, recognized by 78% of them. Despite that the producers perceive the "asoleado" as something traditional, unique, and specific to their area, half of them were unaware about its Protected Designation of Origin (D.O.). Among those producers who knew the D.O. "asoleado," 90% of them considered it important to maintain it and they would be willing to pay for certification, depending on the price it entails.

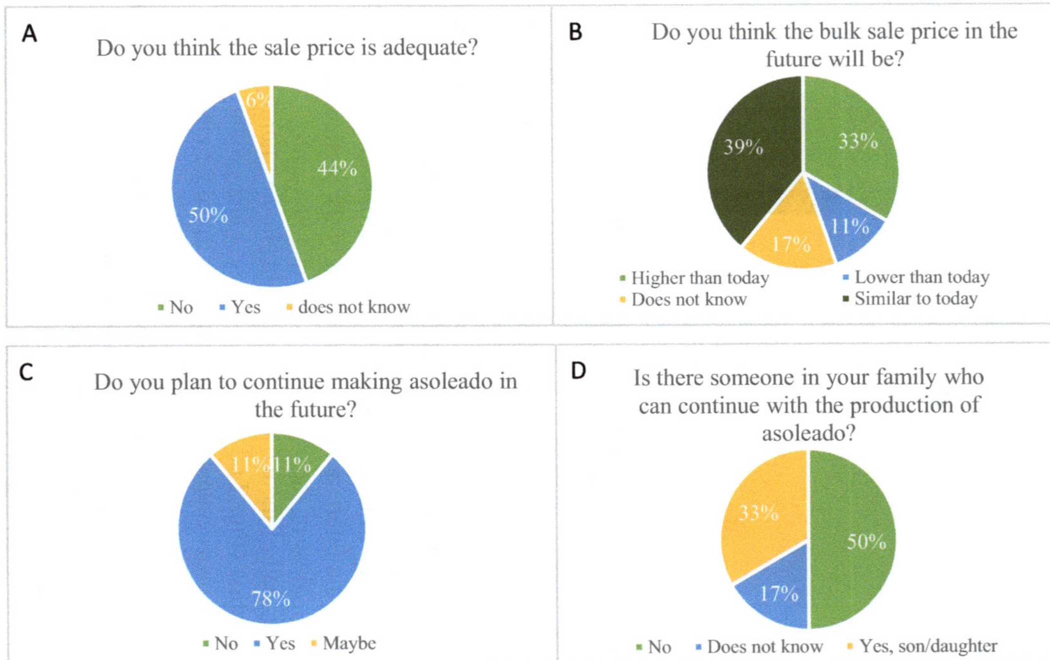

Fig. 7.6 Criteria for future continuity of "asoleado" production. (**A**) Represents the answers of the "Do you think the sale price is adequate?" question. (**B**) Representes the answers of the "Do you think the bulk sale price in the future will be..." question. (**C**) Represents the answers of the "Do you plan to continue making asoleado in the future?" question and (**D**) represents the answers of the "Is there someone in your family who can continue with the production of asoleado?" question

7.4 Conclusions

The "asoleado" continues to be a distinctive product of the Cauquenes commune and a few neighboring communes that were once part of the province. Its production is primarily in the hands of small producers with limited plantations of "País" grapes, and like the producers of the nineteenth century, they see "asoleado" as an alternative to valorize these centuries-old rainfed vineyards. While commercial motivation is the main reason for producing "asoleado," there is also an important component associated with preserving food heritage and family and emotional connection. Despite that most of the producers do not have a direct family heritage and their trajectory is medium to short, it is recognized that the product has been transmitted transgenerationally, thereby connecting past and present generations. However, many of them do not envision the future continuity of family production.

There is a low standardization of processes observed. Some producers follow vinification and aging protocols, and the production method is mainly artisanal and variable. Consistent with what producers have mentioned about their learning in the production of "asoleado," the drying times and methods of the grapes, as well as the production techniques, are diverse. This multiplicity of processes results in a diversity of styles and qualities, with some "asoleados" not reaching the alcohol level to be classified as fortified wines, while others stand out for their quality.

The D.O. "asoleado" establishes territorial production limits and indicates that it is a fortified wine, however, there are multiple questions and gaps regarding production regulations. As mentioned by producers who follow a more artisanal approach, they think that "asoleados" produced with grapes that are not directly sun-dried should not be called "asoleado." Consequently, some producers' products correspond more to late-harvest wines, while others may have been corrected with concentrated must or even artificial sweeteners. This lack of precision and regulation can be highly limiting to repositioning "asoleado" in the major consumer centers of the country, potentially repeating some of the patterns that led to its decline in the national market in the early twentieth century.

In commercial terms, the historical and heritage character of "asoleado," along with its sensory characteristics and organoleptic attributes, could be attractive to the profile of the national consumer, providing an opportunity for revaluation as long as its quality can be guaranteed and maintained over time.

References

Briones, F. (2008). Misiones de estudio y vitivinicultura chilena: 1875–1900. *Tiempo y Espacio, 17*(20), 61–76.

Couyoumdjian, J. (2006). Vinos en Chile desde la independencia hasta el fin de la *Belle époque*. *Historia, 39*(1), 23–64.

Del Pozo, J. (2014). *La historia del vino chileno: desde la época colonial hasta hoy*. Ediciones LOM.

Gay, C. (1855). *Historia física y política de Chile. Tomo II: Agricultura*. Available on: http://www.memoriachilena.gob.cl/602/w3-article-7879.html

Lacoste, P., et al (2016). Asoleado de Cauquenes y Concepción: apogeo y decadencia de un vino chileno con Denominación de Origen. *Idesia 34*(1), 85–99. https://dx.doi.org/10.4067/S0718-34292016000100010

Lomas de Cauquenes. (2023). *Historia*. Available on: https://lomasdecauquenes.cl/content/6-historia

Ministerio de Agricultura. (1938). *Decreto 114*. Available on: http://bcn.cl/2hbpl

Ministerio de Agricultura. (1979). *Decreto Ley 2753*. Available on: https://bcn.cl/2jhre

Mujica, F., Lacoste, M., & Lacoste, P. (2019). Bernardo O'Higgins y el patrimonio del vino en Chile. *Idesia, 37*(4), 109–114.

ODEPA. (2021). *Catastro vitivinícola nacional*. Oficina de Estudios y Políticas Agrarias. Available on: https://www.odepa.gob.cl/rubro/vinos/catastro-viticola-nacional

ODEPA. (2023). *Boletín del vino*. Oficina de Estudios y Políticas Agrarias. Available on: https://bibliotecadigital.odepa.gob.cl/bitstream/handle/20.500.12650/72227/BVino_032023.pdf

Pérez, V. (1886). *Recuerdos del pasado (1814–1860)* (3rd ed., 432 p). Imprenta Gutenberg. Available on: http://www.memoriachilena.gob.cl/archivos2/pdfs/MC0004566.pdf

Pinochet, R. (1941). *El vino asoleado de Cauquenes* (Undergraduated thesis). Universidad de Chile, 117 p.

Reyes, M. (2020). Uvas cultivadas en secano: País y Moscatel. In *Producción vitivinícola en el Secano de Chile Central*. Boletín INIA N 418. Available on: https://biblioteca.inia.cl/bitstream/handle/20.500.14001/6905/NR42181.pdf?sequence=9&isAllowed=y

Reyes, M., & Lavín, A. (2022). Estación Experimental Cauquenes. Historia y aportes a la vitivinicultura nacional. *Revista Iberoamericana de Viticultura, Agroindustria y Ruralidad, 9*(27), 129–142.

Reyes, M., & Salazar, C. (2023). Patrimonial viticulture facing climate change, challenges and opportunities. *BIO Web of Conferences, 56*, 03016.

Rojas, M. (1897). *Tratado de viticultura i vinificación*. Imprenta i encuadernación Barcelona.

Rojas, G. (2021). Chilean vineyards as world heritage. *Revista Iberoamericana de Viticultura, Agroindustria y Ruralidad, 8*(22), 218–225.

Rojas, G., & Espina, N. (2015). *Relación histórica sobre la Vitivinicultura de la Provincia de Cauquenes*. Available on: https://vinifera.cl/editorial/extranet/sistema/informes/documentos/VINOSDECAUQUENES(2015-06-01).pdf

SAG. (2009). *Servicio Agrícola y Ganadero Ley 18.455*. Available on: https://www.sag.gob.cl/sites/default/files/ley_y_dec_78_dec_464_y_dec_521.pdf

Chapter 8
Terroir and Typicity Evolution of Different Uruguayan Wine Regions

Milka Ferrer, Gustavo Pereyra, Ramiro Tachini, Julia Salvarrey, and Mercedes Fourment

8.1 The Concept of Terroir: A Systemic, Dynamic and Constantly Evolving Concept

Terroir is a word of Latin origin derived from *territoire*. The historical evolution of the concept of terroir has its origin in 1200; in 1000, according to other authors in Burgundy, being a linguistic modification of older forms (tieroir, tioroer), originating in the popular Latin "*territorium*" (Ballantyne, 2011; Tonietto, 2007). But it was not until the seventeenth and eighteenth centuries that it was affirmed, understood as "a geographical situation describing the characteristics of the physical environment considered homogeneous". It is from this concept that the first definition, "an extension of land with certain agricultural aptitudes" (Deloire et al., 2008), originates.

Its evolution continued in the twentieth century when soil suitability (Henin, 1957), geology and landscape (Morlat & Jacquet, 1993) and the valorization of a territory/product by incorporating the social and cultural dimension and know-how (Rouquette, 1994) were incorporated.

In a synthesis, Vaudour (2003) indicates that the concept of terroir comprises four dimensions: terroir-physical (relationship with the soil/climate), terroir-space (relationship with the territory), terroir-awareness (relationship with know-how) and terroir-slogan (relationship with communication about an identity). These concepts have in common that they are based on the origin, specificity and typicality of the product and, as a consequence, cannot be reproduced elsewhere. The aim is to add value by differentiating between standard products vs territorial products; for

M. Ferrer (✉) · G. Pereyra · R. Tachini · J. Salvarrey · M. Fourment
Facultad de Agronomía, Universidad de la República, Montevideo, Uruguay
e-mail: mferrer@fagro.edu.uy

G. Gutiérrez Gamboa, M. Fourment (eds.), *Latin American Viticulture Adaptation to Climate Change*, https://doi.org/10.1007/978-3-031-51325-1_8

some consumers, particularly those from the "New World", this message may be a relatively recent construction (Muchnik, 2006).

In 2005, UNESCO, INAO and INRA defined terroir as "A responsible alliance between people and their territory, based on know-how, production, culture, landscape and heritage. As such, they are the source of great human, biological and cultural diversity, expressed in products, typicity, originality and the recognition that goes with them. They create value and wealth. Terroirs are living, innovative spaces where human communities, through their history, build viable, sustainable development. Terroirs help to meet consumer expectations in their search for diversity, authenticity, food cultures, nutritional balance and health".

For its application to wine products, the general assembly of the International Organization of Vine and Wine (OIV) defines vitivinicultural "terroir" as a concept that refers to an area in which collective knowledge of the interactions between the identifiable physical and biological environment and applied vitivinicultural practices develops, providing distinctive characteristics for the products originating from this area. "Terroir includes specific soil, topography, climate, landscape characteristics and biodiversity features" (OIV, 2010). The Resolution OIV-VITI 423-2012 expressed: "that terroir has a spatial dimension, which implies a need for delimitation and zoning and that different aspects of terroir can be zoned, particularly physical environment aspects: soil and climate", decides to adopt the following resolution, concerning the OIV "Guidelines for vitiviniculture zoning methodologies on a soil and on a climate level" (OIV, 2012).

From these definitions/resolutions, it emerges that the various dimensions of terroir require a multidisciplinary approach that allows robust scientific analysis of the social, cultural and environmental characteristics of plant requirements, leading to sustainable grape and wine production, as well as implementing a consumer communication strategy.

This term, which originated in France some 1000 years ago, was adopted by winemakers in many "New World" countries less than 50 years ago (Tonietto, 2007) and is recognized by wine consumers to indicate and appreciate the unique qualities of certain wines. This recognition does not need to be linked to Appellations of Origin, the use of which is restricted. Moreover, this concept, which associates wine with a territory, generates "changing geographies" that are not necessarily the traditional ones, which contributes to incorporating new producers and new consumers.

An economic analysis by several authors on the interest in studying terroir in "New World" countries indicates that more competitive prices are achieved when a wine is linked to its place of origin, as this strategy contributes to consumer recognition (Ballantyne et al., 2019; Cross et al., 2017).

8.2 Methodology for Studying the Different Components of the Terroir

In one of the reports of Resolution OIV-VITI 423-2012 (OIV, 2012), it considers that the "terroir" represents a spatial dimension that implies the need for delimitation and that different aspects of the "terroir" can be zoned the physical elements of the environment: climate and soil (Fig. 8.1). Terroir is a system in which its components interact in a dynamic association. Although they can be studied separately, it is ultimately the analysis of the response of the whole plant that is the synthesis of these interactions (Deloire et al., 2003).

8.2.1 Climate as a Component of Terroir

The climate is a significant factor in the physical environment influencing terroir expression (De Rességuier et al., 2020). It is also a critical component of the system because it limits where grapes can be grown to reach maturity globally and locally (Jones, 2018; van Leeuwen, 2010). For its analysis, bioclimatic indices are used, the purpose of which is to describe and delimit homogeneous climatic zones. The main objective of these zones is to ensure that the variety(ies) to be planted complete oenological maturity.

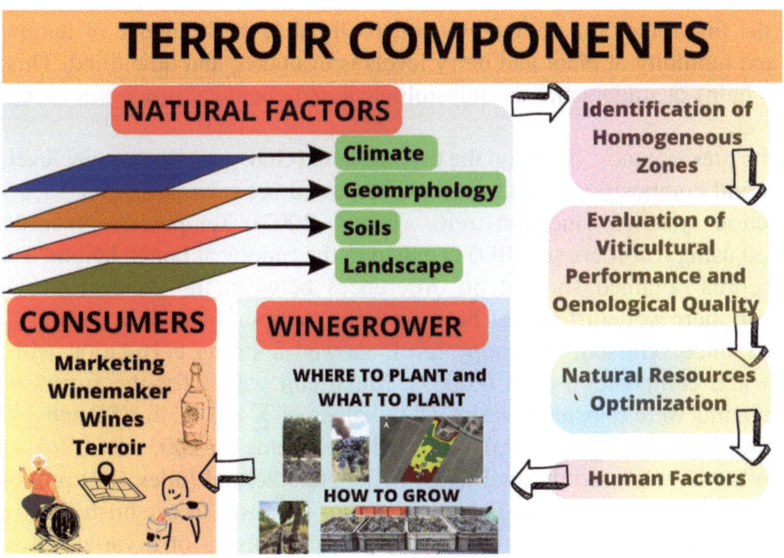

Fig. 8.1 Terroir components

The spatial scales used to study climate are the macroclimate, which is used to define an extensive climatic zone; the mesoclimate for a local scale; and the microclimate, which is quantified at the plot level. These last two scales also allow the analysis of the annual variability of climate and incorporate another component of terroir, the plant variety.

One of the first precedents for macro-scale regionalization was the Winkler Index (WI) (Amerine & Winkler, 1944) in California. This index is based on the sum of active temperatures during the growing season from first April to 31th October for HN conditions. Active temperatures are defined as values above the vegetative zero, set by convention for vines at 10 °C, and are calculated as the difference between the average daily temperature and 10 °C. The definition of this thermal threshold arises from the recording of the average bud break temperature of a significant number of varieties. These authors defined 5 zones with climates classified from cold to warm, according to the thermal range from <1389 °C in the cold region to >2222 °C in the warm region.

Other benefits of this index, particularly in mesoscale studies, are determining the climatic supply to analyse the possibility of acclimatization of varieties to local climatic conditions to maximize the expression of the terroir. In this sense, it is necessary to know the thermal requirements of the different varieties to ensure optimal grape ripening. Using the WI, van Leeuwen et al. (2008) report the thermal needs and ripening order of many varieties. In that study, they report, for example, requirements for Merlot of $\sum 1474$ °C, Tannat of $\sum 1501$ °C and Cabernet Sauvignon of $\sum 1520$ °C, thermal degrees recorded from first January until harvest, for HN conditions.

In the fine-scale climate analysis (microclimate), the influence of temperature, light and humidity at plant and berry levels is evaluated and quantified. This information helps us understand the physiological and chemical processes occurring at that level.

In the present study, based on the thermal sum (GDD_{10}) at the canopy level, berry weight and composition evolution were evaluated between 2002 and 2003 in two conduction systems, vine and trellis (Ferrer, 2007). Temperature records were obtained using recorders (HOBO) installed in the canopy at bunch height. The thermal records obtained showed no differences between the conduction systems. However, there were differences between years (temporal variability). GDD_{10} accumulated since veraison, at canopy level, showed a strong correlation with grape weight and composition ($p < 0.0001$). Maximum values for the sugars and the anthocyanins were registered with GDD_{10} values of $\sum 443$ °C in VSP and $\sum 439$ °C in lyre for 2002, and $\sum 428$ °C in both trelling systems for 2003.

Another widely used thermal index is the Helothermal Index (HI) proposed by Huglin (1978). Based on this index this author proposes to establish a correlation between temperature and the number of potential sugars of a variety in a given region. As the quantity of sugars is a product of the photosynthesis process, the index considers the temperature favourable to this process, as well as the length of the day, depending on the latitude of the region to be evaluated. Therefore, the HI relates the air temperature during the active period of vegetative growth

(October–March in the Southern Hemisphere) and a coefficient of day length that varies according to the latitude. The HI provides information about the local heat summation considering the average and maximum temperatures, weighing the accumulated temperatures to the daytime period. The index establishes ranges of heat summation in base 10 °C with a minimum of ∑1500 °C (cold climate), up to >∑3000 °C (warm climate); outside these ranges, the minimum sugar contents of 180–200 g/L considered for grape ripening would not be reached. The spatial and temporal evolution of this index over long periods is used to analyse climate change, as proposed by Le Roux et al. (2008) for the period 1986–2005 and by Fourment et al. (2013) for Uruguay for the period 1994–2009.

Huglin and Schneiner (1998), to corroborate the relevance of the index and to determine the potential sugars content in grapes of a region, analysed the behaviour of a wide range of varieties and regions and established a correlation of $R = 0.86$ between this index and the sugar content. Ferrer (2007) proposed for Uruguay the viticultural regionalization based on the Multicriteria Climate System (Tonietto, 1999; Tonietto & Carbonneau, 2004), in which the Heliothermal Index is included. For the calculation of the index, thermal information was processed for 30 years from 23 meteorological stations that meet WMO standards and are distributed throughout the territory. The index values were mapped using the programme "SPRING: Integrating remote sensing and GIS by object-oriented data modeling" (Camara et al., 1996) (Fig. 8.1, left). This index ranges from values of ∑2100 °C to ∑2600 °C.

This study aimed to determine the sugar accumulation potential for 3 varieties, Tannat, Cabernet Sauvignon and Merlot, concerning the ranges of the Heliothermal Index (Fig. 8.2). The information on the sugar content of these varieties came from the processing of 3000–4000 forms, which must be filled in by winemakers (sworn declaration), and which were provided by INAVI (National Institute of Viticulture).

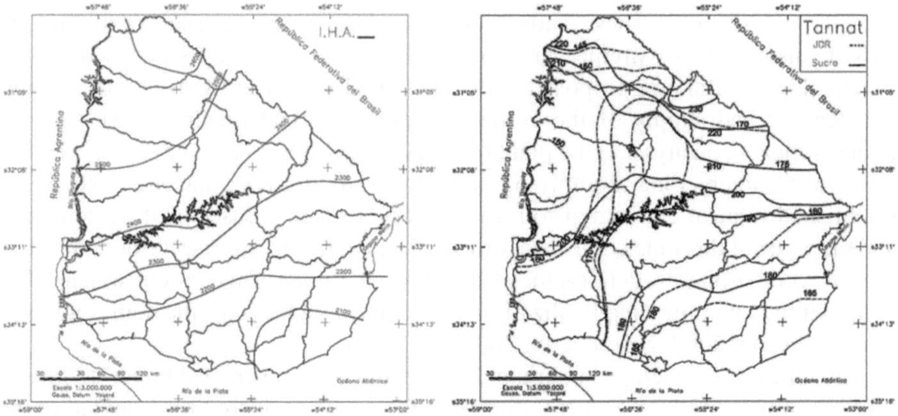

Fig. 8.2 Heliothermal Index calculated from first September to 28th February (HS) (Ferrer, 2007). (left), sugar content (g/L) (in bold) and length of growing cycle from budburst to harvest (days) (in dotted line) for the Tannat grape variety (right) in Uruguay

The information, by variety, includes grape production per hectare, the probable amount of sugar per litre of must at harvest and the probable harvest date. For the calculation of the cycle duration, the number of days from first September (the date of bud break of the HS vine and the start of the HI heat summation) to the date of harvest from the affidavit was counted. The processed data were analysed in ten departments where the three varieties were present and for 2001–2004.

The sugar content and the cycle length were related to the HI values. The established correlation of the index with the sugars content for all three varieties and a wide range of regions was $R = 0.81$ ($p = 0.001$). For each variety, the correlations established were: Merlot $R = 0.85$ ($p < 0.01$), Tannat $R = 0.84$ ($p < 0.01$), Cabernet-Sauvignon, $R = 0.83$ ($p = 0.001$), which is in agreement with that reported by Huglin and Schneider (1998). The maps show that the sugar content increases from south to north as the index increases (Example in Fig. 8.1, right).

The varieties show a different response regarding sugar accumulation at the same or increased thermal supply. In Merlot, there is a transition between IH4 and IH5, where it gains 10–15 g/L of sugars. In Cabernet Sauvignon, the difference between the extremes of the index (IH3-IH5) is 10 g/L. Tannat is more sensitive to the increase in temperature, with extreme values between IH3 and IH5 of 40–50 g/L. The application of this index allowed us to determine that there are no limitations to reaching maturity for these three varieties in the country, that their thermal needs are different, and that according to the thermal offer of the region, they can reach different sugar content (probable alcohol) and therefore wines with different characteristics. Finally, the vegetative cycle of the varieties follows the same trend as sugar accumulation.

Tonietto and Carbonneau (2004) propose to regionalize the multi-criteria climate system (MCC). This system analyses three indices: the Heliothermal Index (HI), the Cold Night Index (CI) and the Drought Index (IS). The CI refers to the average minimum temperature the month before harvest. It is a qualitative index that aims to evaluate the potential of a region in the production of secondary metabolites (polyphenols and aromas). These authors propose four climate classes: warm nights >18 °C and at other extreme very cold nights <11.9 °C. The IS is based on the Potential Water Balance proposed by Riou et al. (1994). The IS takes into account the demand and supply of the climate (temperature and precipitation), the soil (water supply capacity) and the plant (crop evapotranspiration). The index is calculated for 6 months on a monthly basis (October–March HS), starting from a pre-established value of 200 mm of soil water reserve capacity. Based on the calculation and the value of this index, five categories were established, from wet (more than151 mm) to very strong drought (values less than −201 mm). This system was used to carry out the viticulture regionalization of Uruguay as a basis for the delimitation of terroirs (Ferrer, 2007; Ferrer et al., 2007, 2012a, b).

The vine cycle develops in our conditions from September (bud break) to March (harvest of most of the varieties); these months were considered for calculating the indices, proposing an adaptation of these. The sum of temperatures of the HI was carried out from first September to 28th February. In the case of the CI, the average minimum air temperature from 15th February to 15th March was considered. The calculation of the IS was carried out, considering the real initial useful water reserve capacity of the dominant soils of the wine-growing regions, where 82.3% of the soils have values between 40 and 160 mm (Molfino & Califra, 2001). This balance was carried out from 1st September to 28th February.

The climate index data were treated by multivariate analysis: Principal Component Analysis (PCA) and Cluster Analysis (CA), the latter to confirm the correspondence and determine the boundaries of the regions delimited by the PCA. Before PCA the data were standardized. The CA is calculated by applying the Ward Hierarchical Algorithm, and the number of clusters is determined by the pseudo-F (relative maximum), graphically translated into a dendrogram. Using the programme "SPRING: Integrating remote sensing and GIS by object-oriented data modelling" (Camara et al., 1996), the different climatic zones of Uruguay were mapped and delimited. Based on this, six climatic zones were delimited and described in Fig. 8.3.

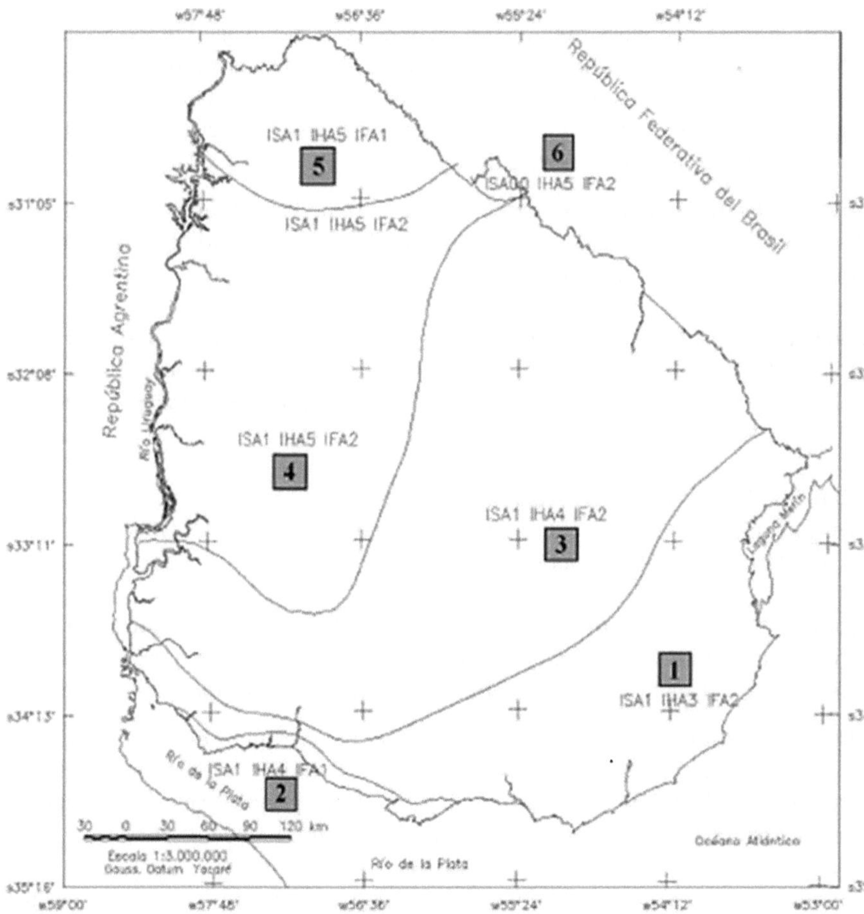

Fig. 8.3 Delimitation of Uruguay's viticultural regions. (Ferrer, 2007)

Region 1 – South East Zone A region with climatic type ISA1 IHA3 IFA2 is delimited, i.e., with a viticultural climate of "moderate drought, temperate, mild nights". This zone represents 60.4% of the total production, and the main varieties in this region are red varieties (73%). According to INAVI (2023) data for 2022, this region and in particular Maldonado department has grown in surface area by 46%, positioning itself as the new emerging region of Uruguay, thanks to being a terroir directly influenced by the Atlantic Ocean and presenting soils.

Region 2 – South West Zone A region with climatic type ISA1 IHA4 IFA1 is delimited, that is to say, with a viticultural climate "moderate drought, warm temperate, with warm nights" is located on the banks of the estuary of the "Río de la Plata", and with a width of approximately 15 km. This region is characterized by the influence exerted by the sea breeze. This area represents 35% of the total production, and the main varieties of this region, 77.4% are red.

According to INAVI (2023) data for 2007, 95.4% of the total production is concentrated in these two areas. As most of the vineyards are located in the South East and South West regions of the country, we can say that they are subject to the influence of bodies of water and are planted in two types of vine-growing climate, one "temperate, with mild nights and moderate drought" and the second "warm temperate, with warm nights and moderate drought".

Region 3 – Central Zone This is the region with the largest territory classified as ISA1 IHA4 IFA2, i.e., with a "warm temperate climate with mild nights and moderate drought". This zone represents 1.3% of the total production.

Region 4 – West Coastal Zone A region spread over territory classified ISA1 IHA5 IFA2, i.e., a "warm wine climate with mild nights and moderate drought", which, being far from the Atlantic Ocean, has a continental character. This zone represents 1.9% of total production.

Region 5 – North Zone It is classified as ISA1 IHA5 IFA1, i.e., with a winegrowing climate of "moderate drought, warm, with warm nights". This zone represents 1.25% of the total production.

Region 6 – North East Zone Classified as ISA00 IHA5 IFA2, with a wine-growing climate "warm, with warm nights and humid", very rainy where a high proportion of sand characterizes the soils. This zone represents 0.25% of the total production.

The publication *Climate, zoning and wine typicity in Ibero-American wine regions* (Ferrer et al., 2012a, b) presents a series of studies on the climatic characterization using the CCM methodology in the following Ibero-American countries: Argentina, Bolivia, Brazil, Chile, Cuba, Mexico, Spain, France, Peru, Portugal and Uruguay.

To corroborate the proposed climatic delimitation, a study was carried out that analysed the effect of the environment using vine response and grape composition as indicators. Four cv. Tannat vineyards in three different climatic regions of Uruguay (a mesoscale) with similar soil conditions were studied in 2008 and 2009. Salto: moderate drought, warm, with warm nights (hereinafter abbreviated as "warm climate"); Colonia: moderate drought, warm temperate, with warm nights (hereinafter abbreviated as "temperate-warm climate"); Canelones: moderate drought, temperate, temperate nights (hereinafter abbreviated as "temperate climate") (Fig. 8.4). Vines were grafted onto SO4 (*Vitis berlandieri* x *V. riparia*) rootstock and were trained on a trellis system. At each vineyard, we recorded: yield per plant, pruning weight, leaf area and pre-dawn leaf water potential. We analysed sugars, total acidity and pH, polyphenolic potential, organic acids and berry weight. Weather information was obtained from weather stations (MMO standards) (Ferrer et al., 2012a, b).

The proposed climatic delimitation is manifested in the plant response and the primary and secondary composition of the grapes. The discriminating factors of the climate are thermal accumulation and rainfall (Table 8.1). From the plant response, it was found that for the same variety, the ripening period was shorter in the warm climate (Salto) than in the other two regions (Colonia and Canelones).

The influence of the terroir is evident in the composition of the grapes, according to the differences in the sugar and anthocyanin content of the must (Table 8.2). In Uruguay, the minimum legal alcohol content for the VCP wines (top category) requires a minimum sugar content of grapes (210 g/L); this value was reached only

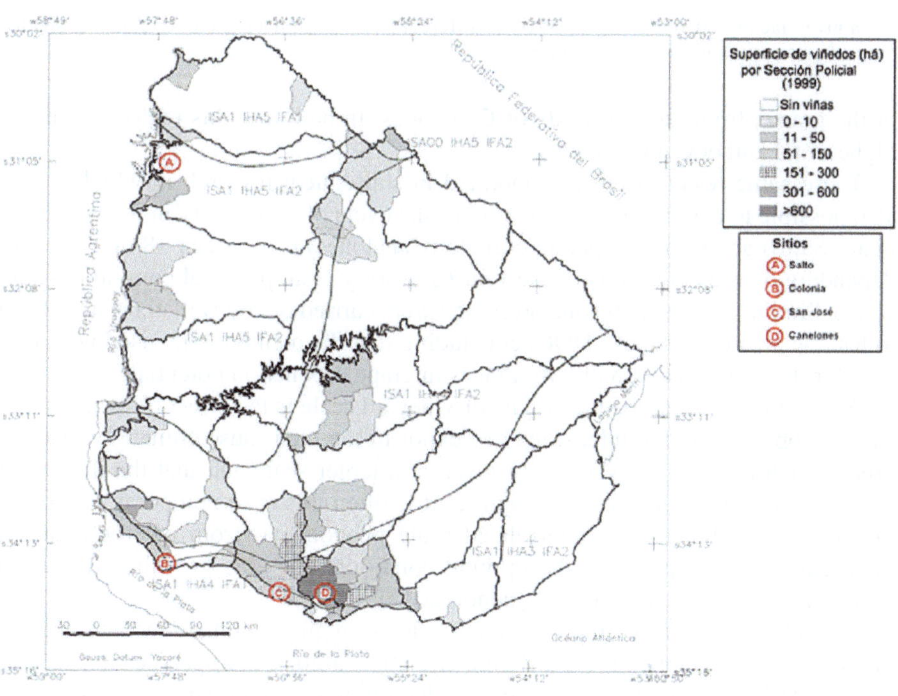

Fig. 8.4 Map with the location of the trials

Table 8.1 Climatic averages for three wine-growing regions of Uruguay (Salto, Colonia and Canelones) in 2008

Variable	Salto (Warm)	Colonia (Temperate-warm)	Canelones (Temperate)
Rainfall during the crop cycle, mm (1972–2000)	722	523	522
Rainfall during the crop cycle, mm (year 2008)	704	450	592
Maximum temperature during January (MTJ), °C (1972–2000)	31.6	27.9	29
Maximum temperature during January (MTJ), °C (year 2008)	31.7	29.7	29
Degree day base 10 °C, veraison at harvest (DD_{10}, V-H) (year 2008)	876	845	936
Number of days veraison at harvest (Julian days), N° J, V-H, (year 2008)	56	65	76

Table 8.2 Tannat berry composition in 2008 harvest

	Salto (Warm)	Colonia (Temperate-warm)	Canelones (Temperate)
Sugars (g/L)	197 b	184 b	224 a
pH	3.57 a	3.32 b	3.40 b
Anthocyanins (mg/L)	1820 c	2002 b	2033 a
Phenols richness (ua)	57.4 b	58.7 b	63.3 b

For a given factor and significance $p < 0.05$, different letters within a same row represent significant differences (Duncan's test, $p < 0.05$)

in the grapes from the vineyards of Canelones. In addition, this region presented higher total anthocyanin contents.

These same results were corroborated in subsequent research in which it was also possible to determine the precocity of the cycle of this variety in the warm climate (Salto) about the temperate climate (Canelones) of between 18 and 42 days depending on the conditions of the year (Salvarrey et al. personal communication, 2023). Similar studies to the one presented were carried out by Bollatia et al. (2015) to test the area of Cesanese DOC, concluding that the methodology used was relevant for different sectors with differences in grape production potential.

Another example at the mesoscale or vineyard scale is the effect of local factors such as topography (altitude, slope and exposure), which cause climatic variations greater than the climatic variability given at a larger scale and that the data from regional weather stations cannot capture. It is often this spatiotemporal variability of climate, combined with interrelated scale phenomena (from macroclimate to microclimate), that provides the optimal conditions for vine growth and gives a viticultural terroir its specificity (Quénol, 2017).

Mesoclimatic analysis improves knowledge of regional climate by studying the influence of topography on the different climatic variables at a specific site, such as a vineyard (terrain on the incidence of radiation, temperature and wind exposure) (De Rességuier et al., 2020; Dumas et al., 1997; Quénol & Bonnardot, 2014).

Mesoclimatic conditions also include the role of proximity to bodies of water such as oceans, lakes or rivers, giving rise to local air circulation. This is due to thermal differences between the air above the water and that on the ground, as in the south-eastern region of Uruguay (Fourment et al., 2017). The arrival of sea breezes can prevent extreme temperatures, avoiding thermal stress during the vine's vegetative cycle and the grapes' ripening period.

Based on the topoclimate, Tachini et al. (2022) analysed the thermal behaviour of a commercial vineyard of the Tannat variety and evaluated the response of the plant. The trial was located in Region 2 South-East of Uruguay, which corresponds to a "temperate with mild nights and moderate drought" type of viticultural climate surrounded by a chain of low-altitude mountain ranges (Cuchilla Grande) and simultaneously influenced by the Atlantic Ocean and the estuary of the "Río de la Plata".

Altitude and exposure to oceanic winds were the main factors causing temperature variability at the mesoscale, while the effect of sun exposure was not significant. Thermal differences were observed in the extreme values (minimum and maximum temperatures, Fig. 8.5), where high areas exposed to sea breezes reach lower temperatures during the day, while low, concave areas protected from oceanic winds reach the highest values. However, during the night, the high plots were the

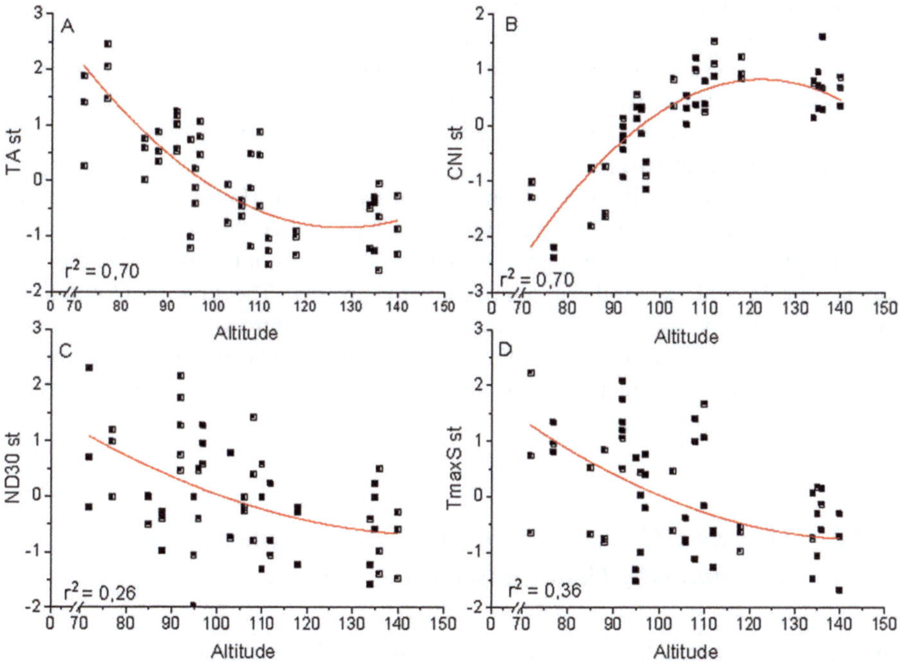

Fig. 8.5 Correlations between plots' topography and standardized bioclimatic indices in 2019, 2020 and 2021 vegetative cycles: correlation between altitude and (**a**) thermal amplitude under ripening period; (**b**) cool night index; (**c**) number of day above 30 °C during ripening period; (**d**) mean maximum temperature during ripening period

warmest as the cool air drained towards the low areas, causing a greater thermal amplitude in these areas.

The combination of topography-soil-climate had an impact on the response of Tannat, where high plots favoured malic acid in 15%. At the same time, lower elevations showed higher amounts of secondary metabolites, mainly the potentially extractable anthocyanins at pH 1 and the actual anthocyanins at pH 3.2 by 21% and 16%, respectively. Consequently, the characterization of the region concerning its topoclimate and its response on the vine will allow agronomic decisions aimed at enhancing the typicity of the terroir.

8.2.2 *Soil as a Component of Terroir*

The effect of the soil on vine behaviour and grape composition is complex because the soil influences vine mineral nutrition and water uptake (van Leeuwen & Seguin, 2006).

The analysis of the "soil-landscape" factor in terroir is carried out in a homogeneous region from the climatic point of view. Therefore, its approach is at the meso or microclimate scale. In this sense, in a mesoscale study carried out in Uruguay, the methodology proposed by Vaudour (2003) was applied, consisting of delimiting homogeneous territorial units by edaphic-landscapes and applying zonal geographic operators (Barbosa et al., 1998); using the Geographic Information System, basic territorial units (UTB) were obtained, which, grouped with the vineyards, made it possible to delimit two viticultural territorial units or "terroirs" (see diagram in Michelazzo et al. (2018)).

The homogeneous climatic region, previously defined (Ferrer, 2007), is South West: corresponding to the climatic type ISA1 IHA4 IFA1, "warm temperate, with warm nights and moderate drought", located on the banks of the estuary of the "Río de la Plata", with a width of approximately 15 km. The city of Colonia del Sacramento (Department of Colonia), whose historic centre was declared a World Heritage Site by UNESCO in 1995, is in this region. The methodology applied made it possible to characterize and separate two potential terroirs in the Colonia del Sacramento area (Fig. 8.6).

8.2.3 *Man as a Component of the Terroir: Winegrowers, Agronomists, Oenologists and Consumers*

At the vineyard scale, there is generally a high spatial heterogeneity of plant vigour due to structural factors (such as soil morphological characteristics as soil composition, topoclimate), functional factors (such as the cultivation practices applied by the vine grower), and the climate of the season (Morlat et al., 2001; van Leeuwen &

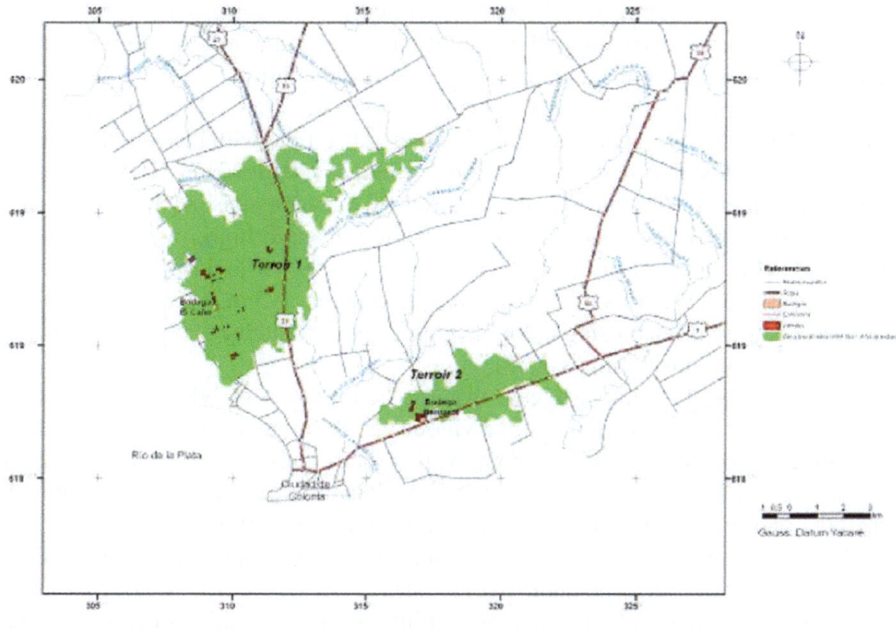

Fig. 8.6 Two potential terroirs in the Colonia del Sacramento area, Uruguay

Seguin, 2006). Often, these cultivation practices result from their experience and are used in a general way in the vineyard without considering the internal variability of the vineyard plot. This has direct negative consequences on the composition of the grape, the yield of the plot and the overall profitability of the company, as well as causing undesirable negative environmental impacts (nitrogen fertilization, phytosanitary applications, etc.), not exploiting the expression of terroir on the composition of the grape or the opportunity for differentiated harvests (Ferrer et al., 2020).

One of the objectives of the terroir study is to provide the winegrower with scientifically proven tools to help him make informed decisions, such as the choice of management practices such as irrigation, fertilization and sanitary control according to the characteristics of the plot. The application of precision viticulture techniques is a set of tools that allows the viticulturist, winemakers and researchers to understand the functioning of terroir components, in particular soil and topoclimate, as well as to verify the existence of spatial variability of these factors at the vineyard scale (Bramley & Hamilton, 2007; Kasimati et al., 2023). The most widespread methodology is the use of the Normalized Difference Vegetative Index (NDVI), obtained by multispectral remote sensing, which refers to the estimation of the proportion of the area covered by photosynthetically active vegetation calculated from reflectance measurements (red and near-infrared). This index provides a broad view of the area studied, is positively correlated with plant vigour (weight of pruned wood, leaf area) and allows a first delimitation of sub-plots based on the values of the index. These sub-plots, of uniform vigour, are the basis for the application of site-specific management, which can be defined as the adaptation of cultivation techniques according to the spatial variability of the plot and the requirements of the plants to increase economic and environmental sustainability.

As an example of the application of precision viticulture, Ferrer et al. (2020) conducted a study in Uruguay in a rainfed vineyard (1.1 ha) with the cultivar Tannat, on a vertical trellis, during three consecutive vintages (2015–2017). NDVI was estimated from high-resolution images acquired by airborne sensors, which allowed differentiating of three vigour levels within the vineyard: high, medium and low (Fig. 8.7). The vineyard of 1.1 ha was planted in 1998 with *Vitis vinifera* L. cv. Tannat, grafted on SO_4 rootstock. The vine spacing was 2.5 m × 1.2 m (3333 vines ha^{-1}). Vines were pruned using a double guyot system, and the shoots were trained to a vertical shoot positioning system.

The results of this study suggest that the delimitation of NDVI zones within vineyards was relevant to characterize vigour variability and had consequences on vegetative growth, yield and berry composition. During the 3 years of study, vigour zones selected by NDVI values within the vineyard remained stable, although weather conditions each year modified the absolute values of NDVI. This indicates that NDVI information allows site-specific management practices to reduce production costs and improve grape quality (Fig. 8.8).

The effect of water and nitrogen availability associated with soil type on plant vigour and grape composition has been demonstrated for Cabernet Sauvignon (Choné et al., 2001) and Merlot (Trégoat et al., 2002). In Uruguay, Pereyra et al. (2022a, b) addressed the effect of cultivation techniques on plant vigour and, thus, on plot terroir. Based on the results obtained in that study and the communications of several authors, a study was installed in the same plot (Ferrer et al., 2020) to evaluate the impact of two components of the terroir, the soil and the intervention of the winegrower (site-specific management) to increase yields and improve berry quality. At plot scale, during three consecutive seasons, contrasting treatments

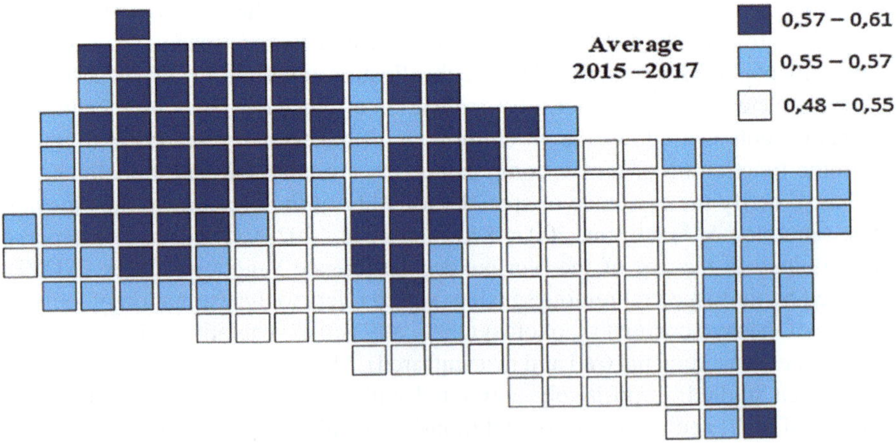

Fig. 8.7 Maps of NDVI values depicting the three vigour zones (white = low, grey = medium, black = high), Ferrer et al. (2020)

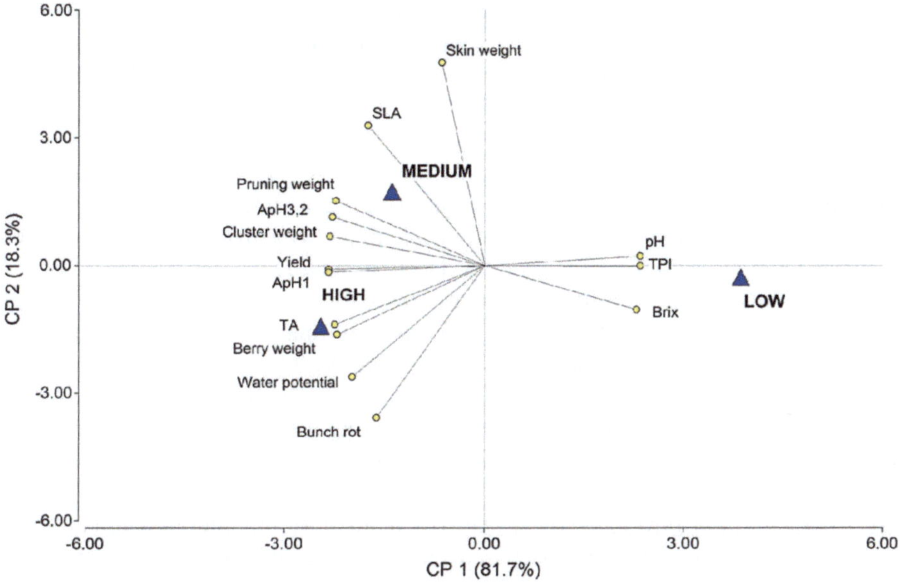

Fig. 8.8 Bi-plot from the principal component analysis discriminating the three vigour categories within the vineyard studied according to vegetative growth, yield, sanitary status and berry compositional traits (*TA* total acidity, *SLA* surface leaf area, *TPI* Total Phenolic Index)

designed ad hoc were tested for two vigour zones pre-established by NDVI: high vigour zone (HV) and low vigour zone (LV). The vineyard was not irrigated and received standard fertilization with urea, distributed half pre-flowering and half post-harvest at a total dose of 140 kg of urea (46% N) fertilizer per ha. Soil physical and chemical characteristics also showed a strong spatial variability, mainly regarding the percentage of clay, clay type and total available water (TAW). The TAW estimated from soil texture and root depth was greater than 180 mm in the HV with a predominance of montmorillonite (expansive clay) and less than 140 mm in the LV with higher content of illite compared to HV (Pereyra et al., 2022a). The grapes are harvested uniformly and the grapes from the entire plot are destined to produce common wine. The treatments aimed to reduce water and nitrogen supply and improve microclimatic conditions in the cluster zone in the HV zone. In the LV zone, treatments were aimed to increase the water and nitrogen supply (Pereyra et al., 2022b) H-L (High Vigor Leaf removal) and L+W (Low Vigor Irrigation) treatments could produce good quality wines even with higher yields and better bunch health than HV and LV. In addition, yield and composition parameters were more homogeneous, a desirable situation for winegrowers. Without intervention using management techniques, wines produced from the control grapes (HV and LV) would be impossible to sell under the category of quality wines according to Uruguayan regulations because of their low alcohol level (less than 12%). Therefore, their sale price would be automatically significantly lower than that of quality wine (USD 2.5 vs USD 7.2; 750 ml bottle), which justifies a differentiated vintage. In rainy years, the water reduction treatment improved bunch health and grape composition. In contrast, nitrogen supply in the LV zone was highly dependent on soil water availability. Winegrowers could use this approach on a larger scale to determine micro-terroirs and thus generate the application of site-specific techniques to obtain the potential for productivity.

Best et al. (2013) of the INIA Precision Viticulture Programme propose to advance in Digital Terroir because highly innovative solutions are required due to the need to automate the processes of capturing, analyzing and interpreting field information in a format and at a cost accessible to the winegrower. This information collected and processed is supported for decision-making.

In that sense, several digital tools are available for smartphones and were used to define when and how much to irrigate. In this trial, images were taken to estimate the leaf area index (LAI) and processed with the VitiCanopy app (De Bei et al., 2016), as well as the Easy Leaf Area® app (Ealson & Bloom, 2014) for estimating leaf area. This information was useful for estimating the Kc of the crop every week and making the irrigation adjustment. The growth rate of the vine and its relationship with the level of water stress can also be determined using digital tools such as the Apex Vigne app developed by the AgroTic team of Montpellier SupAgro (Pichon et al., 2021). This application is based on observing the shoot apex to estimate growth and water status. Jouzier (2020) cite other applications: to facilitate grassing-growing management, the winegrower can rely upon the applications Dicot'ID and PlantNet, which can help establish flower readings between the rows by identifying weeds; Canopeo application that evaluates the quantity of vegetative cover; FlirOne sensor, an infrared camera that connects to a smartphone to estimate the water status of the vines; WineOz SmartGrape application counts the number of berries and estimates their size and colour.

8.2.4 The Retailer and the Consumer Are Other Components of Terroir

Terroir, as a marketing strategy, transfers the image of the region, the vineyard or the winery to the wine product, which the consumer must recognize, puts him in contact with the product, its production, the production environment, etc. (Charters, 2010; Muchnik, 2006; Riviezzo et al., 2017).

These communicators have identified that the complexity of the terroir concept means that, in general, the message is based on one of its components: climate, soil, winegrower, low environmental impact cultivation techniques, local wine identity, etc., as a way of valorizing a wine as opposed to a standard product. Knowing what the word terroir is intended to convey by suppliers and retailers, as well as what it represents for consumers, is as important as knowing how producers translate it into their products and include terroir references in their marketing, particularly when the reference to terroir is often used to represent a promise between producers, sellers and consumers (Ballantyne et al., 2019).

In the case of Uruguay, according to the Bodegas del Uruguay website (Bodegas del Uruguay, 2023), of the 42 wineries with a website, 17 use terroir or terroir in their website communication to affirm the excellence and originality of their wines. The wineries use the terroir concept as a tool to inform consumers about their wines. They rely on the different components of terroir: (1) the influence of the climate, in particular the influence of the ocean breeze or the distance to the Río de la Plata, which allows a greater thermal amplitude; (2) that of the soil (depth, geological material, rocky) of a new region of implantation of vineyards (Maldonado) or the "traditional" area (Canelones); (3) in the winegrower due to the cultivation techniques he applies or his family tradition in the sector or his Italian or Portuguese origin. Some examples of communication from these companies are: "*a sustainable use of natural resources, preserving the unique characteristics of the terroir, with a high maritime influence, very low intervention, no fertilization or irrigation and a very austere pruning, obtaining a very selective harvest per hectare, which allows to achieve the expected typicity and quality*"; "*In our territory, the soil, the climate and the influence of the Río de la Plata converge to create a propitious balance, an accomplice pact between the work of our people and the genuine expression of our land through our wines*"; "*The experience accumulated over decades allows the winery to identify small plots with great potential that, when worked individually, results in wines with more depth, more character and a higher level of focus in a seal of identity in each bottle*"; "*The Terroir of Salto, the combination of soil, climate and the work of the people of this place, makes this wine the best Tannat*"; "*Our vineyards are located on the rocky sierras of Pueblo Eden. The mineral character of its soils and the oceanic influence create the ideal conditions for creating and sustainable production of unique style wines. Steep slopes, stony soils and the influence of the Atlantic Ocean characterize our terroir*"; "*Very close to the Uruguayan Atlantic coast and 160 meters above sea level, the vineyard was planted in a privileged terroir that gives our grapes a unique and unrepeatable character, which added to the mild climate and the organic practices developed in the plantations, allows us to reflect in each wine the authentic character of the Garzón terroir*".

Two of the important functions of the wine bottle label for the consumer are: evocation, which is to convey with text, image and design the wine it contains, and the other is differentiation. The labels of several wineries refer to some of the components of the terroir to identify a wine, detailing the particularities of the area of origin of the wine "Balastro" (geological composition), Colinas del Uruguay (topography) "Atlántico Sur" (proximity to the coast), the place of implantation of the vineyard (Pan de Azúcar, Isla de Lobos, Rocha, Las Brujas), the scale of the plot "Viñedo Único" and "Parcela Única".

Wine tourism, an area that many wineries have started to include in their business, is a strategy that puts the winegrower/winery producer in direct contact with the consumer. The companies offer different activities such as guided tours of the vineyards and winery, catering and sale of their wines. To facilitate future purchases, online sales from the winery or specialized websites are another promotional and consumer loyalty strategy.

References

Amerine, M. A., & Winkler, A. J. (1944). Composition and quality of musts and wines of California grapes. *Hilgardia, 15*(6), 493–675.

Ballantyne, D. (2011). Sustaining the promise of terroir: The case of Central Otago wine region. In *6th AWBR international conference*. Retrieved from http://academyofwinebusiness.com/wp-content/uploads/2011/09/88-AWBR2011_Ballantyne.pdf

Ballantyne, D., Terblanche, N., Lecat, B., & Chapuis, C. (2019). Old world and new world wine concepts of terroir and wine: Perspectives of three renowned non- French wine makers. *Journal of Wine Research, 30*(20), 122–143.

Barbosa, C., Camara, G., Medeiros, J. S., Crepani, E., Novo, E. M. L. M., & Cordeiro, J. P. C. (1998). Operadores zonais em álgebra de mapas e sua aplicação a zoneamento ecológico-econômico. In *Simposio Brasileiro de Sensoriamento Remoto (SBSR)* (pp. 487–500). INPE.

Best, S., León, L., Quintana, R., Aguilera, H., Flores, F., & Concha, V. (2013). Viticultura de Precisión, Terroir Digital. *Boletín Digital INIA, 269*, 139.

Bodegas del Uruguay. (2023). *Bodegas del Uruguay*. Retrieved from https://www.bodegasdeluruguay.com.uy/

Bollati, A., Molin, P., Cifelli, F., Petrangeli, A., Parotto, M., & Mattei, M. (2015). An integrated methodology of viticultural zoning to evaluate terrains suitable for viticulture: The test area of Cesanese DOC (Latium, central Italy). *Journal of Wine Research, 26*(1), 1–17.

Bramley, R. G. V., & Hamilton, R. P. (2007). Terroir and precision viticulture: Are they compatible? *OENO one, 41*(1), 1–8.

Camara, G., Souza, R., Freitas, U., & Garrido, J. (1996). Integrating remote sensing and GIS by object-oriented data modelling. *Computers & Graphics, 20*(3), 395–403.

Charters, S. (2010). Marketing terroir: A conceptual approach. In *Proceedings of the 5th international conference of the academy of wine business research*. Retrieved from http://academyofwinebusiness.com/wp-content/uploads/2010/04/Charters-Marketing-terroir.pdf

Choné, X., Trégoat, O., & van Leeuwen, C. (2001). Fonctionnement hydrique des terroirs, base de l'irrigation raisonnée de la vigne. *Vigne et Vin Publications Internationales*, 47–51.

Cross, R., Plantinga, A., & Stavins, R. (2017). Terroir in the new world: Hedonic estimation of vineyard sale prices in California. *Journal of Wine Economy, 12*(3), 282–301.

De Bei, R., Fuentes, S., Gilliham, M., Tyerman, S., Edwards, E., Bianchini, N., Smith, J., & Collins, C. (2016). VitiCanopy: A free computer app to estimate canopy vigor and porosity for grapevine. *Sensors, 16*(4), 585.

De Rességuier, L., Mary, S., Le Roux, R., Petitjean, T., Quénol, H., & van Leeuwen, C. (2020). Temperature variability at local scale in the Bordeaux area: Relations with environmental factors and impact on vine phenology. *Frontiers in Plant Science, 11*, 515.

Deloire, A., Ferrer, M., & Carbonneau, A. (2003). Respuesta de la viña al terroir: Elementos para un método de estudio. *Agrociencia, 7*(1), 105–113.

Deloire, A., Prévost, P., & Kelly, M. (2008). Le concept de terroir est-il quantifiable à travers. *Revue Francaise d'Oenologie, 231*, 5–11.

Dumas, V., Lebon, E., & Morlat, R. (1997). Differentiation of local climate in the Alsatian vineyard. *OENO one, 31*(1), 1–9.

Easlon, H., & Bloom, A. (2014). Easy leaf area: Automated digital image analysis for rapid and accurate measurement of leaf area. *Applied Plant Science, 2*(7), apps.1400033.

Ferrer, M. (2007). *Étude du climat des régions viticoles de l'Uruguay, des variations climatiques et de l'interaction apportée par le microclimat et l'écophysiologie des systèmes de conduite Espalier et Lyre sur Merlot*. Thesis doctorat Ecole Nationale Supérieure Agronomique de Montpellier.

Ferrer, M., Pedocchi, R., Michelazzo, M., González Neves, G., & Carbonneau, A. (2007). Delimitación y descripción de regiones vitícolas del Uruguay en base al método de clasificación climática multicriterio utilizando índices bioclimáticos adaptados a las condiciones del cultivo. *Agrociencia, 11*(1), 47–56.

Ferrer, M., González-Neves, G., Echeverria, G., & Camussi, G. (2012a). Plant response and grape composition of *Vitis vinifera* L. cv Tannat in different climatic regions. *Journal of Agricultural Science and Technology, 2*, 1252–1261.
Ferrer, M., Pedochi, R., & González-Neves, G. (2012b). Delimitación climática de las regiones vitícolas del Uruguay: Utilización de la metodología de sistemas de clasificación multicriterio (CCM) y adaptación de los Índices Climáticos. In *Clima, zonificación y tipicidad del vino en regiones vitivinícolas iberoamericanas* (pp. 373–387). CYRED. Retrieved from https://oa.upm.es/13203/
Ferrer, M., Echeverria, G., Pereyra, G., González-Neves, G., Pan, D., & Miras-Avalos, J. M. (2020). Mapping vineyard vigor using airborne remote sensing: Relations with yield, berry composition and sanitary status under humid climate conditions. *Precision Agriculture, 21*, 178–197.
Fourment, M., Ferrer, M., & Quénol, H. (2013). *Vitis vinifera* L. cv. Tannat: Reponses to climate variability. *Agrociencia, 17*(2), 45–54.
Fourment, M., Ferrer, M., González-Neves, G., Barbeau, G., Bonnardot, V., & Quénol, H. (2017). Tannat grape composition responses to spatial variability of temperature in an Uruguay's coastal wine region. *International Journal of Biometeorology, 61*, 1617–1628.
Henin, S. (1957). Quelques aspects de la pédologie, l'évolution du concept de sol et ses conséquences. *Comptes Rendus de l'Academie d'Agriculture de France, 2*, 66–69.
Huglin, P. (1978). Nouveau mode d'évaluation des possibilités héliothermiques d'un milieu viticole. *Comptes Rendus de l'Academie d'Agriculture de France, 64*, 1117–1126.
Huglin, P., & Schneider, C. (1998). *Biologie et écologie de la vigne*. Lavoisier.
INAVI. (2023). *Viñedos: Estadísticas*. INAVI, Montevideo. Retrieved from https://www.inavi.com.uy/vinedos/estadisticas/
Jones, G. (2018). The climate component of terroir. *Elements, 14*(3), 167–172.
Jouzier, P. (2020). Smartphone: The winegrowers' Swiss army knife. *IVES Technical Reviews*. Retrieved from https://doi.org/10.20870/IVES-TR.2020.3622
Kasimati, A., Psiroukis, V., Darra, N., Kalogrias, A., Kalivas, D., Taylor, J. A., & Fountas, S. (2023). Investigation of the similarities between NDVI maps from different proximal and remote sensing platforms in explaining vineyard variability. *Precision Agriculture, 24*, 1220–1240.
Le Roux, R., van Leeuwen, C., de Rességuier, L., Neethling, E., Irimia, L., Patriche, C., Santesteban, L.G., Stoll, M., Hofmann, M., Foss, C., Bonnardot, V., Planchon, O., & Quénol, H. (2008). Climate modeling at vineyards scale in a climate change context. In *XIIth international terroir congress*. Zaragoza.
Michelazzo, M., Ferrer, M., & Hernández, J. (2018). Delimitación y caracterización de unidades territoriales vitícolas en la región de Colonia del Sacramento (Uruguay), con Sistemas de Información Geográfica. *Geografía y Sistemas de Información Geográfica, 10*(10), 85–99.
Molfino, J. H., & Califra, A. (2001). *Agua disponible de las tierras del Uruguay*. INIA, Montevideo. Retrieved from http://www.inia.uy/Publicaciones/Documentos%20compartidos/111219220807174344.pdf
Morlat, R., & Jacquet, A. (1993). The soils effects on the grapevine root system in several vineyards in the Loire Valley (France). *Vitis, 32*, 35–42.
Morlat, R., Barbeau, G., & Asselin, C. (2001). Facteurs naturels et humains des terroirs viticoles français: Méthode d'étude et valorisation. *Etudes et Recherches sur les Systèmes Agraires et le Développement, 32*, 111–127.
Muchnik, J. (2006). Identidad territorial y calidad de los alimentos: Procesos de calificación y competencias de los consumidores. *Agroalimentaria, 12*(22), 89–98.
OIV. (2010). *Definition of Vitivinicultural "terroir": Resolution OIV/VITI 333/2010*. OIV, Tbilisi. Retrieved from https://www.oiv.int/public/medias/379/viti-2010-1-en.pdf
OIV. (2012). *OIV guidelines for vitiviniculture zoning methodologies on a soil and climate level: Resolution OIV-VITI 423-2012 REV1*. OIV, Izmir. Retrieved from https://www.oiv.int/public/medias/400/viti-2012-1-en.pdf
Pereyra, G., Ferrer, M., Pellegrino, A., & Gaudin, R. (2022a). Montmorillonite content is an influential soil parameter of grapevine development and yield in South Uruguay. *Agrociencia, 26*(2), e1124.

Pereyra, G., Pellegrino, A., Gaudin, R., & Ferrer, M. (2022b). Evaluation of site-specific management to optimise *Vitis vinifera* L. (cv. Tannat) production in a vineyard with high heterogeneity. *OENO one, 56*(3), 397–412.

Pichon, L., Brunel, G., Payan, J. C., Taylor, J., Bellon-Maurel, V., & Tisseyre, B. (2021). ApeX-Vigne: Experiences in monitoring vine water status from within-feld to regional scales using crowdsourcing data from a free mobile phone application. *Precision Agriculture, 22*, 608–626.

Quénol, H. (2017). Viticulture: Experimentation or adaptation? In *Adapting to climate change* (pp. 333–340).

Quénol, H., & Bonnardot, V. (2014). A multi-scale climatic analysis of viticultural terroirs in the context of climate change: The "teradclim" project. *Journal International of the Sciences de la Vigne et du Vin*, (Spécial Laccave), 25–34.

Riou, C., Pieri, P., & Le Clechz, B. (1994). Consommation d'eau de la vigne en conditions hydriques non limitantes: Formulation simplifies de la transpiration. *Vitis, 33*, 109–115.

Riviezzo, A., Garofano, A., Granata, J., & Kakavand, S. (2017). Using terroir to exploit local identity and cultural heritage in marketing strategies. *Place Branding and Public Diplomacy, 13*, 136–149.

Rouquette, J. L. (1994). Un pays, des hommes, des produits, une race: Chronique du développement agricole et rural de l'Aubrac. *Études et Recherches sur les Systèmes Agraires et le Développement, 28*, 195–204.

Tachini, R., Fourment, M., Bonnardot, V., Fanzone, M., & Ferrer, M. (2022). Study of mesoclimate and its impact on *Vitis vinifera* L. cv. Tannat in the Atlantic winegrowing region of Uruguay. In *XIVth international terroir congress and 2nd ClimWine symposium* (pp. 5). Bordeaux.

Tonietto, J. (1999). *Les macroclimats viticoles mondiaux et l'influence du mesoclimat sur la typicite de la Syrah et du Muscat de Hamburg dans le sud de la France*. Thèse de Doctorat ENSA Montpellier.

Tonietto, J. (2007). Afinal, o que é Terroir? *Bon Vivant, Flores da Cunha, 8*(98), 08.

Tonietto, J., & Carbonneau, A. (2004). A multicriteria climatic classification system for grape-growing regions worldwide. *Agricultural and Forest Meteorology, 124*, 81–97.

Trégoat, O., van Leeuwen, C., Choné, X., & Gaudillère, J. P. (2002). Étude du régime hydrique et de la nutrition azotée de la vigne par des indicateurs physiologiques: Influence sur le comportement de la vigne et la maturation du raisin. *OENO one, 36*, 133–142.

van Leeuwen, C. (2010). Terroir: The effect of the physical environment on vine growth, grape ripening and wine sensory attributes. In *Managing wine quality: volume 1: Viticulture and wine quality* (pp. 341–381). Elsevier.

van Leeuwen, C., & Seguin, G. (2006). The concept of terroir in viticulture. *Journal of Wine Research, 17*(1), 1–10.

van Leeuwen, C., Garnier, C., Agut, C., Baculat, B., Barbeau, G., Besnard, E., Bois, B., Boursiquot, J. M., Chuine, I., Dessup, T., Dufourcq, T., García-Cortazar, I., Marguerit, E., Monamy, C., Koundouras, S., Payan, J. C., Parker, A., Renouf, V., Rodríguez-Lovelle, B., Roby, J. P., Tonietto, J., & Trambouze, W. (2008). Heat requirements for grapevine varieties is essential information to adapt plant material in a changing climate. In *VIIe Congrès International des terroirs viticoles*. ACW, Wädenswil.

Vaudour, E. (2003). *Les terroirs viticoles: Définitions, caractérisation, protection*. Dunod.

Chapter 9
Adaptation to Climate Change and Variability for Viticulturists in Uruguay

Mercedes Fourment, Ramiro Tachini, and Milka Ferrer

9.1 Introduction

Uruguay has a long history of wine production dating back to the early nineteenth century when European immigrants first introduced vineyards to the country's fertile soils. Despite its relatively small size, the wine industry in Uruguay has developed a reputation for producing high-quality wines that are distinct from those made in other parts of the world. Even if the wine industry in Uruguay is relatively small, it has grown in recent years and gained recognition for its high-quality wines. According to the INAVI (2023), wine production in Uruguay has increased from around 6.5 million litres in 2000 to nearly 14 million litres in 2019. In terms of the economy, the wine industry is a significant contributor to the agricultural sector in Uruguay, and it plays an important role in the country's exports. According to the Ministry of Livestock, Agriculture, and Fisheries, the wine sector generates around 1500 direct jobs and 6000 indirect jobs, with a total export value of around USD 10 million in 2020. Viticulture in Uruguay occupies 5.889 hectares under cultivation, exploited by 812 producers (National Institute of Viticulture – INAVI, 2023). 78.7% of the total area is Southern Uruguay (departments of Montevideo and Canelones), the traditional wine region of the country.

The Tannat grape, which is now considered the national grape of Uruguay, was first brought to the country by Basque immigrants in the nineteenth century and has since become a hallmark of the country's wine industry. The unique combination of climate, soil and topography in Uruguay's wine regions contributes to the distinct character of its wines, which are known for their elegance, complexity and balance.

M. Fourment (✉) · R. Tachini · M. Ferrer
Facultad de Agronomía, Universidad de la República, Montevideo, Uruguay
e-mail: mfourment@fagro.edu.uy

G. Gutiérrez Gamboa, M. Fourment (eds.), *Latin American Viticulture Adaptation to Climate Change*, https://doi.org/10.1007/978-3-031-51325-1_9

However, as with many wine-producing regions worldwide, the wine industry in Uruguay is vulnerable to the impacts of climate change, which could have significant economic and social consequences for the country (MGAP-FAO, 2013). This chapter develops three central themes for discussing climate change in Uruguayan viticulture: (1) Climate change impacts in Uruguay. In this section, we describe the general climate trends in Uruguay, including temperature and precipitation changes over time; (2) Climate impact in the Uruguayan vineyards: vulnerability. We provide more detailed information on how climate change affects vineyards in Uruguay, including specific impacts on the vineyards' vulnerability, the adaptive capacity of viticulturists and climate change perceptions; (3) Adaptation strategies: finally, in this section, we discuss various adaptation strategies that viticulturists in Uruguay use to mitigate the impacts of climate change on their vineyards.

9.2 Climate Change Impacts in Uruguay

Climate records have shown that weather patterns have changed significantly over time. According to the latest Intergovernmental Panel of Climate Change – IPCC reports (IPCC, 2001, 2023), there has been a clear increase in the intensity and frequency of hot weather extremes, while cold weather extremes have become less frequent. In the South Eastern South America (SES) region, there is an interesting trend. Rainfall patterns have undergone notable shifts, especially during the warm season. Data analysis from various weather stations reveals a strong positive trend in annual precipitation since the early twentieth century, leading to increased rainfall across the region. This positive trend is further highlighted by a significant increase in heavy rainfall events. The SES region is familiar with the impact of extratropical cyclones, which have been observed to occur more frequently, affecting the area. This serves as a reminder of the ever-changing dynamics of the climate in the region.

In order to analyse the evolution of temperature as an example in the wine-growing region of eastern Uruguay (Departments of Maldonado and Rocha), observed daily temperature data were taken for the period 1991–2023 (last 32 years) from the INUMET Rocha meteorological station (Uruguayan Institute of Meteorology), located in the south-eastern part of the country 18 km from the Atlantic Ocean (34°48′ S; 54°30′ W). The average maximum, minimum and average temperature of each growing season (1 September to 15 March) shows an increase in warmer conditions over the last 4 years. However, what is observed is a statistically significant trend in the increase of days with temperatures above 30 °C (Fig. 9.1). This trend of warmer conditions during the summer has repercussions on the vines, bringing forward the ripening period, altering the composition of the grapes and therefore modifying the typicity of the wines. In that sense, local events of changes on temperature are of importance in coastal wine regions because of the reduction of the impact of heat stress when the ripening process occur (Fourment et al., 2013, 2014, Hunter and Bonnardot, 2011).

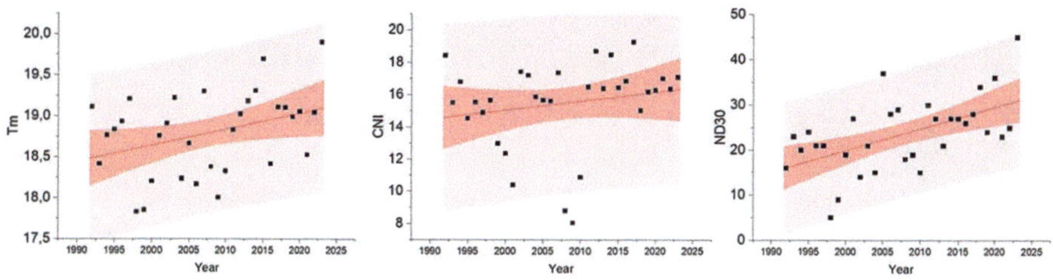

Fig. 9.1 Average temperature (°C) during the growing season (left), Cool Night Index (middle) and number of days with temperatures above 30 °C (ND35) (right) from 1992 to 2023 in Rocha, Uruguay

In two wine-growing regions of Uruguay (Departments of Canelones and Rocha), Tachini et al. (2023) analyse observed daily rainfall data from two WMO-endorsed weather stations for the period 1991–2023 (last 32 years). The first one is the agro-meteorological station of INIA Las Brujas (Instituto Nacional de Investigaciones Agropecuarias), which is located in south-central Uruguay, 13 km from the estuary of the Río de la Plata (34°64′ S; 56°33′ W). The second is the INUMET Rocha (Instituto Uruguayo de Meteorología) weather station (34°48′ S; 54°30′ W). Based on the climatic data, indicators associated with rainfall were calculated: total accumulated rainfall of the vegetative cycle (1-Sep to 15-Mar); the number of days with rainfall during the vegetative cycle and number of dry periods (DryP; 1 period is equivalent to 15 days with accumulated rainfall less than 6 mm). The value of 6 mm for the calculation of dry periods refers to the average value of daily potential transpiration for a summer day in southern Uruguay (Instituto Nacional de Investigaciones Agropecuarias, INIA).

Rainfall during the period of vine growth and development averages 608 mm between the two stations, however Rocha (a station located on the shores of the Atlantic Ocean) records 127 mm more during the cycle compared to Las Brujas (located on the estuary of the Rio de la Plata). The minimum value recorded for a whole season is 189 mm at Las Brujas, which is 111 mm less than the minimum water requirement of 300–600 mm in humid climates (Williams, 2014). The maximum rainfall value was recorded by Rocha with 1154 mm, showing a maximum difference of 940 mm between the driest and wettest seasons on record.

Regarding the number of days with rainfall, the average value between both seasons was 59. Considering an average vegetative cycle of the grapevine in Uruguay of 197 days (1 Sep – 15 Mar), it is presented as a precipitation every 3.3 days. Of the 59 days, 40 are light rains (between 1 and 10 mm), 17 are moderate rains (10–40 mm) and 2 are heavy rains (>40 mm). Among the stations, Rocha has significantly 14 days with more rainfall on average than Las Brujas. The difference lies mainly in light rainfall (1–10 mm), with no difference in moderate and heavy rainfall. Regarding the evaluation of drought periods, the average of both stations is 20 days with this phenomenon. The number of days can vary from 0 days without generating this condition to 57 in Las Brujas and 50 in Rocha. This alternation between years forces the plant and the vine grower to adapt quickly between

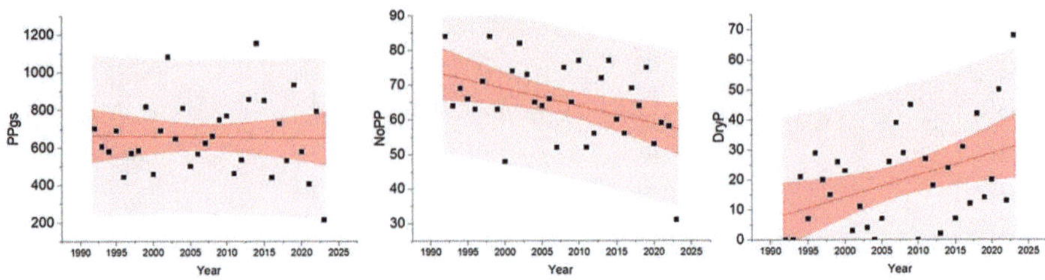

Fig. 9.2 Accumulated precipitation (in mm) (left), number of days with precipitation (middle) and number of dry periods (DryP; 1 period is equivalent to 15 days with accumulated rainfall less than 6 mm) (right) during the growing season from 1992 to 2023 in Rocha, Uruguay

production cycles. Therefore, knowing and describing the variability of rainfall allows for more efficient crop management (Fig. 9.2).

Peering into the future, IPCC's final report on climate projections offers compelling insights. There is a high confidence level in the projected increases in mean air temperature and extreme heat events, indicating a significant decrease in the occurrence of cold spells. Moreover, it is noteworthy to observe that the frequency of warm nights is expected to rise even more prominently than warm days.

As we look into the implications of precipitation, the projections provide interesting findings. There is a strong belief that the amount of rainfall will increase, which means the region can expect more rain on average. Furthermore, there is a moderate belief that both pluvial floods and river floods will become more intense. This information provides valuable insights into the potential challenges and risks associated with changing rainfall patterns. However, when it comes to assessing droughts, the situation becomes less clear. Projections suggest that droughts in the River Plate basin will happen more often in the medium and distant future. Yet, there is less certainty about the severity and duration of these drought events, especially under the more extreme emission scenario (RCP8.5).

9.3 Climate Impact in the Uruguayan Vineyards: Vulnerability

9.3.1 Vulnerability Definition and Its Components

The definition of the vulnerability of an agroecosystem to climate change was made by the IPCC (2007) as the sum of its physical exposure, sensitivity and adaptive capacity, i.e., the impact of exposure or threat to a particular system as viticulture systems (IPCC, 2007). Vulnerability, in the latest report of IPCC (2023), is defined as the propensity or predisposition to be adversely affected. It encompasses a variety of concepts and elements, including sensitivity or susceptibility to harm and lack of capacity to cope and adapt.

Physical exposure (or physical vulnerability) refers to the climatic conditions, often adverse, to which an agroecosystem is subjected: adverse temperatures for its correct development and growth, and deficient or excessive rainfall, among other variables (as discussed in part 1).

The vulnerability of the viticulture systems is analysed based on adverse physical exposure: extreme temperatures (e.g., heat waves with ambient temperatures above 35 °C and cooling due to low temperatures), deficient rainfall (water stress during grape ripening), extreme rainfall, direct radiation (Baló et al., 1986; Dokoozlin & Kliewer, 1996; Schultz, 2000; Van Leeuwen & Vivin, 2008) or combinations such as the effect of temperature extremes and direct radiation, temperature extremes and different water regimes (Bergqvist et al., 2001; Spayd et al., 2002; Goto-Yakamoto et al., 2011).

The sensitivity of an agroecosystem can be assessed by examining the impact of environmental conditions on the system. In the case of grapevines, studies have analysed their sensitivity to climatic variables (Barbeau et al., 2014). Sensitivity is determined by indicators of growth and development (phenological stages, Chuine et al., 2004; Coombe, 1995, Duchêne et al., 2010), production and quality of grapes and wine (Baciocco et al., 2014; Duchêne & Snheider, 2005; Jones et al., 2005; Jones & Davis, 2000; Mozel & Thach, 2014; Quénol & Bonnardot, 2014). Grapevines are known for their phenotypic plasticity, which is the ability to adapt and change their traits in response to environmental cues (Sadras et al., 2009). Compared to other crops like wheat and barley, grapevines exhibit high phenotypic plasticity.

The adaptive capacity of an agroecosystem involves evaluating the ability of producers to cope with threats resulting from physical exposure. This capacity is largely influenced by human factors, such as the type of producer (family or entrepreneurial), access to resources, knowledge of the crop and the ability to gather information (Grothmann & Pratt, 2005; Yaro, 2013). Winegrowers' ability to adapt management practices has been found to mitigate the impacts of climate change on grapevines (Van Leeuwen et al., 2013).

The perception of climate change by viticulturists and advisors within the production sector is an important aspect of the adaptive capacity of the system (Grothmann & Pratt, 2005). Globally, the perception of climate change risk can facilitate effective adaptation strategies (Battaglini et al., 2009; Yaro, 2013). At a local scale, grape growers' awareness of spatial climate variability is crucial for implementing vineyard management practices tailored to specific conditions. For example, delaying pruning in vineyards at risk of late frost (temperatures below 0 °C in September or October). Interviews with winegrowers can provide valuable insights into their knowledge and adaptation practices in vineyard management (Goulet & Morlat, 2011).

As Kelly and Adger (2000) emphasize, adaptation to climate change is not a future endeavour but an ongoing process that requires careful study. Neethling et al. (2016) investigated the evolution of adaptive practices in two regulated winegrowing regions in France (Anjou and Saumur), revealing various levels of adaptive responses, ranging from reactive to anticipatory tactical strategies. In comparison, the capacity to adapt management practices in Uruguay is less restricted, as there are no production or quality regulations apart from the minimum alcohol content nor limitations on grape varieties. Lereboullet et al. (2013) provide an example of specific adaptation responses between countries, comparing traditional France with Australia, which is considered a more resilient system due to its liberal production regulations, weak traditions and effective collective actions enabling large-scale changes like water recycling systems or variety shifts.

In that sense, Neethling et al. (2016) have studied the evolution of practices as adaptive climate measures in two regulated winegrowing regions in France (Anjou and Saumur). Their study has shown that there are several levels of adaptive response, from reactive to anticipatory tactical strategies. The capacity to adapt management measures is much more limited (due to the AOCs' own national regulations) than in Uruguay, where there are no restrictions on production or grape and wine quality (except for the minimum alcohol content), nor on the varieties to be planted. The work of Lereboullet et al. (2013) is an example of specific adaptation responses between countries, such as France (a traditional country) and Australia (from the "new world"). The latter is considered to be a more resilient system in the sense that production regulations are more liberal, have weak traditions and effective collective actions that allow major changes to be implemented on large scales (e.g., water recycling systems or switching varieties).

9.3.2 Perception of Climate Change for Viticulturists in Uruguay

Fourment et al. (2020) conducted 41 semi-direct interviews to understand how Uruguayan viticulturists perceive climate change. The results of the interviews showed that producers in the region perceive the local variability of the climate but, above all, of the annual variation (inter-annual variability). For viticulturists, the climate is responsible for bad years of "vintages" according to the quality of the grapes obtained and the quality of the wine made. Another agent that stood out in the interviews was the extreme events that have occurred in recent years (rains with strong gusts of wind, hail), of which 71% considered an increase in the frequency of extreme phenomena (Fourment et al., 2020).

Bossière et al. (2013) define "local perceptions" as how producers identify and interpret observations and concepts. While climate change can bring conditions beyond previous experience, local knowledge and perception remains the foundation of any local response. In the interviews, perceptions of climate change were not consistent. Despite many being aware of information from the scientific community about climate change in Uruguay, scepticism was found in the responses towards climate change in the region. Most family farmers were the most perceptive, especially with vineyard conditions, while larger farmers were more sceptical about the impacts of climate change. According to Yaro (2013), smallholders perceive the local impact of the observed changes because they relate them to productivity, although larger farmers have a better understanding of the science of climate change (Bossière et al., 2013; Yaro, 2013).

From the interviews, it emerged that there are external climatic hazards in the region that affect plant components and translate into risk when producing grapes for winemaking. These climatic hazards were defined as the determinants of the "bad years" of the series studied.

Grapevine phenology was the most sensitive plant component, as it is driven by climate. By affecting the development of the crop, yield and also the final grape composition are affected. A clear example is the effect of summer rainfall. The availability of water in the vine causes an increase in vegetative growth and development (vigour). By that time, the grape production per plant is already determined, so there is competition between growth and the accumulation of photosynthesis products in the grape (pit). Grape ripening is delayed and can disrupt the correct accumulation and synthesis of primary and secondary compounds. In addition, the larger leaf area modifies the microclimate in the bunches, leading to a possible increase in the incidence and/or severity of diseases (Fourment et al., 2020).

From the interviews, it emerges that there are external non-climatic threats that affect the plant components, such as the cost of labour, the final prices of grapes, restrictions on the marketing of wine in stock in the winery and the competitiveness of a certain product (very recurrent in some specific V.C.P. wines) that also cause greater vulnerability of the viticultural systems studied. Several studies have established that the vulnerability of wine-growing systems results from climatic and non-climatic factors (Belliveau et al., 2006; Hadarits et al., 2010; Pérez-Catalá, 2013). As in the national case analysis, Füssel (2010) identifies these non-climatic agents as factors external to the system that intervene in biophysical and social vulnerability, i.e., they act outside its boundaries (Fourment et al., 2020; Belliveau et al., 2006; Hadarits et al., 2010).

9.4 Adaptation Strategies to Face Climate Change and Variability

Adaptation in the IPCC report (IPCC, 2023) is defined, in human systems, as the process of adjustment to actual or expected climate and its effects to moderate harm or exploit beneficial opportunities. In natural systems, adaptation is the adjustment process to actual climate and its effects; human intervention may facilitate adjustment to expected climate and its effects (IPCC, 2023). Resilience in this report is defined as the capacity of social, economic and environmental systems to cope with a hazardous event or trend or disturbance, responding or reorganising in ways that maintain their essential function, identity and structure, while also maintaining the capacity for adaptation, learning and transformation (IPCC, 2023).

Adaptation measures are identified according to the timing of action (anticipatory or reactive) and their duration (tactical or strategic) (Belliveau et al., 2006; Neethling et al., 2016). Tactical measures are short-term and respond to daily climate variability, some being anticipatory (Belliveau et al., 2006), such as irrigating the vineyard to reduce heat stress during the summer and avoid blocking ripening (Flexas et al., 2010; Fraga et al., 2016), and others reactive, such as applying hydrogen cyanamide to homogenise bud break in situations of mild winters (low accumulation of chilling hours) (Martin & Dunn, 2000).

Long-term strategic adaptation measures can be anticipated, such as choosing rootstocks that are more resistant to root asphyxia and thus excess water, for example (Koundouras et al., 2008), or choosing other planting sites, e.g., high topographic situations to avoid frost damage in sites that are susceptible to these events. As discussed by Belliveau et al. (2006), the choice of varieties is not only determined by climate and is, therefore, not a single measure of adaptation to climate change, as their choice also depends on their competitiveness on the national and/or international market.

A reactive strategic measure occurs when there is an opportunity in a certain year, e.g., when in a "good" year, finances allow investment in the system to reduce risks (Belliveau et al., 2006). For example, from the interviews done by Fourment et al. (2020), the change of conduction system materials (metal poles) improves wind resistance if concrete poles were previously in place. Another example is the possibility of buying better equipment for phytosanitary application (sprayer) to optimise the use of products and thus improve the vineyard's health.

A crucial element in decreasing vulnerability emerged from the Uruguayan interviews, and that was the distribution of vineyards in the region. Belliveau et al. (2006) highlight this capacity as an adaptive measure as a risk reduction strategy that allows the winegrower to have a diversity of grapes to obtain wines of different quality or to carry out trimming. Sixty-seven percent of the winegrowers interviewed have vineyards in different locations within the region, and they were emphatic about the benefits of reducing risk. The extreme weather events suffered in the region have been a clear example. The hail of January 2013 affected several vineyards in the area but did not affect 100% of the winegrowers because they also have vineyards that were not affected by this event. It also happens with events such as rainfall, with great spatial variability, which can greatly affect the final quality of the grapes if they occur close to the moment of technological maturity.

In the medium and long term, adaptation measures also refer to the systematisation of vineyards, especially in terms of variety choice and production objectives (Fourment et al., 2020). In this sense, the producers suggest not taking the risk of producing a single variety since, in adverse conditions, they could lose the entire production. This example is typical for those producers who produce Tannat as a red variety for VCP. They prefer to have a part of the surface with another variety (e.g., Merlot) because they know the risk of losing grapes to rot. Producers with a less diverse, mono-varietal production propose to manage tables differently to reduce the risk of losing grape volumes targeted at a particular type of wine. As with the previous measure, the diversification of grape quality also makes it possible to reduce risks in the winery.

One way to reduce vulnerability in cases of low adaptive capacity is the grouping of producers (e.g., through cooperative groups). This provides them with technical advice, information exchange and access to machinery to use on their farms. The latter is relevant and increasingly used in the region. Many producers are concerned about the increasingly high cost of labour and the lack of adaptation of mechanisation systems for their vineyards, for example, by having a two-plane conduction system (lira). They emphasise the need to improve the mechanised management system in order to adopt it in the vineyards in an appropriate manner and respect the timing of activities (e.g., time for leaf removal).

Figure 9.3 shows the adaptation strategies for Uruguayan viticulture through the interviews done by Fourment et al. (2020) based on the work of Nicholas and Durham (2012) and Neethling et al. (2016). In Uruguay, there are no restrictions to planting at different sites, so that is an opportunity to explore different terroir conditions, taking into account social aspects such as labour availability or capacity for a mechanised management system. Irrigation is a much-discussed adaptation technique when there are water source restrictions. It is essential to manage irrigation promptly with the minimum possible water use when water constraints occur at the early stages of vine development.

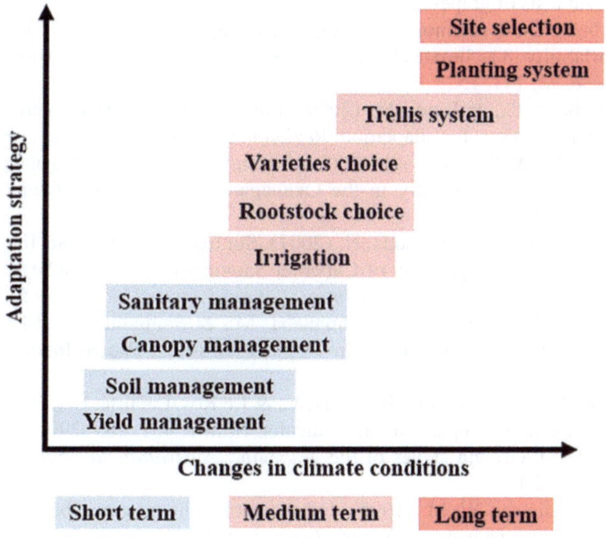

Fig. 9.3 Adaptation strategies for Uruguayan viticulture. (Modified from Nicholas & Durham, 2012; Neethling et al., 2016)

9.5 Perspectives

As the impact of climate change on agroecosystems becomes increasingly evident, it is imperative to develop proactive and effective adaptation measures in the viticultural sector. Building upon the insights gained from research and the perceptions of winegrowers, future adaptation strategies should focus on enhancing the resilience and adaptive capacity of grapevines and the wine industry. These measures may encompass innovative viticultural practices, such as precision irrigation and canopy management, as well as the exploration and selection of climate-adapted grape varieties. Furthermore, the collaboration between researchers, winegrowers and policymakers will be crucial in facilitating knowledge exchange, supporting decision-making processes and implementing adaptive management approaches. By embracing these future adaptation measures, the viticultural sector can navigate the challenges of climate change and ensure the long-term sustainability and success of grape cultivation and wine production.

References

Baciocco, K. A., Davis, R. E., & Jones, G. V. (2014). Climate and Bordeaux wine quality: Identifying the key factors that differentiate vintages based on consensus rankings. *Journal of Wine Research, 25*, 75–90.

Baló, B., Mustárdy, L. A., Hideg, É., & Faludi-Daniel, Á. (1986). Studies on the effect of chilling on the photosynthesis of grapevine. *Vitis, 25*, 1–7.

Barbeau, C., Barbeau, G., & Joannon, A. (2014). Analyzing the sensitivity of viticultural practices to weather variability in a climate change perspective. An application to workable-days modelling. *OENO one, 48*, 141–152.

Battaglini, A., Barbeau, G., & Bindi, M. (2009). European winegrowers' perception to climate change impact and options for adaptation. *Regional Environmental Change, 9*, 61–73.

Belliveau, S., Smit, B., & Bradshaw, B. (2006). Multiple exposures and dynamic vulnerability: Evidence from the grape industry in the Okanagan Valley, Canada. *Global Environmental Change, 16*, 364–378.

Bergqvist, J., Dokoozlian, N., & Ebisuda, N. (2001). Sunlight Exposure and Temperature Effects on Berry Growth and Composition of Cabernet Sauvignon and Grenache in the Central San Joaquin Valley of California. *American Journal of Enology and Viticulture* 52:1–7.

Boissière, M., Locatelli, B., Sheil, D., Padmanaba, M., & Sadjudin, E. (2013). Local perceptions of climate variability and change in Tropical Forests of Papua, Indonesia. *Ecology and Society, 18*, 13.

Chuine, I., Yiou, P., Viovy, N., Seguin, B., Daux, V., & Le Roy-Ladurie, E. (2004). Historical phenology: Grape ripening as a past climate indicator. *Nature, 432*, 289–290.

Coombe, B. G. (1995). Growth stages of the grapevine. *Australian Journal of Grape and Wine Research, 1*, 100–110.

Dokoozlian, N., & Kliewer, W. M. (1996). Influence of light on grape berry growth and composition varies during fruit development. *Journal of the American Society for Horticultural Science, 121*(5), 869–874.

Duchêne, E., & Schneider, C. (2005). Grapevine and climatic changes: A glance at the situation in Alsace. *Agronomy for Sustainable Development, 25*, 93–99.

Duchêne, E., Huard, F., Dumas, V., Schneider, C., & Merdinoglu, D. (2010). The challenge of adapting grapevine varieties to climate change. *Climate Research, 41*, 193–204.

Flexas, J., Galmés, J., Gallé, A., Gulías, J., Pou, A., Ribas-Carbo, M., Tomàs, M., & Medrano, H. (2010). Improving water use efficiency in grapevines: Potential physiological targets for biotechnological improvement. *Australian Journal of Grape and Wine Research, 16*, 106–121.

Fourment, M., Ferrer, M., González-Neves, G., Barbeau, G., Bonnardot, V., & Quénol, H. (2013). Spatial variability of temperature and grape berry composition at terroir scale in Uruguay. In *Proceedings of 18th international symposium GiESCO*, 7th to 11th July, Porto, Portugal.

Fourment, M., Bonnardot, V., Planchon, O., Ferrer, M., & Quénol, H. (2014). Circulation atmosphérique locale et impacts thermiques dans un vignoble côtier: observations dans le sud de l'Uruguay. *Climatologie, 11*, 47–64.

Fourment, M., Ferrer, M., Barbeau, G., & Quénol, H. (2020). Local perceptions, vulnerability and adaptive responses to climate variability in a winegrowing region in Uruguay. Environmental Management. https://doi.org/10.1007/s00267-020-01330-4

Fraga, H., Santos, J. A., Malheiro, A. C., Oliveira, A. A., Moutinho-Pereira, J., & Jones, G. V. (2016). Climatic suitability of Portuguese grapevine varieties and climate change adaptation. *International Journal of Climatology, 36*, 1–12.

Füssel, H. M. (2010). *Review and quantitative analysis of indices of climate change exposure, adaptive capacity, sensitivity and impacts. Background note to the World Development Report 2010. Development and climate change*. Retrieved from http://siteresources.worldbank.org/INTWDR2010/Resources/5287678-255547194560/WDR2010_BG_Note_Fussel.pdf

Goto-Yakamoto, N., Mori, K., Numata, M., Koyama, K., & Kitiyama, M. (2011). Effects of temperature and water regimes on flavonoid contents and composition on the skin of red-wine grapes. *OENO one, 43*, 75–80.

Goulet, E., & Morlat, R. (2011). The use of surveys among wine growers in vineyards of the middle-Loire Valley (France), in relation to terrois studies. *Land Use Policy, 28*, 770–782.

Grothmann, T., & Patt, A. (2005). Adaptive capacity and human cognition: The process of individual adaptation to climate change. *Global Environmental Change, 15*(3): 199–213.

Hadarits, M., Smit, B., & Diaz, H. (2010). Adaptation in viticulture: A case study of producers in the Maule region of Chile. *Journal of Wine Research, 21*, 167–178.

Hunter, J. J., & Bonnardot, V. (2011). Suitability of some climatic parameters for grapevine cultivation in South Africa, with focus on key physiological processes. *South African Journal of Enology and Viticulture, 32*, 137–154.

INAVI. (2023). Instituto Nacional de Vitivinicultura. Retrieved from www.inavi.com.uy. Visited 7th June 2023.

IPCC. (2001). *Climate change 2001: Impacts, adaptation and vulnerability, summary for policymakers*. WMO.

IPCC. (2007). *Climate change 2007: Impacts, adaptation and vulnerability. Contribution of Working Group I to the Fourth Assessment. Report of the Intergovernmental Panel on Climate Change*. Cambridge University Press.

IPCC. (2023). Ara Begum, R., Lempert, R., Ali, E., Benjaminsen, T.A., Bernauer, T., Cramer, W., Cui, W., Mach, K., Nagy, G., Stenseth, N.C., Sukumar, R., & Wester, P. (2022). Point of departure and key concepts. In H.-O. Pörtner, D. C. Roberts, M. Tignor, E. C. Poloczanska, K. Mintenbeck, A. Alegría, M. Craig, S. Langsdorf, S. Löschke, V. Möller, A. Okem, & B. Rama (Eds.), *Climate change 2022: Impacts, adaptation and vulnerability. Contribution of Working Group II to the Sixth Assessment Report of the Intergovernmental Panel on Climate Change* (pp. 121–196). Cambridge University Press. https://doi.org/10.1017/9781009325844.003

Jones, G., & Davis, R. (2000). Using a synoptic climatological approach to understand climate–viticulture relationships. *International Journal of Climatology, 20*, 813–837.

Jones, G. V., White, M. A., Cooper, O. R., & Storchmann, K. (2005). Climate change and global wine quality. *Climate Change, 73*, 319–343.

Kelly, P. M., & Adger, W. N. (2000). Theory and practice in assessing vulnerability to climate change and facilitating adaptation. *Climatic Change, 47*, 325–352.

Koundouras, S., Tsialtas, I. T., Zioziou, E., & Nikolaou, N. (2008). Rootstock effects on the adaptive strategies of grapevine (*Vitis vinifera* L. cv. Cabernet-Sauvignon) under contrasting water status: Leaf physiological and structural responses. *Agriculture, Ecosystems and Environment, 128*, 86–96. https://doi.org/10.1016/j.agee.2008.05.006

Lereboullet, A., Beltrando, G., & Bardsley, D. K. (2013). Socio-ecological adaptation to climate change: A comparative case study from the Mediterranean wine industry in France and Australia. *Agriculture, Ecosystems and Environment, 163*, 273–285.

Martin, S. R., & Dunn, G. M. (2000). Effect of pruning time and hydrogen cyanamide on budburst and subsequent phenology of *Vitis vinifera* L. variety Cabernet Sauvignon in central Victoria. *Australian Journal of Grape and Wine Research, 6*, 31–39.

MGAP-FAO. (2013). *Clima de cambios: Nuevos desafíos de adaptación en Uruguay* (L. Astigarraga, V. Picasso, M. Ferrer, M. Mondelli, G. Cruz, G. Camussi, M. Fourment, V. Varela, G. Pereyra, J. Taks, S. Contreras, & P. Cobas). Resultados del proyecto FAO/TCP/URU 3302, Montevideo, Uruguay.

Mozell, M. R., & Thach, L. (2014). The impact of climate change on the global wine industry: Challenges and solutions. *Wine Economics and Policy, 3*(2), 81–89. https://doi.org/10.1016/j.wep.2014.08.001

Neethling, E., Petitjean, T., Quénol, H., & Barbeau, G. (2016). Assessing local climate vulnerability and winegrowers' adaptive process in the context of climate change. *Mitigation and Adaptation Strategies of Global Change, 22*, 777–803. https://doi.org/10.1007/s11027-015-9698-0

Nicholas, K. A., & Durham, W. H. (2012). Farm-scale adaptation and vulnerability to environmental stresses: Insights from winegrowing in Northern California. *Global Environmental Change, 22*(2), 483–494.

Pérez-Catalá, A. (2013). *Viticulture under climate change: Understanding vulnerability and adaptive capacity of wineries and growers. A case study in El Penedès region, Catalonia* (p. 34). PhD thesis, Universitat Autònoma de Barcelona.

Quénol, H., & Bonnardot, V. (2014). A multi-scale climatic analysis of viticultural terroirs in the context of climate change: The "TERADCLIM" project. *International Journal of Vine and Wine Sciences*, (Special Laccave), 23–32.

Sadras, V. O., Reynolds, M. P., de la Vega, A. J., Petrie, P. R., & Robinson, R. (2009). Phenotypic plasticity of yield and phenology in wheat, sunflower and grapevine. *Field Crops Research, 110*, 242–250.

Schultz, H. (2000). Climate change and viticulture: An European perspective on climatology, carbon dioxide, and UV-B effects. *Australian Journal of Grape and Wine Research, 6*, 2–12.

Spayd, E. S., Tarara, J. M., Mee, D. L., & Ferguson, J. C. (2002). Separation of sunlight and temperature effects on the composition of *Vitis vinifera* cv. Merlot berries. *American Journal of Enology and Viticulture, 53*, 171–182.

Tachini, R., Bonnardot, V., Ferrer, M., & Fourment, M. (2023). Topography interactions with the Atlantic Ocean and its impact on Vitis vinifera L. 'Tannat'. Vitis. 163–177. https://doi.org/10.5073/vitis.2023.62

Van Leeuwen, C., & Vivin, P. (2008). Alimentation hydrique de la vigne et la qualité des raisins. *Innovations Agronomiques, 2*, 159–167.

Van Leeuwen, C., Schultz, H. R., García de Cortázar-Atauri, I., Duchêne, E., Ollat, N., Pieri, P., Bois, B., Goutouly, J. P., Quénol, H., Touzard, J. M., Malheiro, A. C., Bavaresco, L., & Delrot, S. (2013). Why climate change will not dramatically decrease viticultural suitability in main wine-producing areas by 2050. *Biological Sciences, 110*(33), E3051–E3052. https://doi.org/10.1073/pnas.1307927110

Williams, L. E. (2014). Determination of evapotranspiration and crop coefficients for a Chardonnay vineyard located in a cool climate. *American Journal of Enology and Viticulture, 65*, 159–169.

Yaro, J. A. (2013). The perception of and adaptation to climate variability/change in Ghana by small-scale and commercial farmers. *Regional Environmental Change, 13*, 1259–1272.

Chapter 10
Climate Change Adaptations of Argentine Viticulture

J. A. Prieto, M. Bustos Morgani, M. Gomez Tournier, A. Gallo, M. Fanzone, S. Sari, E. Galat, and J. Perez Peña

10.1 Observed and Projected Climatic Evolution in the Region

In Argentina, the main grape-growing region comprises the provinces of Mendoza and San Juan. In this region, commercial viticulture depends on irrigation water that comes from the snow that accumulates in the Andes mountains during winter. In

J. A. Prieto (✉) · M. Fanzone
Estación Experimental Agropecuaria Mendoza, Instituto Nacional de Tecnología Agropecuaria (INTA), Mendoza, Argentina

Centro de Estudios Vitivinícolas y Agroindustriales (CEVA), Universidad Juan Agustín Maza, Mendoza, Argentina
e-mail: prieto.jorge@inta.gob.ar

M. Bustos Morgani
Research & Development, Grupo Peñaflor, Mendoza, Argentina

Consejo Nacional de Investigaciones Científicas y Técnicas (CONICET), Buenos Aires, Argentina

M. Gomez Tournier
Estación Experimental Agropecuaria Mendoza, Instituto Nacional de Tecnología Agropecuaria (INTA), Mendoza, Argentina

Consejo Nacional de Investigaciones Científicas y Técnicas (CONICET), Buenos Aires, Argentina

A. Gallo
Estación Experimental Agropecuaria Junín, Instituto Nacional de Tecnología Agropecuaria (INTA), Mendoza, Argentina

S. Sari · E. Galat · J. Perez Peña
Estación Experimental Agropecuaria Mendoza, Instituto Nacional de Tecnología Agropecuaria (INTA), Mendoza, Argentina

G. Gutiérrez Gamboa, M. Fourment (eds.), *Latin American Viticulture Adaptation to Climate Change*, https://doi.org/10.1007/978-3-031-51325-1_10

spring and summer, the melting snow provides water for the rivers and aquifers, on which agriculture, tourism, hydroelectricity generation, and population consumption depend. Therefore, this region is highly vulnerable to any change in temperature or rainfall/snowfall regimes since it will have a strong impact on several economic and social aspects (Boninsegna, 2014).

In this region, located at the foot of Andes mountains, an increase in mean and minimum spring and summer temperatures has been observed, particularly since the 1970s (Boninsegna, 2014; Deis et al., 2015). This increase has also been observed in the mountains, changing the rivers water regime (Villalba, 2007). Higher spring temperatures cause an earlier melting of snow, advance the runoff peak of the rivers, change their flows, and reduce water availability for the irrigation of crops at the end of summer (Boninsegna, 2014). In addition to these changes in the rivers, a general reduction in their flows has been observed since 1950, with some short periods of recovery (Boninsegna, 2014). In years when snowfall is scarce, rivers maintain their flows with the water from the melting mountain glaciers. This generates a retraction in the glaciers of this region, as observed from the beginning of the twentieth century (Boninsegna, 2014; Villalba, 2007). Although it is not clear if this retraction of glaciers is related to the increase in temperature, less snowfall, or both (Boninsegna, 2014), the consequences of this retraction imply for the future less water availability for agricultural, human, industrial, and recreational uses.

The climatic projections for the near (present-2039) and far future (2075–2099) indicate an increase in average annual temperature between 1.5 and 4 °C depending on the scenario, with a higher increase in the mountain's areas than in the irrigated valleys (Cabré & Nuñez, 2020; Cabré et al., 2016). This variation would have a stronger impact in the southern basins of Mendoza due to its lower altitude (Boninsegna, 2014). Concerning precipitation, two contrasting trends will occur. During summer, storms with high intensity accompanied by hail will increase, and during winter, snow precipitation in the Andes mountains will decrease. These projections are consistent with the trends observed during the twenty-first century (Villalba, 2007). Considering this scenario of higher temperatures and less water for irrigation, in the following points, we will discuss some adaptation and mitigation strategies at the vineyard level. And taking into account that an extensive literature exists concerning the impacts of water deficit on grapevine responses, regulated deficit irrigation and other irrigation strategies, here we will focus mainly on strategies to decrease the negative effects of high temperatures and their impact on water use. In the last point, we will briefly discuss some oenological alternatives that might also be applicable.

10.2 Adaptation Strategies

Grape production aims to achieve the maximum yield of a given composition, ultimately determining the wine that can be obtained. Environmental conditions directly influence not only yield and quality but also production costs. The choice of plant material, planting site, trellis system, viticultural and enological practices, and wine style are some of the factors that must be taken into account to optimize the relationship between yield and quality (Leeuwen et al., 2019). In the context of climate change, characterized by the increase of temperature and lower water availability, it is necessary to adapt and adopt viticultural practices to maintain sustainable production in terms of quantity and quality. Consequently, a better understanding of the effects of an increase in temperature on photosynthesis, water use efficiency, ripeness, and berry and wine chemical composition is essential to assure vineyard sustainability in the future environmental conditions (Naulleau et al., 2021). The effects of high temperature on photosynthesis acclimation (Galat et al., 2019; Gallo et al., 2020), stomatal conductance and water use efficiency (Gallo et al., 2022), hydraulic conductivity (Galat et al., 2020), berry and wine phenolic (De Rosas et al., 2017) have already been tackled in our region, under arid and warm conditions and out of the scope of this chapter.

Adaptation strategies can generate effects in the short, medium, or long term. The different adaptation strategies include changes in vineyard establishment, selection of plant material, and viticultural techniques (Palliotti et al., 2014). While the first two strategies (i.e., vineyard establishment and plant material) have long-term effects, viticultural techniques have short term effects and may apply to established vineyards (Fig. 10.1). They are based on delaying the phenological stages,

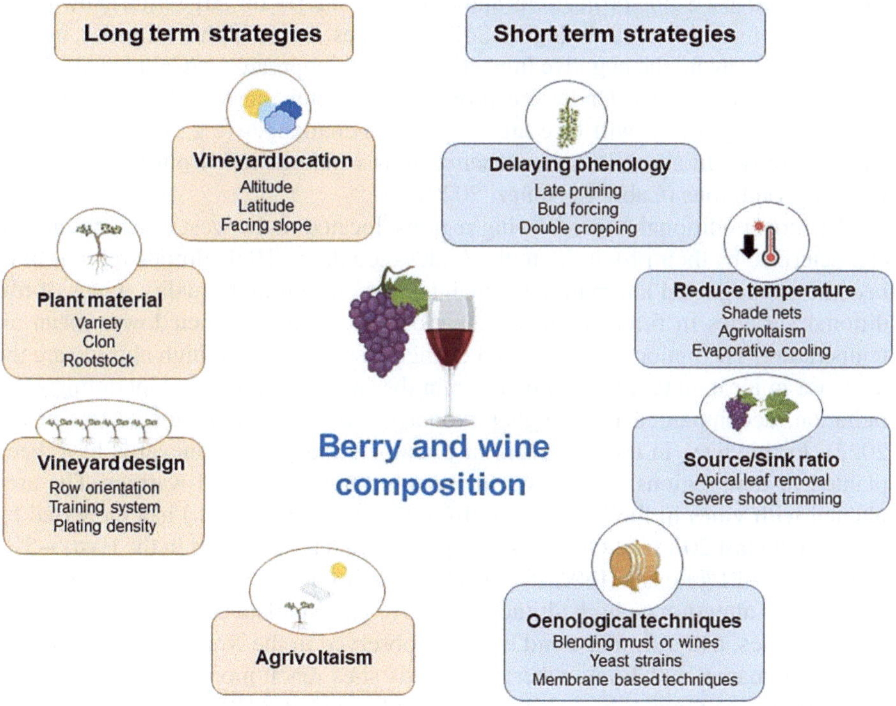

Fig. 10.1 Schematic representation of the different adaptation strategies at the vineyard and oenological level considering long- and short-term effects

altering the leaf/fruit ratio to limit sugar accumulation in the berry, or lowering the canopy temperature (Gutierrez-Gamboa et al., 2021). It is unrealistic to think that a single strategy will allow adaptation, on the contrary, a combination of factors or strategies must be adjusted and fine-tuned to adapt to the new conditions. It is important to note that, first, the adaptation strategies proposed by researchers or government entities will succeed if they are applicable and accepted by grapegrowers. Second, they must consider the final product and its acceptance by consumers (Naulleau et al., 2021).

10.2.1 Vineyard Location: Exploring New Wine Regions

Vineyard plantation sites are particularly important in terms of the available solar energy that is necessary for optimal and complete grape ripening (Gutiérrez-Gamboa et al., 2021). Currently, 18 provinces of Argentina produce grapes and wine (INV, 2020), many of them showing large increases in their planted area these last years. In this sense, the increase in temperature may be negative for some areas and beneficial for others given that the most limiting environmental factor constraining viticulture differs considerably from one region to another (Schultz & Hofmann, 2016). Some projections and modeling studies for Argentina show a displacement of suitable areas to higher South latitudes, and to higher altitudes, mainly to the West, to avoid the negative impacts of higher temperatures (Cabré et al., 2016; Cabré & Nuñez, 2020). Under the projected scenarios, current Argentina's main wine producing region will face high adaptation challenges (e.g., severe irrigation water scarcity). In contrast, some other regions would benefit from new favorable growing conditions (Cabré & Nuñez, 2020).

The most traditional wine-growing regions, located in the west of Argentina, are characterized by their proximity to the Andes mountains. High altitude regions have become a recognized alternative for maintaining current high-quality wines of traditional varieties in future climate scenarios, mostly due to their lower mean air temperature. The major environmental changes associated with high altitude are the decrease in air temperature, the increase in thermal amplitude, and incoming solar radiation, accompanied by a higher proportion of UV-B radiation (Arias et al., 2022). In this way, in the last years in Mendoza, many new vineyards have been planted in new regions, away from those more traditional and warmer. The area planted with vines in the Uco Valley (from 900 to 1200 m a.s.l.) has increased by 117% in the last 20 years. On the contrary, the area has decreased in the East (−5%) and South (−21%) oasis (INV, 2020). This search for new grape production areas, with special attention on high altitude valleys (e.g., Uspallata at 2039 m a.s.l.), actually continues, and a similar trend has been observed in the Northwest provinces of Salta, Catamarca, and Jujuy, where vineyards sites reach more than 3000 m a.s.l. The two highest vineyards in our country are located at 3229 m a.s.l. in Quebrada de Humahuaca (Jujuy), and at 3111 m a.s.l. in Los Valles Calchaquíes (Salta). The huge potential of these vineyards at extreme altitude is under study, and it is evidenced in recently published studies, showing the characteristics of the grapes and the wines (Barroso et al., 2019; Arias et al., 2022). However, planting vineyards at high altitude regions is an adaptation strategy only feasible in certain regions and for certain companies and big growers with financial resources but more difficult for small ones. Furthermore, the environmental conditions imposed by high-altitude vineyards imply new challenges in terms of vineyard management, grapevine physiology, and yield potential associated with those conditions.

As already mentioned, the cultivated area in non-traditional grapegrowing regions of the country has increased these last years. Even if it represents a small percentage in terms of area (around 1%), it illustrates the search for new suitable terroirs by the wine industry. This search for new wine regions in Buenos Aires, Chubut, and La Pampa provinces and the revalorization of ancient wine regions, like those in Entre Ríos and Córdoba provinces, is just beginning. Each zone must carry out a process of reflection and study to make the appropriate decision on what to plant (variety, rootstock, clone) and how (training system, irrigation, planting distances). It undoubtedly represents an opportunity to increase the diversity of products made in the country by exploring new climates, soils and varieties.

10.2.2 Plant Material

Grapevine presents a high genetic diversity. Worldwide, there are between 6000 and 10,000 different *Vitis vinifera* genotypes (Wolkovich et al., 2018). This high genetic diversity of varieties with different characteristics (e.g., phenology, cold, heat and drought tolerance, among others) could contribute to a better adaptation to climate change and the creation of new wines and styles (Gutiérrez-Gamboa et al., 2020; Morales-Castilla et al., 2020). However, the world wine market is limited to a few varieties. This has caused a loss of local or minor varieties as they were considered by the international markets as difficult to commercialize and some of them with poor oenological potential. Currently, only 13 varieties represent 30% of the cultivated area and 33 varieties represent 50% of the cultivated area worldwide (OIV, 2017). These 13 varieties (representing less than 1% of the total diversity) occupy around 80% or more of the cultivated area with vines in Australia, New Zealand, and Chile (Wolkovich et al., 2018). In Argentina, the situation is a little different since there is still 30% of the cultivated area with local varieties, and the most cultivated red varieties, Malbec and Bonarda, are not within the group of these 13 varieties. Efforts to rescue and value minority varieties have been carried out in Spain, Greece, France, and Argentina, looking for local genotypes more adapted to climate change conditions. It has recently been shown that maintaining and/or increasing the diversity of cultivated varieties can reduce the loss of areas suitable for viticulture globally (Morales-Castilla et al., 2020). The same study highlights that if the increase in global temperature is higher than 2 °C by the end of the century, more than half of the current wine-growing area will be lost. For a given environmental condition, there is great diversity in terms of phenology, so it is especially important to choose the right variety for each plantation site (Parker et al., 2013). With the increase in temperature, varieties that were adapted to certain site will become unsuitable as they will not have an ideal temperature for ripening, which would affect wine quality and typicity. Although progress has been made in rescuing and characterizing minority varieties, exploiting this high genetic diversity within *Vitis vinifera* to adapt to climate change poses some challenges. On the one hand, conservation policies are necessary to maintain and increase the germplasm banks and to increase our knowledge of the varieties conserved in terms of phenology and oenological potential to be able to propose the most suitable varieties for a given climate. On the other hand, it is necessary to work together with grapegrowers and winemakers to choose varieties that are well adapted to a certain site, and finally, that their wines are aligned with consumers' preferences (Wolkovich et al., 2018).

In addition, within each variety, there is also a certain degree of variability among clones. This clonal variability has previously been used to obtain certain differential characteristics regarding grape composition. Clonal selection in the past was focused on yield, berry composition, and soluble solid accumulation. Currently, there is a growing interest in evaluating the behavior of different clones under stressful conditions, like those produced by global warming. While in some old varieties (e.g., Pinot Noir) this variability is better known, in others, the diversity and potential remain unknown. Some recent studies with Tempranillo have shown that certain clones maintained the same sugar per berry vs anthocyanin ratio under higher temperatures combined with different environmental factors, such as drought or high CO_2, as observed in lower temperature conditions (Arrizabagala et al., 2021). In the same way, under water deficit, differences in water use efficiency among Tempranillo clones can be up to 80%, even greater than what is frequently observed between varieties (Tortosa et al., 2020). Variability has also been found in soluble solids accumulation pattern that can lead to differences of 3.5 °Brix at harvest in clones of Sangiovese or Chardonnay and differences up to 17 g/L of sugar (1% alcohol) in Cabernet Franc clones (Van Leeuwen et al., 2013). In Mendoza, Malbec clones differ in the date of the fruitset, length of the fruitset-maturity period, days between flowering and veraison, and yield components, such as bunch weight and number of berries per bunch (van Houten et al., 2020). The selection goals should be oriented in the coming years to tolerate drought and higher temperatures and to identify those more adapted for different situations and regions. In this sense, it is important to preserve intra-varietal variability, as well as the use of molecular markers to characterize it and assist selection. The non-regulated urban advance over ancient vineyards in traditional areas of Mendoza (e.g., Maipú and Luján de Cuyo counties) threatens the conservation of genetic material of more than 100 years of traditional varieties, especially Malbec, of great economic importance. Finally, although almost 90% of the planted vineyards in Argentina are own-rooted, the use of rootstocks is essential where biotic or abiotic limitations are present. The evaluation of different rootstocks at the local level showed that some local varieties and other *Vitis* genotypes present a high potential tolerance to salinity (Lucero et al., 2017; Martin et al., 2020).

10.2.3 Vineyard Design

10.2.3.1 Row Orientation

Row orientation is a planting decision that depends on several issues like plot shape and slope, row length, prevailing wind direction, and irrigation design, among others. In Argentina, the predominant row orientation is North-South, which allows a similar light interception on both sides of the canopy. However, in warm regions with high radiation intensity, exacerbated by this context of higher temperatures, rows could be orientated to minimize the incident radiation on the canopy, especially on the bunches during the hours of high temperatures (i.e., afternoon). A study conducted at INTA Mendoza Experimental Station on a Malbec vineyard (see Fig. 10.4 for a picture of the whole experiment) showed that East-West and Northwest-Southeast orientations (scarcely planted in the region) favored the higher concentration of phenolic compounds in grapes and wines. Less exposure of bunches to radiation prevented the degradation of chemical compounds responsible for wine color and mouthfeel attributes. On the other hand, regardless of the maturity of the berries, wines from the different row orientations showed similar levels of acidity and pH.

10.2.3.2 Training System

The choice of the training system aims to achieve an optimal distribution of canopy organs (i.e., bunches and leaves) in the space, generating an adequate microclimate for the bunches to ensure berry maturity and chemical composition. Besides these general premises for training systems, some modifications can be introduced to mitigate the effects of the increase in temperature. These modifications are intended to: (i) delay maturity, (ii) reduce berry sugar accumulation, (iii) reduce radiation in the bunches zone, and (iv) improve water use efficiency (Santos et al., 2020). The interaction between the training system, canopy management, variety, and the environment determine the canopy architecture, which plays a key role in plant functioning. Studies at the whole plant level have shown that water use efficiency of different training systems is largely determined by the number of leaves at the inner layers of the canopy, and its relationship with the exposed leaf area (Prieto et al., 2013). In that sense, training systems with divided or open canopies present higher efficiencies than constraint and dense canopies with high percentage of leaves in the inner layers of the canopy.

Minimal pruning is a training system that delays maturity, decreases berry sugar accumulation, increases acidity and decreases pH, and in addition reduces operational costs (Clingeleffer, 2009). However, this system presents some drawbacks, such as premature plant aging, an alternation in production, overcropping, higher water consumption at the beginning of the season, and the need of some green interventions to allow vineyard operations (Intrieri et al., 2011). Alternatively, free-canopy training systems have been developed, such as free cordon, box pruning, or semi-minimal Pruned Hedge (Intrieri et al., 2011). This last system is an adaptation of the minimal pruning, it allows a higher control of yield, and it generates less compact bunches, with a lower incidence of botrytis and other diseases (Intrieri et al., 2011). The results regarding the delay in the accumulation of sugars are more variable and seem to depend on the variety. The free cordon is currently being evaluated in Mendoza, to determine the optimal leaf-fruit ratio for different situations and its impact on the yield and berry and wine chemical composition (Ahumada et al., 2021). The overhead (i.e., "parral" or "pergola") is a system that protects the bunches from direct radiation. Still, since it presents other operational drawbacks in terms of mechanization and costs, it has been gradually displaced by VSP (even though it still occupies about 50% of the cultivated surface) and more recently by free cordon implantation. The interaction between canopy architecture with irrigation strategies for different varieties should be more explored to better understand their contribution to mitigate the effects of climate change on the production and quality of grapes.

10.2.4 *Canopy Management Practices to Delay Phenology*

10.2.4.1 Late Pruning

Winter pruning is a routine practice in almost all deciduous fruit trees. One of its objectives is to maintain the balance between vegetative and reproductive growth and to adapt plant growth to the training system and to facilitate harvest. Grapevine is a liana that presents a marked correlative inhibition where the distal buds of the shoot are the first to burst while the basal ones remain dormant. Winter pruning is conducted when buds are still in their ecodormancy phase and to break this correlative inhibition, some practices such as short pruning or cane arching are applied. In recent years, late pruning has been applied to retard basal bud development. This practice has been used to escape late frosts (Howell & Wolpert, 1978), and to increase fruitset in varieties with a tendency to shatter (Friend & Trought, 2007). In recent years, due to climate change, late pruning has been studied in different regions of the world as a technique to delay phenology and bring berry maturity to a cooler period (Allegro et al., 2020; Frioni et al., 2019; Gatti et al., 2018; Moran et al., 2018; Palliotti et al., 2017; Petrie et al., 2017).

In our region, late pruning is applied by some growers to escape spring frosts, but without an objective criterion on when to prune. In an experiment carried out at INTA Mendoza Experimental Station on Malbec vines, the effect of late pruning at three phenological stages were compared with winter pruning: budburst, two to three separate leaves and eight separate leaves. Results after 3 years of study showed that it is possible to delay the phenological stages until veraison, the later the pruning, the greater was the delay in phenology (Bustos Morgani et al., 2023). In this way, plants pruned at budburst (i.e., when control plants were at this stage) delayed

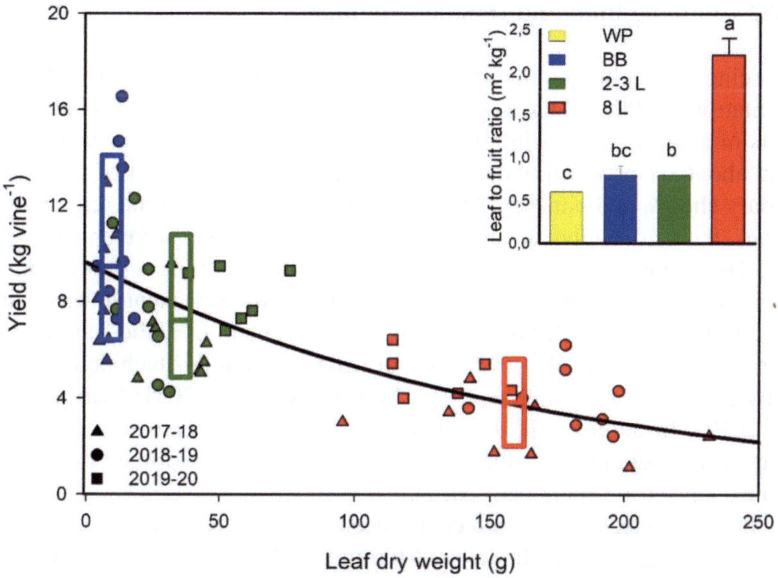

Fig. 10.2 Relationship between yield (kg vine^{-1}) and leaf dry weight removed for Malbec plants pruned during winter (WP), at budburst (BB), at 2–3 L and at 8 L. Upper and lower limits of the boxes represent the first and third quartile for yield and the interior horizontal line represents the median yield for each treatment. Insert shows the leaf to fruit ratio

budburst of remaining buds by 11 days, plants pruned at two to three separated leaves delayed it 15 days, and plants pruned at eight separate leaves delayed it almost a month. However, berry sugar accumulation occurred faster in those plants pruned at eight leaves, and consequently, the harvest date could not always be delayed. The higher rate of sugar accumulation was due to a higher leaf/fruit ratio (Fig. 10.2) caused by a decrease in yield (Bustos Morgani et al., 2022). The decrease in yield relative to winter pruned plants was 18% on those pruned at two to three separate leaves and 60% on those pruned at eight separate leaves, and it was related to the leaf dry weight eliminated at pruning (Fig. 10.2). Therefore, with Malbec plants it was not possible to delay harvest (i.e., considering °Brix as an indicator of harvest time). However, beyond the fact that the harvest date did not change, some effects of late pruning on berry and wine chemical composition were observed (Bustos Morgani et al., 2022). In general, late pruning treatments maintained or improved the quality of Malbec wines, although they increased their pH. These changes in the oenological composition were mainly due to changes in the thermal regime during the early stages of berry development and the yield reduction that changed the leaf:fruit ratio. If the goal is to delay budbreak between 10 and 15 days to escape late spring frosts, late pruning should be conducted between budbreak and two to three separate leaves to avoid significant yield losses.

10.2.4.2 Forcing Buds and Double Cropping

Bud forcing is a technique that delays berry development, especially soluble solid accumulation, and thus delays maturity in extremely hot areas (Gu et al., 2012). It consists of a severe shoot trimming above the last bunch a few days after fruitset, leaving about six or seven nodes, and the simultaneous removal of all the leaves, secondary shoots, and bunches. This releases the inhibition of the remaining buds, "forcing" them to budbreak and delaying the subsequent development of their shoots. This technique delays maturity between 30 and 60 days depending on the moment in which the forcing is carried out, with an improvement in berry enological quality (i.e., lower pH, higher acidity, higher content of anthocyanins, tannins, and total phenols; Gu et al., 2012). One drawback is that this technique can reduce yield by 20% or more (Gu et al., 2012; Martinez de Toda, 2021; Martinez Moreno et al., 2019). In addition, it is difficult to convince grapegrowers to remove all bunches at the beginning of the season, expecting that the forced buds will then generate a new crop (Poni et al., 2021).

An alternative technique has recently been proposed (called double-cropping or double harvesting) that facilitates bud-forcing application and would additionally allow an increase in yield. It has been demonstrated that it is possible to force bud development by only removing the secondary shoots, without removing the bunches (Martinez de Toda, 2021; Poni et al., 2021). Thus, it is possible to obtain a second harvest from the forced buds with a lower pH, higher acidity, and higher content of anthocyanins and phenols, which is added to the harvest of the primary bunches (Martinez de Toda, 2021; Poni et al., 2021). These authors showed that this second harvest could represent between 30% (Martinez de Toda, 2021) and 40–50% of the primary harvest (Poni et al., 2021). Some experiments are currently being carried out in Malbec with promising but contrasting results. In the first season of bud forcing (Fig. 10.3), a second harvest, representing 10% of the primary harvest, was

Fig. 10.3 Crop forcing to delay maturity and to obtain a second harvest in the same season. The red arrow indicates a forced shoot carrying two new bunches at pea size stage. In the lower part of the canopy, the bunches of the primary shoots left during winter pruning are visible at veraison. Malbec, Mendoza

achieved 1 month later with improved berry and wine characteristics (i.e., higher phenolic content, higher acidity, lower pH). By contrast, in the second season, the forced buds had no bunches, suggesting that no bunch induction occurred (Prieto et al. unpublished data, in progress). Anyway, in both seasons, the primary harvest was delayed between 10 and 15 days compared to the control. Obtaining a second harvest might be dependent on environmental conditions during spring and the phenological stage at which it is applied.

10.2.5 Strategies to Decrease Temperature

10.2.5.1 Shade Nets

In the past, several experiments carried out in cold or humid areas highlighted the importance of adequate bunch exposure to achieve the desired ripeness and improve berry chemical composition (Smart & Robinson, 1991). In addition, developing training systems pursued the main objective of maximizing canopy radiation interception, especially in regions with low sun radiation. However, given the current and future conditions, it is necessary to develop and evaluate different management strategies adapted for warm areas to decrease the negative effects of extreme events such as heat waves. In this sense, it is necessary to consider the different effects of bunch exposure on warm or cool regions.

Reducing vine organs' temperature by reducing radiation can be implemented by different techniques. Shade nets over the canopy have been evaluated in recent years in different places. Shade reduces solar radiation and decreases canopy temperature and also canopy photosynthesis which may slow down berry ripening. Reducing radiation between 35% and 60% using semi-shade nets in the bunch zone when berries were at pepper–corns size decreased berry temperature by 4 °C, preventing the loss of acidity and phenolic compounds (Lobos et al., 2015; Martinez-Lüscher et al., 2017). Similar results were obtained by Caravia et al. (2016) through a chapel-type shading (60% reduction) above the vineyard without reducing yield or photosynthesis. Using different colors of anti-hail net on different trellis structures in Malbec improved vine water status (Nahuel et al. in progress). However, excessive bunch shading (between 1% and 5% relative to ambient radiation) decreases berry and wine color and increases malic acid content (Chorti et al., 2010; Palliotti et al., 2014). In our area, using anti-hail nets could have some complementary applications in the future. For example, nets with different thread density to lower the temperature or the use of photoselective nets that absorb a specific wavelength (Martinez-Lüscher et al., 2017) could generate some benefits in terms of berry chemical composition, plant hydraulic architecture, or water use efficiency (Gonzalez et al., 2021).

10.2.5.2 Agrivoltaism: Production of Grapes and Energy from Renewable Sources

Agrovoltaism is a system that combines agricultural production with the generation of renewable energy in the same space to improve the efficiency of land use. It involves installing photovoltaic panels above the crop (Abidin et al., 2021). This agricultural-energy system is one of the innovations in agriculture of the last 10 years that has attracted the development of numerous investigations in different crops and in different countries (Abidin et al., 2021). It appears as another alternative strategy to face climate change and the challenges of energy availability in the future (Elamri et al., 2018). It may represent a diversification alternative for agricultural producers (Cuppari et al., 2021). An agrivoltaic system modifies the environmental conditions for the crop. Generally, the air ambient temperature at the crop level and that of the soil is lower due to the solar radiation intercepted by the panels (Marrou et al., 2013a), and water consumption is also lower (Marrou et al., 2013b). It is important to consider the rapid and constant advances in the innovations of photovoltaic solar panels that will also modify the effects of these systems on the crops.

Currently, our team is conducting a study that combines two mitigation techniques for the increase in temperature (Fig. 10.4): solar panels and row orientation (Mariana Gomez Tournier et al., in progress).

Our objective is to evaluate the effects of shading caused by agrovoltaic panels on different aspects of grapevine functioning, yield, berry, and wine composition with two different row orientations. The experiment is carried out in a Malbec vineyard where simulated solar panels generating 47% of shade to the soil and the

Fig. 10.4 Experimental agrivoltaic Malbec vineyard at INTA Mendoza Experimental Station planted at two different row orientations where the impact of the shade produced by simulated agrivoltaic panels is studied on grapevine physiology, water use, yield, and wine composition

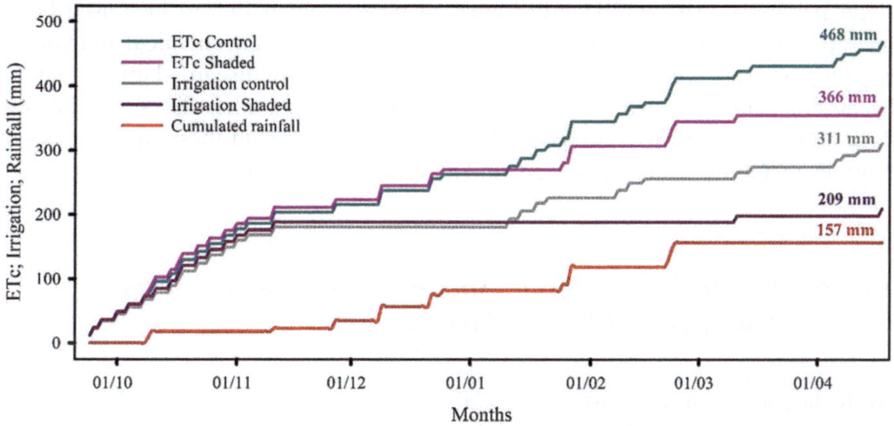

Fig. 10.5 Crop evapotranspiration (ETc), irrigation, and accumulated rainfall under simulated solar panels (Shaeded) and under direct radiation (Control) of Malbec experimental vineyard at INTA Mendoza Experimental Station, Mendoza, Argentina

canopy were installed over two row orientations, East-West and North-South. In each row orientation, plants under the agrivoltaic system were compared with control plants (without simulated solar panels). Our preliminary results, after two seasons of experience, showed that vines under partial shade required 31.7% less irrigation water than control vines (Fig. 10.5, Gomez Tournier et al., 2022). At harvest, no difference in total berry polyphenols was found between control plants and shade-grown plants. Consistent with previous works (see the point row orientation), wines from the E–W orientation showed higher levels of anthocyanins. On the other hand, regardless of the row orientation, wines from the agrivoltaic system showed higher total acidity and lower volatile acidity, pH, and alcohol content than those from control plants (Gomez Tournier et al., 2022). These results must be confirmed in the long-term. Still, our results show that agrivoltaism is a production system that reduces vineyard irrigation requirements throughout the season without significant changes in the accumulation in berry polyphenols, reducing wine pH and alcohol content and increasing total acidity.

10.2.5.3 Water Sprinkling (Evaporative Cooling)

Evaporative cooling is another technique to reduce canopy temperature. It consists of spraying water over the canopy during the warmest hours of the day. The evaporation of the sprayed water requires latent heat, part of which comes from the plant organs, thus lowering their temperature. There are some precedents with variable results. The first studies were carried out by spraying water above the canopy, while recently, a misting system that sprayed low water volumes inside the canopy has been evaluated (Caravia et al., 2017). The system can turn on when the temperature exceeds a certain preset threshold. Although no differences in Brix were detected, sugar per berry was significantly higher in treated plants due to increased berry mass (Caravia et al., 2017). On the other hand, berry and wine chemical and phenolic composition were not affected. Some studies in the past discarded this technique because it generates favorable conditions for botrytis and other diseases. The new development shown by Caravia et al. (2017) seems to avoid these drawbacks but requires installing a misting system and additional management costs.

10.2.6 Limiting the Source: Sink Ratio to Balance Berry Maturity and Delay Harvest

At the beginning of the season, vine carbohydrate sources are the perennial organs (i.e., cordons, trunk, and roots), and thereafter, the leaves. Berry quality is largely determined by the leaf area/fruit ratio, in particular by the percentage of leaves that are exposed to the sun and its relation to yield (Kliewer & Dokoozlian, 2005). It has been proposed that a leaf:fruit ratio between 0.6 and 1.2 $m^2\ kg^{-1}$ indicates balanced vines that produce grapes and wines of high oenological quality (Kliewer & Dokoozlian, 2005). Less than 0.6 m^2/kg leaf:fruit ratio would slow down ripening. Therefore, decreasing leaf:fruit ratio during ripening could slow down berry sugar accumulation and provide more time for a higher synthesis and accumulation of phenolic compounds (Palliotti et al., 2014). In any case, a delay in maturity does not always imply an improvement in grape chemical composition in terms of phenolic compounds or total acidity (Caccavello et al., 2017; Gatti et al., 2019).

10.2.6.1 "Apical" Leaf Removal

Leaf removal is a widespread management practice that can be applied in different ways, at different stages of the vegetative cycle, and pursuing different objectives. Traditional leaf thinning removes basal leaves in the cluster zone between pea-size and veraison to improve microclimate and berry chemical composition, and to avoid fungal diseases. In our area, it is carried out only on the east side of the canopy in VSP systems to avoid excessive exposure of the west-side bunches during the afternoon when temperature is high. When leaf removal is conducted between flowering and fruitset, called "early leaf removal", its main objective is to reduce fruitset and therefore number of berries in varieties with very compact bunches to avoid botrytis and to favor the oenological quality of the grapes. In this practice, all the leaves below the inflorescences are removed after flowering or fruitset. This practice strongly alters the carbohydrate source and, therefore, affects the final number of berries per bunch. In both types of leaf removal (traditional and early) the basal leaves, or up to the first third of the shoots, are removed.

By contrast, an "apical" leaf removal has been evaluated and proposed these last years to reduce the leaf:fruit ratio and slow-down ripening. It consists of the removal of the leaves of about five or six nodes just above the last bunch between veraison and harvest when the berry is around 12–15 °Brix (Palliotti et al., 2014). At this time, leaves located in the two thirds of the canopy have already reached their maximum expansion, are far from their senescence, and they constitute the most photosynthetically active leaves (Palliotti et al., 2014). Because they are far from the bunches, apical leaf removal is easy to mechanize (Palliotti et al., 2013). It has shown to be effective to delay harvest by about 2 weeks, slow down sugar accumulation, and decrease wine alcohol (Caccavello et al., 2017). However, results related to phenolic compounds (and sugar: anthocyanin ratio) are variable. It has been observed that post veraison leaf removal can decrease, increase, or not affect the concentration of phenolic compounds (Caccavello et al., 2017; Palliotti et al., 2013). The results seem to depend on the intensity and moment of leaf removal, being necessary to avoid excessively severe defoliation, eliminating less than 30% of the total leaf area (Pallioti et al., 2013; Caccavello et al., 2017). Some current experiments with Malbec in our region, have shown that when performing this practice at 12° and 14° Brix, no delay in maturity was observed over two seasons.

10.2.6.2 Severe Shoot Trimming

Shoot trimming is a management technique that is commonly performed to facilitate machinery transit, to improve the effectiveness of phytosanitary treatments, and to adapt the shape of the canopy to the training system. It eliminates the portion of shoots above 15–20 cm from the top wire. The number of trimming operations depends mainly on vine vigor. By contrast, a severe shoot trimming consists of removing a larger portion of the shoot, leaving between 7 and 15 nodes per shoot. This technique was evaluated post-setting (Martinez de Toda et al., 2014) or after veraison when the berry was close to 15 °Brix (Filippetti et al., 2015; Caccavello et al., 2017). This practice proved to be effective to delay sugar accumulation, increase or maintain acidity, and increase anthocyanins concentration (Martinez de Toda et al., 2014) when it was applied at post fruitset. Our group has obtained similar preliminary results in Malbec over two seasons, where harvest was delayed up to 2 weeks with consequences on berry and wine composition (Prieto et al., in progress). When carried out in veraison it can also favor or maintain the chemical composition of the grape and reduce the concentration of soluble solids at harvest (Filippetti et al., 2015; Caccavello et al., 2017).

10.2.7 Some Enological Alternatives

In the last decades, the alcohol levels of wines have increased in most of the producing wine regions (Longo et al., 2017). This can be associated with consumer preferences for well-structured, full-bodied wines, rich in ripe-fruit flavors. To achieve this purpose, it is required to efficiently extract phenolic and aromatic compounds from the skins during maceration, grape maturity being a key factor. Under this scenario, winemakers tend to delay harvest, searching a high concentration of these compounds, leading simultaneously to the increase of berry soluble solid concentration and, therefore, increasing wines alcohol concentration (Rolle et al., 2018). This trend has been exacerbated these last years by the observed climatic changes that promote a higher berry sugar concentration. In this context, several possible oenological approaches to reduce the alcohol levels while preserving high phenolics compounds in wines have been proposed. Pre-fermentation and post-fermentation practices can be applied to lower the alcohol levels. Pre-fermentation practices include dilution or blending with juice from low-sugar grapes early harvested, microbiological strategies (including the use of yeast strains that reduce the efficiency of ethanol production), and post-fermentation technologies include blending of high and low alcohol wines, and membrane-based techniques with physical removal of alcohol after fermentation.

In several experiments carried out in Mendoza, a technique easy-to-adopt, flexible, and cost-effective, for simultaneously reducing the alcohol content and the pH of red wines was tested on Malbec, Bonarda, and Syrah varieties. This technique, previously proposed by other authors (Piccardo et al., 2019), consisted of blending wines made from grapes of different ripeness. Specifically, using a low-ethanol and highly acidic wine to replace an equal volume of well-ripened grape juice, prior to fermentation, without altering the chemical and sensory quality of the wines. In another study (Fanzone et al., 2020), harvest blending was combined with inoculation of native yeasts to reduce simultaneously alcohol content and pH of Malbec wines. For this, Sauvignon blanc grapes harvested at the beginning of ripening were employed to obtain a wine with high acidity and low alcohol content (LW) that was blended with Malbec grapes of full phenolic maturity. Three treatments were established: Malbec grapes were harvested at 13.0% (1H) and 14.5% (2H) v/v of probable alcohol and elaborated following a standard protocol. A third treatment was obtained from 2H, where a part of the grape juice was removed and replaced with LW (2RW) as a strategy for alcohol reduction. Each treatment (1H, 2H, 2RW) was elaborated by triplicate. These wines were also inoculated with three kinds of yeasts, including native strains (NS; S. cerevisiae BSc114 native yeast), commercial yeasts (CS; S. cerevisiae EC 1118) and co-inoculation of different strains (CI; Hanseniaspora uvarum BHu9/Saccharomyces cerevisiae BSc114). Results showed that 2RW wines, fermented with different yeast strains, showed similar levels of total phenols, tannins, anthocyanins, and polymeric pigments concerning control wines (2H). In all cases, those wines presented greater phenolic potential compared to wines from 1H. At the same time, the strategy of vintages blending (2RW) produced wines with 1.4% v/v less alcohol than 2H wines, also achieving a pH decrease of 0.3 units. Combined treatments with native strains, especially as single inoculum (2RW–NS), were the most efficient in reducing both parameters, showing higher tannins, anthocyanins levels, and color saturation, without affecting the sensory quality, in terms of aromas and mouthfeel aspects (Fig. 10.6). These strategies proposed could be simple and economic tools for red wine production with low alcohol content, preserving phenolics concentration and organoleptic perception.

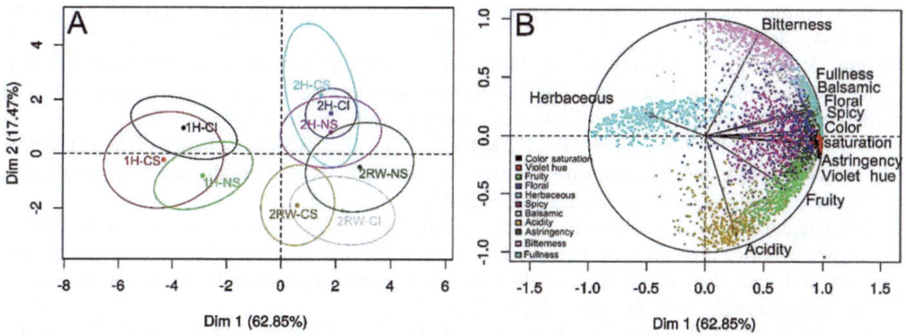

Fig. 10.6 Principal component analysis (PCA) of descriptive sensory data of Malbec wines evaluated by a trained panel ($n = 12$). (**a**) wine factor map. (**b**) sensory attributes loadings with 95% confidence ellipses based on the multivariate distribution of Hotelling's test ($p < 0.05$)

10.3 Conclusions

Climate change in Argentina will modify, on the one hand, the ways and varieties used to produce grapes in traditional wine regions, and on the other hand, it will move the planting of new vineyards to regions where no viticulture existed before. While some companies may relocate their vineyards to more suitable areas in the future, climate change imposes the challenge of maintaining the sustainability of current wine-growing areas. Although some information already exists, more knowledge about the short- and long-term effects of single and combined adaptation strategies to climate change must still be developed. For instance, it is necessary to develop adaptation strategies and study their long-term impacts in different training systems, varieties, and areas. It is also necessary to evaluate the combination of different techniques and their potential cumulative effects (for example, late pruning and subsequent severe shoot thinning, or leaf removal) or to understand their interactions. Generating this information would allow to make more solid recommendations for each situation. It's worth noting that strategies must be sustainable, meaning that they must be economically beneficial for the grapegrower and winery, environmentally friendly, and socially accepted. To achieve this, it will be necessary to provide grapegrowers opportunely with climatic information of different potential viticultural regions and with the necessary tools for canopy management, irrigation, nutrition, and use of adapted plant material.

References

Abidin, M. A., Mahyuddin, M. N., & Zainuri, M. A. (2021). Solar photovoltaic architecture and agronomic management in agrivoltaic system: A review. *Sustainability, 13*, 18–21.

Ahumada, G. E., Catania, A., Fanzone, M. L., Belmonte, M. J., Giordano, C. V., & González, C. V. (2021). Effect of leaf-to-fruit ratios on phenolic and sensory profiles of Malbec wines from single high-wire-trellised vineyards. *Journal of the Science of Food and Agriculture, 101*(4), 1467–1478.

Allegro, G., Pastore, C., Valentini, G., & Filippetti, I. (2020). Post-budburst hand finishing of winter spur pruning can delay technological ripening without altering phenolic maturity of Merlot berries. *Australian Journal of Grape and Wine Research, 26*, 139–147.

Arias, L., Berli, F., Fontana, A., Bottini, R., & Piccoli, P. (2022). Climate change effects on grapevine physiology and biochemistry: Benefits and challenges of high altitude as an adaptation strategy. *Frontiers in Plant Science, 13*, 835425.

Arrizabalaga-Arriazu, M., Gomès, E., Morales, F., Irigoyen, J. J., Pascual, I., & Hilbert, G. (2021). Impact of 2100-projected air temperature, carbon dioxide, and water scarcity on grape primary and secondary metabolites of different *Vitis vinifera* cv. Tempranillo clones. *Journal of Agricultural and Food Chemistry, 22*, 6172–6185.

Barroso, R., Carbajal, H., Ortiz, H., Malaniuk, M., Quenol, H., Murgo, M., Coria, C., Videla, R., Prieto, S., Manzano, H., Quini, C., & Aruani, C. (2019). High-altitude wines from northwest Argentina-physical-chemical and sensory characteristics. *BIO Web of Conferences, 15*, 01002.

Boninsegna, J. A. (2014). Impacto del Cambio Climático en los oasis del oeste argentino. *Ciencia e Investigación, 64*(1), 45–58.

Bustos Morgani, M., Perez Peña, J., Fanzone, M., & Prieto, J. A. (2022). Pruning after budburst delays phenology and affects yield components, crop coefficient, and total evapotranspiration in *Vitis Vinifera* L. cv. Malbec in Mendoza, Argentina. *Scientia Horticulturae, 296*, 110886.

Bustos Morgani, M., Fanzone, M., Perez Peña, J., Sari, S., Gallo, A., Gomez Tournier, M., & Prieto, J. A. (2023). Late pruning modifies leaf to fruit ratio and shifts maturity period, affecting berry and wine composition in *Vitis vinifera* L. cv. Malbec in Mendoza, Argentina. *Scientia Horticulturae, 313*, 111861.

Cabré, F., & Nuñez, M. (2020). Impacts of climate change on viticulture in Argentina. *Regional Environmental Change, 20*, 1–12.

Cabré, M. F., Quénol, H., & Nuñez, M. (2016). Regional climate change scenarios applied to viticultural zoning in Mendoza, Argentina. *International Journal of Biometeorology, 60*, 1325–1340.

Caccavello, G., Giaccone, M., Scognamiglio, P., Forlani, M., & Basile, B. (2017). Influence of intensity of post-veraison defoliation or shoot trimming on vine physiology, yield components, berry and wine composition in Aglianico grapevines. *Australian Journal of Grape and Wine Research, 23*(2), 226–239.

Caravia, L., Collins, C., Petrie, P. R., & Tyerman, S. D. (2016). Application of shade treatments during Shiraz berry ripening to reduce the impact of high temperature. *Australian Journal of Grape and Wine Research, 22*, 422–437.

Caravia, L., Pagay, V., Collins, C., & Tyerman, S. D. (2017). Application of sprinkler cooling within the bunch zone during ripening of Cabernet Sauvignon berries to reduce the impact of high temperature. *Australian Journal of Grape and Wine Research, 23*, 48–57.

Chorti, E., Guidoni, S., Ferrandino, A., & Novello, V. (2010). Effect of different cluster sunlight exposure levels on ripening and anthocyanin accumulation in Nebbiolo grapes. *American Journal of Enology and Viticulture, 61*, 23–30.

Clingeleffer, P. R. (2009). Influence of canopy management systems on vine productivity and fruit composition. In *Recent advances in grapevine canopy management* (pp. 13–18). University of California.

Cuppari, R. I., Higgins, C. W., & Characklis, G. W. (2021). Agrivoltaics and weather risk: A diversification strategy for landowners. *Applied Energy, 291*, 116809.

de Rosas, I., Ponce, M. T., Malovini, E., Deis, L., Cavagnaro, B., & Cavagnaro, P. (2017). Loss of anthocyanins and modification of the anthocyanin profiles in grape berries of Malbec and Bonarda grown under high temperature conditions. *Plant Science, 258*, 137–145.

Deis, L., de Rosas, I., Malovini, E., Cavagnaro, M., & Cavagnaro, J. B. (2015). Impacto del cambio climático en Mendoza. Variación climática en los últimos 50 años. Mirada desde la fisiología de la vid. *Revista de la Facultad de Ciencias Agrarias, 47*, 67–92.

Elamri, Y., Cheviron, B., Lopez, J. M., Dejean, C., & Belaud, G. (2018). Water budget and crop modelling for agrivoltaic systems: Application to irrigated lettuces. *Agricultural Water Management, 208*, 440–453.

Fanzone, M., Sari, S., Mestre Furlani, M., Catania, A., Catelén, M., Jofre, V., González-Miret, M., Combina, M., Vazquez, F., & Maturano, P. (2020). Combination of pre-fermentative and fermentative strategies to produce Malbec wines of lower alcohol and pH, with high chemical and sensory quality. *OENO one, 54*(4), 1041–1058.

Filippetti, I., Movahed, N., Allegro, G., Valentini, G., Pastore, C., Colucci, E., & Intrieri, C. (2015). Effect of post-veraison source limitation on the accumulation of sugar, anthocyanins and seed tannins in *Vitis vinifera* cv. Sangiovese berries. *Australian Journal of Grape and Wine Research, 21*(1), 90–100.

Friend, A. P., & Trought, M. C. T. (2007). Delayed winter spur-pruning in New Zealand can alter yield components of Merlot grapevines. *Australian Journal of Grape and Wine Research, 13*, 157–164.

Frioni, T., Pirez, F. J., Diti, I., Ronney, L., Poni, S., & Gatti, M. (2019). Post-budbreak pruning changes intra-spur phenology dynamics, vine productivity and berry ripening parameters in *Vitis vinifera* L. cv. 'Pinot Noir'. *Scientia Horticulturae, 256*, 108584.

Galat Giorgi, E., Sadras, V. O., Keller, M., & Perez Peña, J. E. (2019). Interactive effects of high temperature and water deficit on Malbec grapevines. *Australian Journal of Grape and Wine Research, 25*, 345–356.

Galat Giorgi, E. G., Keller, M., Sadras, V., Roig, F. A., & Peña, J. P. (2020). High temperature during the budswell phase of grapevines increases shoot water transport capacity. *Agricultural and Forest Meteorology, 295*, 108173.

Gallo, A. E., Perez Peña, J. E., & Prieto, J. A. (2020). Mechanisms underlying photosynthetic acclimation to high temperature are different between *Vitis vinifera* cv. Syrah and Grenache. *Functional Plant Biology, 48*(3), 342–357.

Gallo, A. E., Perez Peña, J. E., Gonzalez, C. V., & Prieto, J. A. (2022). Syrah and Grenache (*Vitis vinifera*) revealed different strategies to cope with high temperature. *Australian Journal of Grape and Wine Research, 28*(3), 383–394.

Gatti, M., Pirez, F. J., Frioni, T., Squeri, C., & Poni, S. (2018). Calibrated, delayed-cane winter pruning controls yield and significantly postpones berry ripening parameters in Vitis vinifera L. cv. Pinot Noir. *Australian Journal of Grape and Wine Research, 24*, 305–316.

Gatti, M., Garavani, A., Krajecz, K., Ughini, V., Parisi, M., Frioni, T., & Poni, S. (2019). Mechanical mid-shoot leaf removal on Ortrugo (*Vitis vinifera* L.) at pre- or mid-veraison alters fruit growth and maturation. *American Journal of Enology and Viticulture, 70*, 88–97.

Gómez Tournier, M., Perez Peña, J., Prieto, J. A., Fanzone, M., Sari, S., & Segovia, B. (2022). *Effect of the shade generated by simulated solar panels on the physiology and productivity of Vitis vinifera L. cv. Malbec*. Agrivoltaic conference.

González, C. V., Prieto, J. A., Mazza, C., Jeréz, D. N., Biruk, L. N., Jofre, M. F., & Giordano, C. V. (2021). Grapevine morphological shade acclimation is mediated by light quality whereas hydraulic shade acclimation is mediated by light intensity. *Plant Science, 307*, 110893.

Gu, S., Jacobs, S. D., McCarthy, B. S., & Gohil, H. L. (2012). Forcing vine regrowth and shifting fruit ripening in a warm region to enhance fruit quality in 'Cabernet Sauvignon' grapevine (*Vitis vinifera* L.). *The Journal of Horticultural Science and Biotechnology, 87*, 287–292.

Gutiérrez-Gamboa, G., Liu, S. Y., & Pszczólkowski, P. (2020). Resurgence of minority and autochthonous grapevine varieties in South America: A review of their oenological potential. *Journal of the Science of Food and Agriculture, 100*, 465–482.

Gutiérrez-Gamboa, G., Zheng, W., & Martínez de Toda, F. (2021). Current viticultural techniques to mitigate the effects of global warming on grape and wine quality: A comprehensive review. *Food Research International, 139*, 109946.

Howell, G. S., & Wolpert, J. A. (1978). Nodes per cane, primary bud phenology, and spring freeze damage to concord grapevines. A preliminary note. *American Journal of Enology and Viticulture, 29*, 5–8.

Intrieri, C., Filippetti, I., Allegro, G., Valentini, G., Pastore, C., & Colucci, E. (2011). The semi-minimal-pruned hedge: A novel mechanized grapevine training system. *American Journal of Enology and Viticulture, 62*, 312–318.

INV. (2020). *Informe anual de superficie*. Instituto Nacional de Vitivinicultura.

Kliewer, W. M., & Dokoozlian, N. (2005). Leaf area/crop weight ratios of grapevines: Influence on fruit composition and wine quality. *American Journal of Enology and Viticulture, 56*, 170–181.

Lobos, G. A., Acevedo-Opazo, C., Guajardo-Moreno, A., Valdés-Gómez, H., Taylor, J. A., & Laurie, V. F. (2015). Effects of kaolin-based particle film and fruit zone netting on Cabernet Sauvignon grapevine physiology and fruit quality. *OENO one, 49*, 137–144.

Longo, R., Blackman, J. W., Torley, P. J., Rogiers, S. Y., & Schmidtke, L. M. (2017). Changes in volatile composition and sensory attributes of wines during alcohol content reduction. *Journal of the Science of Food and Agriculture, 97*(1), 8–16.

Lucero, C., Filippo, M., Vila, H., & Venier, M. (2017). Comparing water deficit and saline stress between 1103P and 101-14 Mgt rootstocks, grafted with Cabernet Sauvignon. *Revista de la Facultad de Ciencias Agrarias, 49*(1), 33–43.

Marrou, H., Guilioni, L., Dufour, L., Dupraz, C., & Wery, J. (2013a). Microclimate under agrivoltaic systems: Is crop growth rate affected in the partial shade of solar panels? *Agricultural and Forest Meteorology, 177*, 117–132.

Marrou, H., Dufour, L., & Wery, J. (2013b). How does a shelter of solar panels influence water flows in a soil-crop system? *European Journal of Agronomy, 50*, 38–51.

Martin, L., Vila, H., Bottini, R., & Berli, F. (2020). Rootstocks increase grapevine tolerance to NaCl through ion compartmentalization and exclusion. *Acta Physiologiae Plantarum, 42*, 1–11.

Martinez De Toda, F. (2021). Global warming allows two grape crops a year, with about two months apart in ripening dates and with very different grape composition-the forcing vine regrowth to obtain two crops a year. *Vitis, 60*, 119–124.

Martínez de Toda, F., Sancha, J. C., Zheng, W., & Balda, P. (2014). Leaf area reduction by trimming, a growing technique to restore the anthocyanins: Sugars ratio decoupled by the warming climate. *Vitis, 53*(4), 189–192.

Martinez-Luscher, J., Chen, C. C. L., Brillante, L., & Kurtural, S. K. (2017). Partial solar radiation exclusion with color shade nets reduces the degradation of organic acids and flavonoids of grape berry (*Vitis vinifera* L.). *Journal of Agricultural and Food Chemistry, 65*(49), 10693–10702.

Martínez-Moreno, A., Sanz, F., Yeves, A., Gil-Muñoz, R., Martínez, V., & Intrigliolo, D. S. (2019). Forcing bud growth by double-pruning as a technique to improve grape composition of *Vitis vinifera* L. cv. Tempranillo in a semi-arid Mediterranean climate. *Scientia Horticulturae, 256*, 108614.

Morales-Castilla, I., de Cortázar-Atauri, I. G., Cook, B. I., Lacombe, T., Parker, A., Van Leeuwen, C., et al. (2020). Diversity buffers winegrowing regions from climate change losses. *Proceedings of the National Academy of Sciences, 117*(6), 2864–2869.

Moran, M. A., Bastian, S. E., Petrie, P. R., & Sadras, V. O. (2018). Late pruning impacts on chemical and sensory attributes of Shiraz wine. *Australian Journal of Grape and Wine Research, 24*, 469–477.

Naulleau, A., Gary, C., Prévot, L., & Hossard, L. (2021). Evaluating strategies for adaptation to climate change in grapevine production – A systematic review. *Frontiers in Plant Science, 11*, 2154.

O.I.V. (2017). *Distribution of the world's grapevine varieties*. Focus OIV. https://www.oiv.int/public/medias/5888/en-distribution-of-the-worlds-grapevine-varieties.pdf

Palliotti, A., Panara, F., Silvestroni, O., Lanari, V., Sabbatini, P., Howell, G. S., et al. (2013). Influence of mechanical postveraison leaf removal apical to the cluster zone on delay of fruit ripening in Sangiovese (*Vitis vinifera* L.) grapevines. *Australian Journal of Grape and Wine Research, 19*(3), 369–377.

Palliotti, A., Tombesi, S., Silvestroni, O., Lanari, V., Gatti, M., & Poni, S. (2014). Changes in vineyard establishment and canopy management urged by earlier climate-related grape ripening: A review. *Scientia Horticulturae, 178*, 43–54.

Palliotti, A., Frioni, T., Tombesi, S., Sabbatini, P., Cruz-Castillo, J. G., Lanari, V., Silvestroni, O., Gatti, M., & Poni, S. (2017). Double-pruning grapevines as a management tool to delay berry ripening and control yield. *American Journal of Enology and Viticulture, 68*, 412–421.

Parker, A., de Cortázar-Atauri, I. G., Chuine, I., et al. (2013). Classification of varieties for their timing of flowering and veraison using a modelling approach: A case study for the grapevine species *Vitis vinifera* L. *Agricultural and Forest Meteorology, 180*, 249–264.

Petrie, P. R., Brooke, S. J., Moran, M. A., & Sadras, V. O. (2017). Pruning after budburst to delay and spread grape maturity. *Australian Journal of Grape and Wine Research, 23*, 378–389.

Piccardo, D., Favre, G., Pascual, O., Canals, J. M., Zamora, F., & González-Neves, G. (2019). Influence of the use of unripe grapes to reduce ethanol content and pH on the color, polyphenol and polysaccharide composition of conventional and hot macerated Pinot Noir and Tannat wines. *European Food Research and Technology, 245*, 1321–1335.

Poni, S., Del Zozzo, F., Santelli, S., Gatti, M., Magnanini, E., Sabbatini, P., & Frioni, T. (2021). Double cropping in *Vitis vinifera* L. cv. Pinot Noir: Agronomical and physiological validation. *Australian Journal of Grape and Wine Research, 27*(4), 508–518.

Prieto, J. A., Louarn, G., Perez Peña, J., Ojeda, H., Simonneau, T., & Lebon, E. (2013). Impact of training system on gas exchanges and water use efficiency: A 3D modeling study with Topvine. In *18th international symposium GiESCO* (pp. 563–567). Porto, Portugal.

Rolle, L., Englezos, V., Torchio, F., Cravero, F., Río Segade, S., Rantsiou, K., & Cocolin, L. (2018). Alcohol reduction in red wines by technological and microbiological approaches: A comparative study. *Australian Journal of Grape and Wine Research, 24*(1), 62–74.

Santos, J., Fraga, H., Malheiro, A., Moutinho-Pereira, J., Dinis, L., Correia, C., Moriondo, M., Leolini, L., Dibari, C., Costafreda-Aumedes, S., Karstchall, T., Menz, C., Molitor, D., Junk, J., Beyer, M., & Schultz, H. (2020). A review of the potential climate change impacts and adaptation options for European viticulture. *Applied Sciences, 10*(9), 3092.

Schultz, H. R., & Hofmann, M. (2016). The ups and downs of environmental impact on grapevines: Future challenges in temperate viticulture. In H. Géros, H. Medrano, S. Delrot, and M. M. Chaves (eds) *Grapevine and environmental stress* (pp. 18–37). John Wiley & Sons Ltd, Chichester, UK.

Smart, R. E., & Robinson, M. (1991). *Sunlight into wine-a handbook for grapevine canopy management*. Adelaide, Winetitles.

Tortosa, I., Escalona, J. M., Toro, G., Douthe, C., & Medrano, H. (2020). Clonal behavior in response to soil water availability in Tempranillo grapevine cv: From plant growth to water use efficiency. *Agronomy, 10*(6), 862.

van Houten, S., Muñoz, C., Bree, L., Bergamín, D., Sola, C., & Lijavetzky, D. (2020). Natural genetic variation for grapevine phenology as a tool for climate change adaptation. *Applied Sciences, 10*(16), 5573.

Van Leeuwen, C., Roby, J. P., Alonso-Villaverde, V., & Gindro, K. (2013). Impact of clonal variability in Vitis vinifera cabernet franc on grape composition, wine quality, leaf blade stilbene content, and downy mildew resistance. *Journal of Agricultural and Food Chemistry, 61*, 19–24.

Van Leeuwen, C., Destrac-irvine, A., Dubernet, M., Duchêne, E., Gowdy, M., Marguerit, E., Pieri, P., Parker, A. K., de Rességuier, L., & Ollat, N. (2019). An update on the impact of climate change in viticulture and potential adaptations. *Agronomy, 10*, 1–20.

Villalba, R. (2007). Cambios Climáticos Regionales en el Contexto del Calentamiento Global. *Director, 6*.

Wolkovich, E. M., de Cortázar-Atauri, I. G., Morales-Castilla, I., Nicholas, K. A., & Lacombe, T. (2018). From Pinot to Xinomavro in the world's future wine-growing regions. *Nature Climate Change, 8*(1), 29–37.

Chapter 11
Autochthonous Grapevine Varieties From Argentina

Jorge Alejandro Prieto, Rocio Torres, Gustavo Alberto Aliquó, Santiago Sari, Simón Tornello, María Elena Palazzo, Anibal Catania, and Martín Fanzone

11.1 General Overview

Currently, between 6000 and 10,000 cultivars of *Vitis vinifera* L. exist worldwide, which are cultivated for different purposes, wine production, table grapes, raisins, or juice (Lacombe et al., 2013). These existing cultivars have originated since grapevine cultivation and domestication commenced 11,000 years ago (Dong et al., 2023). The main sources of variability are crosses between different cultivars and somatic mutations selected and conserved by vegetative reproduction (Myles et al., 2011). While this considerable genetic diversity is mainly conserved in germplasm collections (This et al., 2006), the global wine market is limited to a small group of cultivars. The 13 most planted cultivars around the world, Kyoho, Cabernet Sauvignon, Sultanina, Merlot, Tempranillo, Airen, Chardonnay, Syrah, Red Globe,

J. A. Prieto (✉) · M. Fanzone
Estación Experimental Agropecuaria Mendoza, Instituto Nacional de Tecnología Agropecuaria (INTA), Mendoza, Argentina

Centro de Estudios Vitivinícolas y Agroindustriales (CEVA), Universidad Juan Agustín Maza, Mendoza, Argentina
e-mail: prieto.jorge@inta.gob.ar

R. Torres · G. A. Aliquó · S. Sari · A. Catania
Estación Experimental Agropecuaria Mendoza, Instituto Nacional de Tecnología Agropecuaria (INTA), Mendoza, Argentina

S. Tornello
Agencia Extensión Rural Calingasta, Instituto Nacional de Tecnología Agropecuaria (INTA), San Juan, Argentina

M. E. Palazzo
Secretaria de Agricultura, Ganadería y Pesca, Ministerio de Economía de la Nación, Buenos Aires, Argentina

G. Gutiérrez Gamboa, M. Fourment (eds.), *Latin American Viticulture Adaptation to Climate Change*, https://doi.org/10.1007/978-3-031-51325-1_11

Garnacha Tinta, Sauvignon Blanc, Pinot Noir, and Trebbiano Toscano, cover more than one-third of the total cultivated surface area. The 33 most planted cultivars cover more than 50% of the global cultivated area (OIV, 2017), which represents less than 1% of the total existing genetic diversity (Wolkovich et al., 2018). This global trend to cultivate some international cultivars has led to the disappearance of numerous minor and local genotypes (Muñoz Organero et al., 2015; Zinelabidine et al., 2015). Furthermore, these 13 international cultivars cover more than 60% of the cultivated surface area of the main producer countries (Anderson & Nelgen, 2020), and even more (up to 80%) in some countries, such as Australia, New Zealand, or Chile (Wolkovich et al., 2018). In Argentina the situation is somewhat different since the two more cultivated varieties (Malbec and Bonarda) are not within these 13 most important varieties, and almost 30% of the cultivated area corresponds to local cultivars or *criollas* (INV, 2020). The main cultivated *criollas* in our country are Cereza and Criolla Grande, two varieties widely spread during the 1960s and 1970s because of their high yield potential. Other *criollas* varieties widely cultivated are Pedro Giménez, Torrontés Riojano, and Moscatel Rosado and a large number of minority varieties spread over different regions.

In recent years, however, on a global scale, this situation has slowly begun to change, and producers, wineries, and consumers are looking for new products, elaborated from local resources (García-Muñoz et al., 2014). Minor and local cultivars are thus an opportunity to offer product diversification in the global market (Antolín et al., 2021), and probably to adapt to climate change (Florez-Sarasa et al., 2020). Yet, there is still a long way to evaluate these "new ancient" varieties in different environmental conditions, to develop the most appropriate cultivation and elaboration methods, and especially to convince growers and consumers about their qualities. Here we briefly present our results corresponding to our project conducted for more than 10 years, covering different aspects such as the rescue, identification, conservation, characterization, and valorization of these genotypes.

11.2 A Brief History of Grapevine in America

The history of grapevine (*Vitis vinifera* L.) in America dates to the fifteenth century, when it was first introduced to the Antilles during the Spanish colonization where it did not prosper due to climatic conditions (Maurín-Navarro, 1967; Martínez et al., 2006). However, it was cultivated with success in Mexico a few years later, and then in Peru in the early sixteenth century (Martínez et al., 2006) expanding subsequently to the rest of the South American colonies (Milla Tapia et al., 2007). Some evidence suggest that it was brought from the Canary Islands (Agüero et al., 2003), and others that it arrived from Spain (Toro-Lira, 2018), but it is still unclear if it was introduced in the form of seeds, cuttings, plants, or buds. The introduction to Argentina would have been in 1557, when the father Juan Cidrón introduced some plants to Santiago del Estero (North of Argentina) from La Serena (Chile). A few years later, at the founding of Mendoza in 1561, Pedro del Castillo implanted a plot of plants brought from Santiago de Chile and in 1562 more plants were brought by Juan Jufré who was an experienced grapegrower. From that moment, its cultivation began to grow (Maurín-Navarro, 1967). During these first steps of South American viticulture and for more than 300 years, the Spanish variety Listán Prieto was the predominant variety (Lacoste et al., 2010). It was grown under different names such as Criolla Chica or Criolla de vino in Argentina, Uva País in Chile, Negra Corriente or Negra Criolla in Peru, Misión in Mexico, and Mission in USA (Agüero et al., 2003; Martínez et al., 2006). Then, during the early eighteenth century, Muscat of Alexandria was also highly cultivated in Argentina for longtime, where it was called Uva de Italia (Alcalde, 1989). It was introduced from Spain to Mendoza by Jesuit missionaries at the end of the seventeenth century. It was one of the most cultivated white varieties until the end of the twentieth century (Lacoste, 2013). During the eighteenth century, the vast majority of the *criollas* were originated. Several recent studies have demonstrated that the two most planted varieties at that time (Criolla Chica and Muscat of Alexandria) were the main progenitors of South American varieties, including the group of the Torrontés, Criolla grande, Cereza, and Pedro Giménez, for instance (Agüero et al., 2003; Martínez et al., 2006; Milla-Tapia et al., 2007; Durán et al., 2011; Boursiquot et al., 2014). The Peruvian variety Quebranta, a crossing between Criolla chica and Mollar Cano (This et al., 2006), another Spanish variety, was an exception to this group. The most probable processes leading to the origin of these genotypes may probably be the unintentional sowing of raisin seeds which spontaneously generated from marcs that were mixed and incorporated as manure to fertilize the vineyards (Aliquó et al., 2021).

In 1853, the foundation of the Quinta Normal of Mendoza took place. At this moment, the paradigm shifted towards French viticulture since Michel Pouget introduced French varieties and a gradual abandonment of *criollas* varieties began. Among the varieties introduced, growers rapidly adopted Malbec, which became the most cultivated variety in the country some years later. Between 1948 and 1978 the engineer José Vega pioneered the rescue, identification, and evaluation of native varieties collected in Mendoza, San Juan, and the Northwest provinces of Argentina. In 1949, he implanted the first collection of *criollas* located at the INTA EEA Mendoza. Between 1970 and 1990 the industry was more focused on yield over wine quality. For that purpose, some *criollas* were well adapted, such as Criolla Grande or Cereza, highly productive varieties with more characteristics of table grapes than wine grapes. During those years, they were widely cultivated to make common wines, and from that moment, *criollas* began to be related to low oenological potential and the interest in them declined progressively. In 2011, our team resumed the work concerning the identification, rescue, ampelographic description, and agronomic and oenological characterization of Argentine varieties. Our objective was to identify, conserve, and valorize this genetic diversity that was previously dismissed. Local varieties represent an opportunity to elaborate unique products, related to the identity, history, and culture of a given region. A new collection is implanted nowadays conserving all the genetic diversity found in ancient vineyards. Our collection holds 71 accessions collected from ancient vineyards located in Patagonia, Cuyo, Valles Calchaquies, and Cordoba.

11.3 Genetic Diversity Within Argentinian Varieties

The earliest study on some of the main *criollas* cultivars was made by Storni (1927), while Vega made the first full description of these cultivars in the 1950s and 1960s (Vega, 1950, 1976, 1977; Vega et al., 1962). Many years later, molecular studies characterizing the genetic origin of some of the most known *criollas* (e.g., Torrontés Riojano) appeared (Agüero et al., 2003; Martínez et al., 2006; Milla-Tapia et al., 2007), and a renewed interest has been observed lately in different countries (Prieto et al., 2021).

During these last 10 years, we have analyzed more than 100 accessions. We first analyzed accessions conserved at the Grapevine Collection, located at the INTA EEA Mendoza experimental campus (lat. 33°S; long. 68°51′W), Argentina. Those accessions were collected by Gonzalez and Vega (1949) in ancient vineyards in the western provinces of the country. From that date, they were conserved in the collection without identification. We also collected samples found in ancient vineyards located in different provinces of Argentina (Torres et al., 2022). A first morphological analysis was performed in situ to determine the putative origin of specific plants identified within the vineyards. Then, a sample was extracted and analyzed in the lab. For each collected sample or accession, a set of 19 microsatellites was analyzed (Aliquó et al., 2017) and the identity and parental analysis were performed with CERVUS v3.0.7. The genetic profiles obtained were compared with the *Vitis* International Variety Catalogue (VIVC, http://www.vivc.de/) to verify the identity and check for synonyms and homonyms. We also compared our results with the INRAe database, which comprises more than 4000 accessions genotyped (Lacombe et al., 2013).

Our studies have revealed that (i) the diversity among criollas varieties is higher than previously supposed (Aliquó et al., 2017), and (ii) this diversity is conserved in ancient, isolated, small vineyards where growers have played a key role in conserving this genetic diversity (Torres et al., 2022). Summarizing, we have found 70 different varieties, including 21 European, 49 *criollas* of which 34 of them were previously unknown *criollas* (Aliquó et al., 2017; Torres et al., 2022). As expected, a big group corresponded to the progeny originated from Listan Prieto × Muscat of Alexandria (Fig. 11.1). Still, many other varieties have also participated as parents, such as Muscat a Petits Grains Blanc, Mollar Cano, and many backcrossing. Surprisingly, Malbec also participated in the origin of two red cultivars. This finding suggests that the process leading to the origin of *criollas* varieties continued after the introduction of French cultivars by Pouget in 1853.

11.3.1 Crossings Between Muscat of Alexandria and Listán Prieto

Listán Prieto and Muscat of Alexandria are considered the major founders of South American grapevine varieties (Milla-Tapia et al., 2007). Around 29 direct descendants from this crossing have been reported until now. This group include the widely

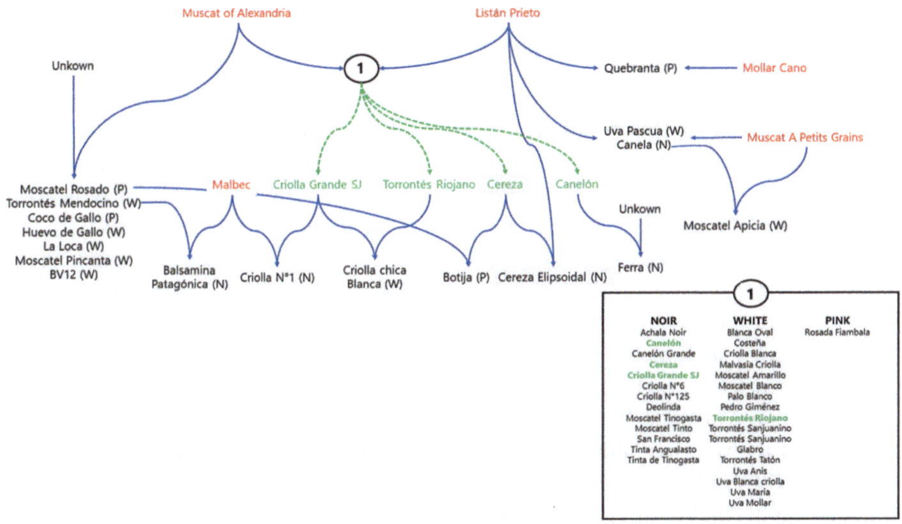

Fig. 11.1 Schematic pedigree and parental relationship of *criollas* varieties showing European varieties (in red). The circle with 1 in the center indicates the group of varieties originated by the crossing between Muscat of Alexandría and Listán Prieto. In the table, a detail is shown of this group of varieties classified by their berry skin color. In the scheme, in green appear varieties belonging to group 1 that gave origin to other variety through backcrossing either with another *criolla* or with other European cultivars

spread Cereza, Criolla Grande Sanjuanina, Torrontés Riojano, Torrontés Sanjuanino, Pedro Gimenez and some minor varieties such as Moscatel Amarillo, Uva Anís, Malvasia Criolla, or San Francisco. It's interesting to note that within this group, we found varieties collected from ancient vineyards located in different regions of our country with long distances among them. Some were collected from isolated valleys where no varietal replacement has been carried out for many years. The main and accepted hypothesis until this moment indicates that all these genotypes originated by spontaneous crossing for around 300 years. Another recent hypothesis is that they proceed from a very ancient, pioneer "breeding" program performed somewhere during the colonial period (Mejía et al., 2021). During the vineyard prospection, we could verify that even if some genotypes are found along the different regions of our country, there are some trends. For example, in the North (e.g., Salta), we found predominantly Torrontes Riojano and Criolla Chica, while Criolla Grande Sanjuanina, Cereza, or Pedro Giménez are scarce. These later varieties are more frequently found in Mendoza and San Juan. Besides those major varieties, we found several minor crossings that are not always the same among the regions. We may suppose that a group of them, the most spread nowadays, originated in a single place and then transported to the other regions where several other minor varieties originated. The number of varieties originated by this crossing is probably larger, considering the preliminary results obtained in different countries of the region such as Peru (Mendoza et al., 2022), Bolivia, or Chile (Prieto et al., 2021).

11.3.2 Crossing With Other Varieties

Besides the crossing between Listán Prieto and Muscat of Alexandria, other cultivars participated in less frequent crosses and originated some *criollas*. In some cases, these crosses involved varieties that the Spanish also introduced during the colony, while in others, some *criollas* have been identified as parental indicating backcrosses. The extended Peruvian variety, Quebranta, derives from Mollar cano and Listan Prieto (Lacombe et al., 2013). In the same way, Uva Pascua and Canela, both cultivars already reported by Vega (1950) and Alcalde (1989) as *criollas*, derived from a crossing between Listán Prieto and Muscat à Petits Grains Blanc. Other varieties such as Breval Negro and Moscatel Rosado also participated, leading to new varieties, such as Fintendo and Botija, respectively. The historical records point out that Muscat à Petits Grains Blanc, Mollar Cano, and Breval Negro were introduced to America and grown in the colonial period mixed in the same plots (Hudson, 1867; Storni, 1927). Moscatel Rosado has been described since the first ampelography of South America (Suarez, 1911; Storni, 1927; Vega et al., 1962; Alcalde, 1989), whereas it is not mentioned in European literature. The smaller number of progenies related to these cultivars might be explained by the fact that they were less spread than Listán Prieto and Muscat of Alexandria. Finally, a group of varieties (e.g., Cereza Ellipsoidal, Blanca Chica, Ferra, Moscatel Apicia) are varieties that originated from other *criollas*, revealing even more clearly the dynamism of the population (Fig. 11.1).

11.3.3 Crossings With Malbec

Interestingly, within the family of *criollas*, there are two genotypes, Criolla N°1 and Balsamina Patagónica, where Malbec participated as a parent. The presence of Malbec in these crossings indicates that these genotypes were generated after 1853, when French varieties arrived to Mendoza brought by Pouget. We can confirm that they are *criollas* since the other parents (i.e., Criolla Grande Sanjuanina and Torrontés Mendocino) are local genotypes. The evaluation carried out these last years confirms the oenological potential of these varieties for red winemaking, since the characteristics of the grapes, berry size, and color intensity resemble those of Malbec more than those of *criollas* (see next point concerning the enological potential). The relationship between Malbec and Criolla N°1 could also be demonstrated through the similar phenolic profile. Some recent studies have demonstrated that Malbec wines and berries have a high concentration of dihydroquercetin-3-glucoside (Fanzone, 2012). This compound has not been found in other varieties such as Merlot or Cabernet Sauvignon. Our analysis showed that Criolla N°1 presented a moderate concentration of dihydroquercetin-3-glucoside whereas the other *criollas* analyzed do not (Table 11.1). Furthermore, we analyzed the berries of the varieties Prunelard and Magdeleine Noire de Charentes, two ancient cultivars which are the

Table 11.1 Berry phenolic composition of some red *criollas* varieties conserved at the Grapevine Collection at INTA Mendoza, Argentina

Compound (mg/kg skin)	CaberINTA	Canela	Cereza	Criolla N°6	Criolla N°1	Valenci
Hydroxibenzoics acids	49.16 ± 5.3	15.93 ± 2.0	11.98 ± 2.4	33.74 ± 6.8	62.21 ± 4.6	27.11 ± 7.9
Hydroxicynamics acids	78.56 ± 8.0	69.29 ± 2.6	32.32 ± 9.8	79.32 ± 7.4	118.67 ± 4.7	66.61 ± 8.0
Stilbenes	6.70 ± 0.4	8.11 ± 4.4	24.58 ± 6.9	18.10 ± 0.4	18.39 ± 2.5	11.17 ± 2.4
Non flavonoids	134.43 ± 13.8	93.32 ± 3.9	68.88 ± 15.0	131.15 ± 13.8	199.27 ± 7.6	104.90 ± 16.5
Flavanols	194.29 ± 8.7	67.57 ± 39.3	17.27 ± 5.8	165.83 ± 19.6	239.30 ± 29.2	47.84 ± 12.4
Flavonols	209.79 ± 23.7	359.79 ± 24.6	283.47 ± 49.0	293.54 ± 15.1	523.32 ± 21.3	323.51 ± 123.2
Dihydroquercetin-3-glucoside	0	0	0	0	48.09 ± 23.6	0
Dihydroflavonols	0	0	0	0	48.09 ± 23.6	0
Flavonoids	404.08 ± 32.4	427.36 ± 57.4	300.74 ± 43.1	459.37 ± 4.5	810.71 ± 57.3	371.35 ± 118.2
Total	538.51 ± 46.3	520.68 ± 60.8	369.62 ± 28.0	590.53 ± 16.1	1009.98 ± 61.0	476.25 ± 119.3

progenitors of Malbec (Boursiquot et al., 2009). This compound was also present in both cultivars, but it was three times higher (170 μg/g of skin) in Magdeleine Noire de Charentes than in Prunelard (67 μg/g of skin). Apparently, Malbec inherited this berry and wine composition related trait from these varieties, and Criolla N°1 from Malbec.

11.3.4 Importance of Ancient Vineyards as Genetic Reservoirs

Most of the prospected vineyards were trained in an overhead system (i.e., "*parral*"), a traditional training system in Argentina, that still represents around 50% of the cultivated surface (INV, 2019). These family vineyards, with less than 1 ha of surface, are mostly cultivated for local market and self-consumption. They are planted with the most cultivated *criollas* (e.g., Pedro Gimenez, Criolla Grande Sanjuanina, Cereza), but they present a high diversity of other varieties planted in the same block, many of them not still identified or confounded with others. The role played by growers in maintaining the vineyards has been crucial to avoid the genetic erosion of the species. Since the first vineyard census carried out in our country in 1936, the cultivated surface has increased by 30% while the number of vineyards decreased by 13% (INV, 2020). In 1936, 11,331 small vineyards up to 1 ha were registered, representing 15% of the total number of vineyards, and concentrating 3.7% of the cultivated hectares. These numbers remained relatively stable until the year 2000 when the number of these small vineyards decreased 60%. In 2019, 4088 small vineyards still existed, representing 8.9% of the total vineyards and corresponding only to 1% of the cultivated hectares. This decline was accompanied by the globalization of international varieties, together with the loss of local, minor varieties. Small vineyards located in isolated Andes valleys, that have not been submitted to varietal changes, represent a reservoir of genetic diversity and growers have played a key role in preserving it.

11.4 Agronomic and Oenological Potential of *Criollas* Varieties

We evaluated the general characteristics of some of these varieties implanted in the Grapevine collection at INTA EEA Mendoza. We evaluated a group of more than 20 different varieties over 8 years (from 2012 to 2019). Not all varieties were evaluated every year. For each variety a total of five plants were evaluated and results presented here correspond to varieties that were studied for at least 2 years. Plants were own-rooted, planting distance was 2.5 m between rows and 1.5 m between plants. The plants were trained in a vertically shoot-positioned system, spur pruned, and drip irrigated. During winter 2012, 2013, and 2016, we measured pruning weight and calculated Ravaz Index (RI). During 2013 and 2016 seasons, berry weight, sugar accumulation, and solids soluble concentration (°Brix) were measured every 10 days from berry set until harvest. For those years, maturity dynamics and sugar accumulation rate were evaluated by the model proposed by Sadras and Petrie (2011). Based on this model, for each variety we calculated the sum of degree days value at which the onset of sugar accumulation occurred, and that at which berry sugar accumulation stopped. The sugar accumulation rate (S_A) was calculated as the slope of the regression between Brix accumulation and the sum of degree days between these two points. White and pink varieties were harvested at 21–22 °Brix while red varieties were harvested at 23–24 °Brix. At harvest, we registered yield (kg/plant), and we analyzed berry maturity (°Brix, pH, and total acidity). After harvest, grapes were transported to the experimental winery at INTA EEA Mendoza for winemaking in 25 L plastic tanks following a standardized protocol. Wine composition was analyzed on red cultivars.

A high variability was found among cultivars in terms of yield level, from 15 kg/plant in Canelón to around 1 kg/plant in Criolla Chica N°2. This latter accession corresponds to a clone of Listan Prieto (syn. Criolla chica) present also in the Vassal collection (http://bioweb.supagro.inra.fr/collections_vigne; Montpellier, France). This accession presents inflorescences with male flowers type 2 (atrophied gynoecium present) and very few berries per bunch, which explains the low yield levels. The same variability was observed in berry weight, which ranges from around 1.3 g (Criolla Chica) to around 5.0 g in Cereza ellipsoidal, Huevo de Gallo or Cereza (Fig. 11.2). It is interesting to observe that most spread *criollas* in Argentina correspond to those with higher berry weight and yield (e.g., Criolla Grande SJ, Cereza). However, there is a huge diversity within this group of varieties and many of them are not cultivated. Sugar content per berry followed the trend in berry weight (i.e., higher berries, higher sugar content per berry). This variability is higher than commonly thought and may be related to the process of natural crossings over 500 years

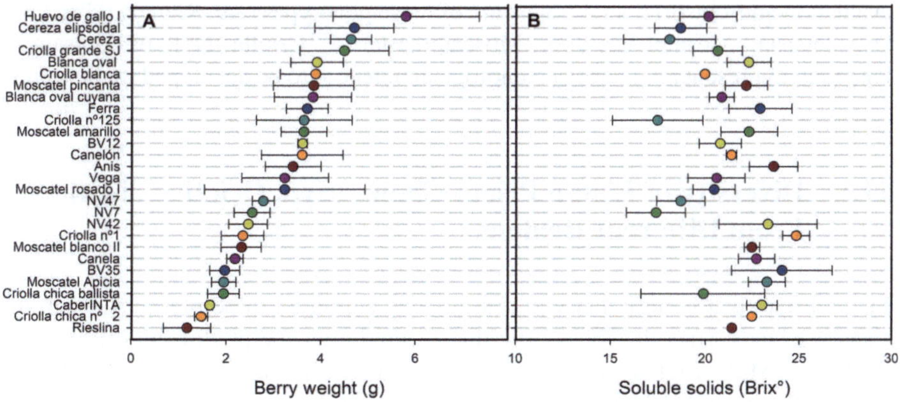

Fig. 11.2 Berry weight (**a**) and soluble solid concentration (°Brix, **b**) of different grapevine cultivars cultivated at the INTA germplasm collection, located at Mendoza, Argentina. Each point is the mean of at least 3 years of measurements ± standard deviation

of cultivation. Beside the two main progenitors of South American varieties (i.e., Criolla Chica and Muscat of Alexandria), many other cultivars have contributed to the origin of *criollas* varieties, leading to a higher diversity. For instance, Muscat à Petit Grains Blanc, Malbec, Criolla Grande Sanjuanina, Canela, and Mollar Cano are also progenitors of some *criollas* (Aliquó et al., 2017) adding more diversity to these agronomic traits.

Sugar accumulation rate (S_A) was calculated as the slope of the first part of the ripening curve (Sadras & Petrie, 2011). A regression analysis showed that the varieties presented statistical differences for the slope between Brix accumulation and accumulated degree days. Some varieties (e.g., Criolla N°1, Canela) presented higher S_A compared to other varieties with lower rates (e.g., Cereza Ellipsoidal). We compared the regressions between Brix and accumulated degree days for two contrasting years: a typical season (2013) with 295 mm of precipitation and 2140 °Cd of accumulated growing degree days and a humid season (2016) with 500 mm of rainfall and 1760 °Cd. We identified some varieties that presented a more stable relationship such as Canela, Anis, or Criolla N°1 while others such as Cereza, Cereza Elipsoidal, or Criolla Chica Ballista showed a more variable relationship, related to climatic conditions. A more stable relationship would be beneficial regarding climate change, given that it would be possible to obtain the same berry characteristics independently of seasonal conditions (at least in terms of soluble solids accumulations).

Color intensity was weak in most of the red or pink *criollas* varieties studied, except Criolla N°1 that presented similar phenolic content than some European ones (Table 11.1). As already explained, Criolla N°1 is a progeny of Criolla Grande Sanjuanina and Malbec. Actually, berry size, color intensity, uniformity, and wine composition of Criolla N°1 are much closer to Malbec than to the wines coming from the other traditional *criollas* varieties (e.g., Criolla Grande Sanjuanina, Cereza). Our analysis suggests that also Balasamina Patagonica presents this color potential similar to Malbec, but no data is presented here. After several years of evaluation and sensorial analysis, some of these genotypes (e.g., Canela, Uva Anis, Uva Pascua, Mosatel Rosado, Moscatel Blanco, Criolla N°1 or Balsamina Patagonica) present an interesting enological potential to be developed in the future.

11.5 Diversity of Listan Prieto Clones (Syn. Criolla Chica) From Different Regions

Listán Prieto is a variety that was traditionally supposed to be present only in the Canary Islands, since it would have disappeared after the phylloxera crisis in the Peninsula (Milla-Tapia et al., 2007). However, recent prospections in ancient vineyards in Spain have detected the presence of Listán Prieto in many vineyards in the region of Castilla La Mancha, Extremadura, and Andalucía (Muñoz Organero et al., 2015). Such evidence supports the hypothesis that this cultivar is not extinct in the continental part of Spain and maybe native of this region. This variety was introduced in America in the early sixteenth century, and as already signaled in the precedent points has a prominent importance in South American genetic diversity. It's actually known as País in Chile, Criolla Chica in Argentina, Negra Corriente in Peru, Misionera in Bolivia, Misión in Mexico, Mission in the United States. Its long history of cultivation in different environmental conditions has raised to different "ecotypes" or clones, which present differences that are evident at a visual level in terms of berry size, color, and cluster size. The oldest vines that are still in production in Argentina, and that are formally registered by National Institute of Vitiviniculture (INV), were implanted between 1862 and 1869 with this variety (INV, 2020) These vines probably represent the oldest formally registered vineyard in South America (Fig. 11.3).

Furthermore, our team has visited and prospected small vineyards in the Calchaquies valleys (Salta, North of Argentina) that are not formally registered and most probably are older (Fig. 11.4).

Including the material collected during the prospection of ancient vineyards, we have attempted to perform a first, preliminary characterization of Criolla Chica accessions conserved in our collection. We evaluated the concentration of phenolics compounds in six accessions of Listan Prieto collecting three samples of 20 berries each per accession during two seasons (2018 and 2019). Berry and skin weight was registered, and the concentration of total phenols (TP), total tannins (TT), and total anthocyanins (TA) were analyzed. The results showed that berry and skin weight were different among the accessions. The accession Criolla Chica N8 presented consistently over the two seasons a higher berry weight and lower phenolic

Fig. 11.3 Vineyard implanted with Listán Prieto between 1862 and 1869, the most ancient vineyard formally registered in Argentina (INV, 2020) located in Cafayate, Salta

Fig. 11.4 Vineyard implanted with Listán Prieto, not formally registered and probably with more than 150 years considering the morphological aspect and sizes of trunk, arms, and whole plant size (e.g., see the trunk in the floor giving rise to a new plant). These kinds of vineyards are maintained by very small, isolated grapegrowers in the middle of Calchaquies valleys, Salta (Argentina)

compounds either expressed per mg of skin or per berry (Table 11.2) indicating a lower phenolic potential. No interaction between accession and season was found for berry weight, and therefore, when expressing the values in mg per berry, no interaction was observed either (except for anthocyanins). Criolla Chica N3 showed the lowest berries weight of the two seasons, and higher concentration (mg/g of skin) but also higher values expressed per berry, indicating a high synthesis. Criolla Chica N4 presented also high values of skin and berry weight but lower AT per berry compared to N3. The accessions N5 presented similar values to N3 while N6 and N7 presented intermediate values (Table 11.2). This variability among the accessions must be more explored including more material in different environments to be able to identify clones with higher concentration of phenolic compounds. These preliminary results open the avenue to interesting questions concerning the physiological response of each accession to different environmental factors (e.g., drought), and the possible adaptation mechanisms developed during long-term cultivation under different environments, specially related to epigenetic modifications. The high variability observed in the accessions of Criolla Chica, in our country and others from South America, deserves certainly more attention, regarding the physiological, sanitary, and molecular basis that would explain the variability in Criolla Chica.

Table 11.2 Analysis of berry total phenols (TP), tannins (TT), and anthocyanins (TA) on different Criolla Chica accessions conserved in the Grapevine Collection at INTA Mendoza, collected from different regions of Argentina

	TP (mg/g skin)		TP (mg/g berry)		TT (mg/g skin)		TT (mg/g berry)		TA (mg/g skin)		TA (mg/g berry)		Berry weight (g)		Skin weight (g)	
2018																
Cch N3	25.97 ± 3.60	a	1.92 ± 0.14	a	15.74 ± 1.39	a	1.17 ± 0.02	a	3.26 ± 0.28	a	0.24 ± 0.02	a	1.77 ± 0.06	a	0.13 ± 0.01	a
Cch N4	21.29 ± 1.98	a	1.42 ± 0.06	a	14.92 ± 1.98	ab	0.99 ± 0.1	a	2.55 ± 0.29	a	0.17 ± 0.03	a	2.55 ± 0.02	b	0.17 ± 0.01	b
Cch N5	17.65 ± 1.27	ab	1.19 ± 0.13	ab	11.05 ± 1.05	ab	0.74 ± 0.03	ab	s/d		s/d		1.87 ± 0.06	a	0.12 ± 0.01	a
Cch N6	22.9 ± 4.31	abc	1.41 ± 0.16	ab	15.92 ± 3.03	bc	0.98 ± 0.11	b	3.4 ± 0.44	a	0.21 ± 0.02	a	2.56 ± 0.05	b	0.16 ± 0.01	bc
Cch N7	17.52 ± 1.08	bc	1.27 ± 0.04	b	11.96 ± 1.42	c	0.87 ± 0.09	b	3.01 ± 0.76	a	0.22 ± 0.05	a	2.57 ± 0.04	b	0.19 ± 0.01	cd
Cch N8	18.57 ± 2.43	c	1.19 ± 0.11	c	12.17 ± 1.55	c	0.78 ± 0.08	c	2.76 ± 0.49	b	0.18 ± 0.02	b	3.09 ± 0.09	c	0.2 ± 0.02	d
p-value	0.0138		<0.0001		0.0223		0.0002		<0.0001		<0.0001		<0.0001		<0.0001	
2019																
Cch N3	17.65 ± 3.85	a	1.82 ± 0.25	a	10.92 ± 1.54	a	1.13 ± 0.08	a	1.88 ± 0.35	a	0.2 ± 0.04	a	1.86 ± 0.04	a	0.19 ± 0.02	a
Cch N4	10.97 ± 2.30	a	1.09 ± 0.2	ab	7.96 ± 0.95	ab	0.79 ± 0.09	ab	0.71 ± 0.07	a	0.07 ± 0.01	ab	2.79 ± 0.1	b	0.18 ± 0.02	ab
Cch N5	22.71 ± 3.38	a	2.08 ± 0.09	ab	12.44 ± 1.71	ab	1.14 ± 0.04	ab	1.93 ± 0.23	a	0.18 ± 0.01	ab	2.20 ± 0.05	c	0.15 ± 0.02	bc
Cch N6	14.38 ± 3.38	a	1.29 ± 0.23	bc	10.21 ± 1.23	abc	0.92 ± 0.06	b	0.89 ± 0.2	a	0.08 ± 0.02	b	2.54 ± 0.06	cd	0.23 ± 0.01	bc
Cch N7	15.64 ± 3.5	ab	1.52 ± 0.26	cd	9.65 ± 1.49	bc	0.94 ± 0.09	c	1.12 ± 0.11	b	0.11 ± 0.01	c	2.68 ± 0.05	cd	0.26 ± 0.02	bc
Cch N8	13.65 ± 4.21	b	1.37 ± 0.24	d	9.06 ± 1.6	c	0.92 ± 0.05	c	0.86 ± 0.29	b	0.09 ± 0.02	c	2.66 ± 0.08	d	0.27 ± 0.04	c
p-value	0.0226		0.0013		0.0367		0.0004		<0.0001		<0.0001		<0.0001		0.009	
$Y \times C$	0.0013		0.16		<0.0001		0.2		<0.0001		<0.0001		0.07		<0.0001	

The analyses were carried out in the mode of berry diameter for each accession

Each value is the mean (± standard deviation) of three samples. Different letters in the same column indicate statistical differences among accessions according to the Duncan test a ($p \leq 0.05$). *p*-values are presented for each season separately, $Y \times C$ indicates the *p*-value for the interaction between accession and the year

11.6 Varietal Potential for Sparkling Wine Production

In order to evaluate the potential of some varieties for sparkling wine elaboration, we selected a group of them that in previous analyses presented high acidity levels. We analyzed the base wine and the final sparkling wine in terms of chemical composition and sensorial perception at different moments of aging (3, 6, 9, and 12 months). The selected varieties were Listán Prieto (syn. Criolla chica), Moscatel rosado, Pedro Gimenez, Blanca Oval, Canelón, while Chardonnay was used as a control. The harvest was performed when berries soluble solids achieved between 17 and 18 °Brix. Three wine base replicates per variety were elaborated and foam intake was performed by the traditional method (i.e., "champenoise"). We present here the results concerning the sensorial analysis carried out using the flash profile technique at 9 and 12 months after contact with the lees (Fig. 11.5). Only the 15 main descriptors that contributed the most to multiple factor analysis were plotted.

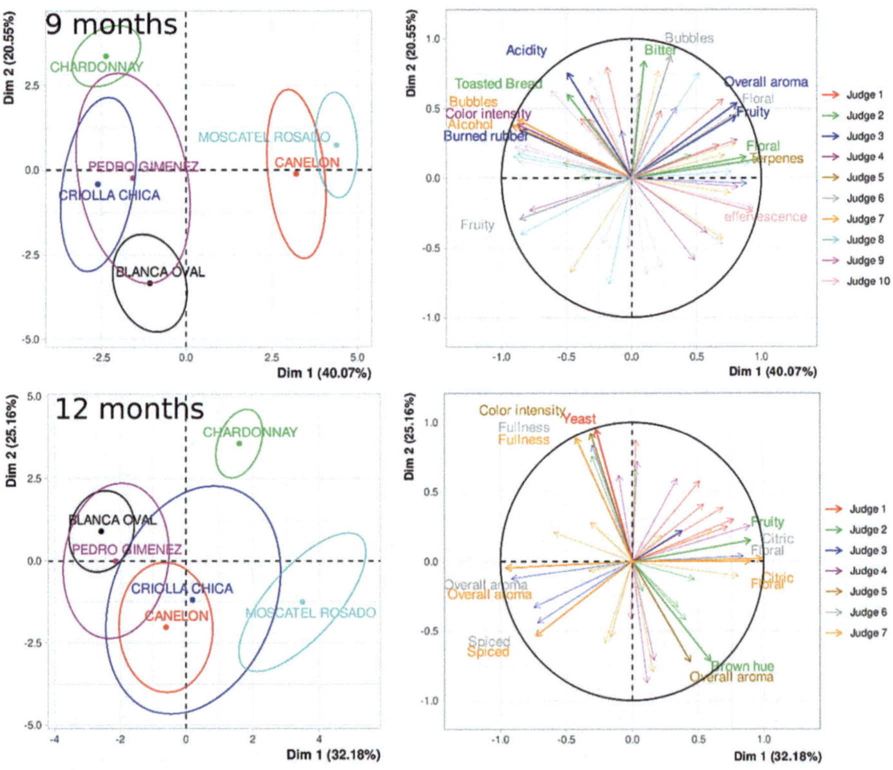

Fig. 11.5 Sensory analysis performed after 9 and 12 months after aging in bottle in contact with the lees for sparkling wines elaborated with Criolla Chica, Canelón, Moscatel Rosado, Blanca Oval, Pedro Gimenez, and Chardonnay. The evaluation was carried out using the Flash Profile technique with 20 panelists

Ellipses representing a 95% confidence interval were used, indicating that wines were perceived differently by panelists if the ellipses did not overlap.

At 9 months of bottle aging, the first two dimensions explained 60% of the existing variability. Sparkling wines made from Moscatel Rosado and Canelón grapes were grouped at the right of Dimension 1. These wines were perceived with floral, terpenic, and fruity aromas, and exhibited high aromatic intensity. To the left of Dimension 1, Blanca Oval, Pedro Gimenez, and Criolla Chica wines were grouped, characterized by descriptors such as color intensity, alcohol, bubbles, and even hints of burned rubber. The Chardonnay wine exhibited similar characteristics to Pedro Gimenez, with descriptors such as acidity and toasted bread. At 12 months of bottle aging, the situation remained similar. The first two dimensions accounted for 57% of the variability. Once again, Blanca Oval, Pedro Gimenez, and Criolla Chica wines were grouped. However, in this case, the Criolla Chica wine exhibited much more variability and even overlapped with Moscatel Rosado. This increased variability may have been due to the smaller number of judges at the 12-month analysis session. The Canelon wine, which was previously close to Moscatel Rosado at 9 months, moved to the left side of Dimension 1, showing similarities to Blanca Oval, Criolla Chica, and Pedro Gimenez wines. These wines were characterized by descriptors such as high aromatic intensity and spiciness. Moscatel Rosado, which remained to the right of Dimension 1, was perceived with fruity, floral, and citrus descriptors. Chardonnay was placed apart from the rest, with descriptors such as yeast, fullness, and higher color intensity.

This study highlights the potential of criollas varieties for sparkling wine elaboration and it will be completed soon with the rest of the sensorial data and chemical analysis such as variety foamability. Our preliminary results revealed the technological potential of autochthonous genetic material to diversify the production of sparkling wines, either as varietal wines or blends, providing products with a strong regional identity.

11.7 Perspectives and Conclusions

Grapevine genotypic diversity in Argentina, and probably also in South America, is higher than previously thought. The identification through molecular markers has allowed us to highlight this diversity. Small grapegrowers distributed in different locations in the western counties of Argentina and the Grapevine Collection at INTA (which conserves material brought to our country more than 100 years ago) has played a key role in conserving this diversity that otherwise would have been lost. This collection has now been enriched with a *criollas* section, including around 71 accessions of different autochthonous varieties and their related genotypes (e.g., Listan Prieto accessions).

Some attempts have been made to characterize the agronomic and enological potential of these varieties under our environmental conditions. However, the small quantities of plants per some of these genotypes make their characterization difficult at a higher level. More efforts should be invested in reproducing free-virus material to generate the information necessary to carry out field experiments. A small group of selected varieties are now being planted in different regions of Argentina to evaluate their characteristics and potential in different environmental conditions. It is also necessary to get a consensus among the different actors of the industry about the potential use of each variety, and the most appropriate blendings for different kinds of wines and/or sparkling wine. A key challenge remains about the communication of this diversity, its origins, and its potential to avoid confounding the consumers and more efforts should be made in this sense.

References

Agüero, C., Rodriguez, J., Martínez, L. E., Dangl, G., & Meredith, C. (2003). Identity and parentage of Torrontés cultivars in Argentina. *American Journal of Enology and Viticulture, 54*, 318–321.

Alcalde, A. J. (1989). *Cultivares vitícolas argentinas* (INTA ed.). EEA.

Aliquó, G., Torres, R., Lacombe, T., Boursiquot, J.-M., Laucou, V., Gualpa, J., Fanzone, M., Sari, S., Perez Peña, J., & Prieto, J. A. (2017). Identity and parentage of some South American grapevine cultivars present in Argentina. *Australian Journal of Grape and Wine Research, 23*, 452–460.

Aliquó, G., Palazzo, M. E., Torres, M. R., Prieto, J., Sari, S., Fanzone, M., Tornello, S., De Biazi, F., & Bolcato, E. (2021). Criollas argentinas. Una línea de tiempo desde sus orígenes hasta nuestros días. In J. A. Prieto (Ed.), *Libro de Resúmenes* (INTA ed.). EEA.

Anderson, K., & Nelgen, S. (2020). *Which winegrape varieties are grown where? A global empirical picture* (Revised ed.). University of Adelaide Press.

Antolín, M. C., Toledo, M., Pascual, I., Irigoyen, J. J., & Goicoechea, N. (2021). The exploitation of local *Vitis vinifera* L. biodiversity as a valuable tool to cope with climate change maintaining berry quality. *Plants (Basel), 10*, 71.

Boursiquot, J.-M., Lacombe, T., Laucou, V., Julliard, S., Perrin, F.-X., Lanier, N., Legrand, D., Meredith, C., & This, P. (2009). Parentage of Merlot and related winegrape cultivars of southwestern France: Discovery of the missing link. *Australian Journal of Grape and Wine Research, 15*, 144–155.

Boursiquot, J.-M., Laucou, V., Llorente, A., & Lacombe, T. (2014). Identification of grapevine accessions from Argentina introduced in the ampelographic collection of Domaine de Vassal. *BIO Web of Conferences, 3*, 01019.

Dong, Y., Duan, S., Xia, Q., Lian, Z., Dong, X., et al. (2023). Dual domestication and origin of traits in grapevine evolution. *Science, 379*, 892–901.

Durán, M., Agüero, C. B., & Martínez, L. E. (2011). Assessing the identity of the variety "Pedro Giménez" grown in Argentina through the use of microsatellite markers. *Revista de la Facultad de Ciencias Agrarias de la Universidad Nacional del Cuyo, 43*, 193–202.

Fanzone, M. (2012). *Caracterización de la composición fenólica de uvas y vinos de la variedad Malbec (*Vitis vinifera *L.): su relación con el origen geográfico, factores vitivinícolas y valor comercial* (PhD thesis doctoral). University Rovira I Virgili, Tarragona, Spain, 357 pp.

Florez-Sarasa, I., Clemente-Moreno, M. J., Cifre, J., Capó, M., Llompart, M., Fernie, A. R., & Bota, J. (2020). Differences in metabolic and physiological responses between local and wide-spread grapevine cultivars under water deficit stress. *Agronomy, 10*, 1052.

García-Muñoz, S., Muñoz-Organero, G., Fernández-Fernández, E., & Cabello, F. (2014). Sensory characterisation and factors influencing quality of wines made from 18 minor varieties (*Vitis vinifera* L.). *Food Quality and Preference, 32*, 241–252.

Gonzalez, F., & Vega, J. (1949). *La viticultura en el Noroeste Argentino* (pp. 5–6). IDIA.
Hudson, D. (1867). La viticultura en Cuyo. *Anales de la Sociedad Rural Argentina, 15*, 477–490. Buenos Aires, SRA.
INV. (2019). *Informe anual de superficie 2019*. Instituto Nacional de Vitivinicultura, Mendoza, Argentina. Retrieved from: https://www.argentina.gob.ar/inv/vinos/estadisticas/superficie/anuarios
INV. (2020). *Informe anual de superficie 2020*. Instituto Nacional de Vitivinicultura, Mendoza, Argentina. Retrieved from: https://www.argentina.gob.ar/inv/vinos/estadisticas/superficie/anuarios
Lacombe, T., Boursiquot, J.-M., Laucou, V., Di Vecchi-Staraz, M., Péros, J.-P., & This, P. (2013). Large-scale parentage analysis in an extended set of grapevine cultivars (*Vitis vinifera* L.). *Theoretical and Applied Genetics, 126*, 401–414.
Lacoste, P. (2013). *Vinos de capa y espada*. Antucura Editorial.
Lacoste, P., Yuri, J. A., Aranda, M., Castro, A., Quinteros, K., Solar, M., Soto, N., Gaete, J., & Rivas, J. (2010). Variedades de uva en Chile y Argentina (1550–1850). Geneaología del torrontés. *Mundo Agrario, 10*, 1–36.
Martínez, L. E., Cavagnaro, P. F., Masuelli, R. W., & Zúñiga, M. (2006). SSR-based assessment of genetic diversity in South American *Vitis vinifera* varieties. *Plant Science, 170*, 1036–1044.
Maurín-Navarro, E. (1967). *Contribución al estudio de la historia vitivinícola argentina. Producción, comercio e industrias de San Juan, desde su fundación hasta principios del siglo XX*. Instituto Nacional de Vitivinicultura.
Mejía, N., Oliva, O., & Ocarez, N. (2021). Caracterización genética de las principales variedades patrimoniales y criollas del Valle de los Cintis, Bolivia. In J. A. Prieto (Ed.), *Libro de Resúmenes* (INTA ed.). EEA.
Mendoza, K., Aliquó, G., Prieto, J. A., Blas, R., Torres, J., Casas, A., Grados, M., Aybar, L., & Torres, M. (2022). Prospection and identification of traditional-heritage Peruvian grapevine cultivars (*Vitis vinifera* L.) from Ica and Cañete valleys. *Vitis, 61*, 47–51.
Milla-Tapia, A., Cabezas, J. A., Cabello, F., Lacombe, T., Martínez-Zapater, J. M., Hinrichsen, P., & Cervera, M. T. (2007). Determining the Spanish origin of representative ancient American grapevine varieties. *American Journal of Enology and Viticulture, 58*, 242–251.
Muñoz Organero, G., Cabello, F., Serrano, M. J., Cretazzo, E., Pérez, J. A., & Gogorcena, Y. (2015). Nuevas prospecciones y recuperación de variedades antiguas de vid en España. In *30ª Reunión del grupo de trabajo de experimentación en viticultura y enología; 25–26 March 2015; Vizcaya, España* (pp. 25–32). Ministerio de Medio Ambiente y Medio Rural y Marino.
Myles, S., Boyko, A., Owens, C., Brown, P., Grassi, F., Arandhya, M., Prins, B., Reynolds, A., Chia, J.-M., Ware, D., Bustamante, C., & Buckler, E. (2011). Genetic structure and domestication history of the grape. *Proceedings of the National Academy of Sciences of the United States of America, 108*, 3530–3535.
OIV. (2017). *Distribution of the world's grapevine varieties*. Organisation Internationale de la Vigne et du Vin.
Prieto, J. A., Hinrichsen, P., Palazzo, M. E., Grinbergs, D., Chilian, J., Fanzone, M., & Tornello, S. (2021). Primeras jornadas Latinoamericanas de vinos y variedades patrimoniales. In J. A. Prieto (Ed.), *Libro de Resúmenes* (INTA ed.). EEA.
Sadras, V., & Petrie, P. (2011). Quantifying the onset, rate and duration of sugar accumulation in berries from commercial vineyards in contrasting climates of Australia. *Australian Journal of Grape and Wine Research, 17*(2), 190–198.
Storni, C. (1927). *Descripción de vidueños que se cultivan en Argentina desde la época colonial*. Tamburini.
Suarez, L. (1911). *Contribución a los estudios ampelográficos*. Sociedad de Vitivinicultores.
This, P., Lacombe, T., & Thomas, M. (2006). Historical origins and genetic diversity of wine grapes. *Trends in Genetics, 22*, 511–519.
Toro-Lira, G. (2018). *Las viñas de Lima: Inicios de la vitivinicultura sudamericana: 1539–1551*. Universidad Ricardo Palma, Editorial Universitaria.

Torres, R., Aliquó, G., Toro, A., Fernández, F., Tornello, S., Palazzo, M. E., Sari, S., Fanzone, M., De Biazi, F., Oviedo, H. J., Segura, R., Laucou, V., Lacombe, T., & Prieto, J. A. (2022). Identification and recovery of local Vitis vinifera L. varieties collected in ancient vineyards in different locations of Argentina. *Australian Journal of Grape and Wine Research, 28*(4), 581–589.

Vega, J. (1950). Identificación de clones de uvas "Criollas". *Informatico De Investigaciones Agrícolas (IDIA),* (Enero), 297–300.

Vega, J. (1976). Experiencia comparativa con 50 variedades de vid. *Informatico De Investigaciones Agrícolas (IDIA),* (Julio–Diciembre), 105–109.

Vega, J. (1977). Experiencia comparativa con 50 variedades de vid. *Informativo De Investigaciones Agrícolas (IDIA),* (Julio–Diciembre), 13–40.

Vega, J., Alcalde, A. J., & Cinta, W. (1962). *Variedades de vid que se cultivan en la región de Cuyo, República Argentina* (INTA ed.). EEA.

Wolkovich, E. M., García de Cortázar-Atauri, I., Morales-Castilla, I., Nicholas, K., & Lacombe, T. (2018). From Pinot to Xinomavro in the world's future wine-growing regions. *Nature Climate Change, 8*, 29–37.

Zinelabidine, L., Cunha, J., Eiras-Dias, J., Cabello, F., Martínez-Zapater, J., & Ibáñez, J. (2015). Pedigree analysis of the Spanish grapevine cultivar Hebén. *Vitis, 54*, 81–86.

Chapter 12
Impact of Climate Change on Argentine Viticulture: As It Moves South, What May Be the Effect of Wind?

Rodrigo Alonso, Rubén Bottini, Patricia Piccoli, and Federico J. Berli

12.1 Introduction

Argentina has diverse wine-producing regions, each one with a unique climate and topography that significantly affects grape-grow and wine characteristics. Argentinian viticulture extends from 22° to 45° South latitude, and mainly along the piedmont of the Andes Mountains under a semi-arid continental climate. The precipitations mostly occur during spring and summer, nevertheless insufficient to cover the vine plant cycle, thus requiring irrigation. The country has vineyards that range from sea level to over more than 3000 m a.s.l., making it one of the highest wine-growing countries in the world (Arias et al., 2022). Mendoza, the most important wine province with 71% of the total vineyards, has five winegrowing zones: North, East, South, First Zone, and Uco Valley. San Juan, the second most important province, with 21% of the total area planted (INV, 2021), presents valleys spread across the centre-west of the province: Pedernal, Calingasta, Zonda, Ullum, Iglesia, and Jáchal. Salta and La Rioja are important wine provinces in northern Argentina, located mainly in the Calchaquí Valleys and Famatina Valleys, respectively. The establishment of vineyards in the mentioned regions has been primarily motivated by climatic conditions, as well as traditional factors. Generally, wine regions located at high altitudes have cooler night temperatures, more thermal amplitude, and lower

R. Alonso (✉) · P. Piccoli · F. J. Berli
Grupo de Bioquímica Vegetal, Instituto de Biología Agrícola de Mendoza (IBAM), Consejo Nacional de Investigaciones Científicas y Técnicas-Universidad Nacional de Cuyo, Facultad de Ciencias Agrarias, Chacras de Coria, Mendoza, Argentina
e-mail: realonso@mendoza-conicet.gob.ar

R. Bottini
Instituto Argentino de Veterinaria, Ambiente y Salud (IAVAS), Universidad Juan Agustín Maza, Guaymallen, Mendoza, Argentina

G. Gutiérrez Gamboa, M. Fourment (eds.), *Latin American Viticulture Adaptation to Climate Change*, https://doi.org/10.1007/978-3-031-51325-1_12

humidity levels than those at lower elevations, all of which create "ideal" growing conditions for wide grape varieties. These cooler temperatures slow down the ripening process, allowing grapes to retain acidity, and to develop more complex aromas and flavours (Gutiérrez-Gamboa et al., 2021). In addition, vineyards located at higher altitudes receive more direct solar UV-B radiation, which increases the levels of polyphenols in red wines (Alonso et al., 2021). On the other hand, regions located at lower altitudes tend to have higher temperatures, which accelerate the ripening process, leading to grapes with higher sugar levels, lower acidity, and less complexity in their flavours and aromas (Van Leeuwen & Destrac-Irvine, 2017).

In the last decades, several publications have highlighted from different perspectives the impact of climate change on viticulture, either worldwide or in specific regions (Arias et al., 2022; Cabré et al., 2016; Caffarra & Eccel, 2011; Dunn et al., 2015; Gutierrez-Gamboa et al., 2021; Hannah et al., 2013; Jones, 2006; Mira de Orduña, 2010; Pomarici & Seccia, 2016; Santillán et al., 2020; Schultz, 2000; Van Leeuwen et al., 2019). Due to global warming, the climatic projections model for the next 50–75 years, Argentine viticulture may have a displacement of winegrowing regions towards the southwest and higher altitudes (Cabré & Nuñez, 2020). Arias et al. (2022) suggest that high altitude viticulture may be an adaptation strategy for Argentina, as it offers cooler temperatures and increased solar radiation, which can positively impact grapevine physiology and biochemistry, and the quality of the wines. Therefore, high altitude wine regions like Uco Valley, Pedernal, Famatina, Cafayate, among others, have the present potential to continue increasing the cultivated area with vineyards.

Patagonia, the southernmost wine-growing region in the world, nowadays represents less than 3% of the area with grapevines in Argentina and includes mainly zones of the banks of the Negro and Colorado rivers, in the provinces of Neuquén, La Pampa, and Río Negro, but also a small oasis in the Chubut province (INV, 2018). In general, the climate of this wine region is continental, temperate, desertic, with a great thermal amplitude, and strong winds that exert their influence on the crops (Villarreal et al., 2007). Taking into account that Patagonia is one of the windiest regions in the world, particularly during spring and summer (Palese et al., 2000), and the potential shift of winegrowing regions in Argentina towards the southwest, it is crucial to investigate the impact of wind on grapevine plants. Therefore, the aim of this paper is to provide a comprehensive review of the existing literature on the effects of wind on vine plants, emphasizing its significance in ensuring the long-term sustainability of Argentine viticulture.

12.2 The Effect of Wind on Plants

One of the major sources of mechanical loading on plants is the wind, which, in turn, has a major impact on plant growth, morphology, physiology, and ecology (Gardiner et al., 2016). Plant responses, ranging from changes in morphology to lodging, are more dependent on the intermittent and turbulent nature of the wind rather than its mean velocity (Van Gardingen & Grace, 1991). Wind can physically damage plants through abrasion (when leaves and other plant parts rub together, disturbing the surface layers of the tissues), leaf stripping (physical removal of plant material), and sandblasting (particles suspended in the air to strike the plant, causing both abrasion and removal of tissue) (Cleugh et al., 1998). Furthermore, plants exposed to wind experience physiological responses that impact the leaf boundary layer, consequently influencing vital gas exchange processes such as photosynthesis and transpiration.

The influence of wind on gas exchange occurs through convection, as the leaf surface is swept by the wind. Additionally, wind indirectly affects photosynthesis by causing light shedding. The motion induced by the wind on leaves and plants plays a significant role in the dispersion of light within a plant canopy (De Langre, 2008). Mechanically-induced stress (MIS) occurs as a natural consequence of environmental conditions as the aerial parts of the plant are moved, usually by wind, but also by such agents as rain, irrigation, animals, or machinery (Biddington, 1986). The most apparent effect of MIS is a reduction in the shoot length, often accompanied by increased radial growth. This response has been named "thigmomorphogenesis" (Jaffe, 1973), and has been described in several species (Gardiner et al., 2016), including grapevine (Dry et al., 1989). Even if the evidence is incomplete, it suggests that endogenous plant hormones elicit MIS responses. Ethylene has been a favourite candidate in causing shoot length decrease, although abscisic acid (ABA) and auxins may also play a role in the response to mechanical stress (Mitchell, 1996).

Most studies on the effects of wind on plants have focused on food crops, given their commercial significance, specifically on cereals, oilseed rape, and sunflower, and less attention has been paid to fruit and hort crops (Gardiner et al., 2016).

12.2.1 The Effect of Wind on Grapevine

Relatively little is known about the effect of wind on grapevines (Kobriger et al., 1984). Although wind is not present in most wine-growing regions of the world, constant winds of moderate (3–6 m/s) to strong (>6 m/s) intensities are components of terroir in viticultural regions such as the Salinas Valley in California (Freeman et al., 1982), the Swan Valley in Australia (Campbell-Clause, 1998), and the Western Cape Province in South Africa (Pienaar, 2005), among others.

Strong daily winds characterize spring and summer months in the Salinas Valley. In this region, Freeman et al. (1982) and Bettiga et al. (1997) evaluated on plants of *Vitis vinifera* cv. Chardonnay the importance of windbreaks on vine vegetative growth, yield components, fruit composition, and wine quality. Freeman et al. (1982) observed that after 13:00 h, when wind speed markedly increased, vines grown without a windbreak had markedly lower stomatal conductance than vines protected by a windbreak. Bettiga et al. (1997) found that sheltered vines had larger primary and lateral leaves, greater total leaf area, and increased pruning weight as compared to the non-sheltered. Similarly, like Freeman et al. (1982), they also observed greater stomatal conductance, compared to non-sheltered vines. Furthermore, their study revealed that wind shelters increased vine yield by 15% over a 5-year period, correlated with an increased number of clusters per vine and higher cluster weight.

Yet, sensory analyses performed on the wines produced indicate no preferences between the treatments. Also in the Salinas Valley, but under controlled conditions in a phytotron, Kobriger et al. (1984) observed in several grape cultivars (Carignane, Chardonnay, Chenin Blanc) that stomatal conductance and transpiration rates decreased in response to moderate (3.6 m/s) and strong (10.7 m/s) winds, but these responses disappeared within 1 day when the moderate wind stopped, while they continued for longer periods when the strong wind stopped. Additionally, leaf water potentials were not affected by either wind treatment. According to Kobriger et al. (1984), exposure to wind can initially have detrimental effects on grapevine growth and water relations. However, these negative impacts are temporary and can be alleviated by the influence of prior wind conditions. It is worth noting that higher plants possess a remarkable ability to induce stress "memory" or "imprinting" (Borges et al., 2014). As defined by Bruce et al. (2007), stress imprinting refers to the genetic or biochemical modifications triggered by an initial stress exposure, which subsequently enhance the plant's resistance to future stressors (phenomenon, also known as "priming"). According to various studies (Bettiga et al., 1997; Freeman et al., 1982; Kobriger et al., 1984), grapevine growers in windy regions may consider implementing windbreaks or other measures to protect their vines from excessive wind exposure.

In the Swan Valley, the main table grape growing area in Western Australia, one of the management problems is wind, where 23% of the long-term wind speed assessments are over 5.5 m/s. To appraise the effect of wind on stomatal resistance, Campbell-Clause (1998) performed a field experiment with *Vitis vinifera* cv. Italia and cv. Ribier. It was found that wind caused some stomatal closure in both cultivars due to an exponential increase in stomatal resistance. In addition, wind speeds above 4 m/s were found to reduce estimated evapotranspiration compared to less windy conditions. Gokbayrak et al. (2008) observed that high wind speeds led to an increase in stomatal density in grapevine leaves, suggesting that this increase may contribute to enhanced gas exchange regulation under high wind speeds. These findings offer a partial explanation for the lower stomatal conductance and transpiration rates previously observed (Campbell-Clause, 1998; Freeman et al., 1982; Kobriger et al., 1984) in windy conditions.

Grapevines situated in the Barossa Hills experience consistent winds of low to moderate intensity throughout the growing season, primarily originating from the south to west. Additionally, there are occasional episodes of strong winds blowing from the southwest during the spring. In this Australian region, Dry et al. (1989) evaluated the effect of wind on the performance of *V. vinifera* cv. Cabernet Franc through the use of an artificial windbreak (shadecloth) in the vineyard. In the two study seasons, sheltered vines exhibited higher pruning weights and longer shoots than non-sheltered vines. Furthermore, sheltered vines produced heavier berries that ripened earlier, showing elevated levels of anthocyanins and phenolics (measured in milligrams per berry). In the second season, no significant effects were observed on bunch weight, berry weight, or ripening time, but the yield was 13% higher in sheltered vines due to an increased number of bunches per vine.

The work of Pienaar (2005) was conducted on *Vitis vinifera* cv. Merlot over a period of 2 years in a vineyard located in the windy South Western Cape region, which is one of the main wine regions of South Africa. The observed effects of wind on vegetative parameters were similar to those obtained by Campbell-Clause (1998), Dry et al. (1989), Freeman et al. (1982), and Kobriger et al. (1984), i.e., sheltered vines exhibited significantly longer shoots, higher pruning weights, and greater stomatal conductance. Regarding yield components, dissimilar results can be observed between Pienaar (2005) and Dry et al. (1989), possibly due to the use of different varietals and/or different experimental systems. The study conducted by Pienaar (2005) revealed that sheltered Merlot plants exhibited more bunches per vine as compared to the exposed vines, but this did not result in a significantly higher crop yield. The reason behind this observation was attributed to the lower mass of individual bunches in the sheltered vines, primarily caused by a reduced number of berries per bunch. It is well-known that the formation of cluster primordia is favoured by high-intensity light on the bud and elevated temperatures (Williams, 2000). Thus, the author speculated that the less favourable microclimatic environment within the denser canopy of the sheltered treatment during the formation of flower cluster primordia might be responsible for this outcome.

Regarding enological quality indicators, the grape skin colour index for sheltered vines was lower compared to exposed vines. In explaining this result, Pienaar (2005) attributed it to the positive response of two important enzymes involved in regulating anthocyanin biosynthesis to UV light and mechanical damage (Seymour et al., 2012). Given that the exposed vines experienced more shoot breakage and defoliation, it was anticipated that these enzymes would be present in higher concentrations, leading to increased anthocyanin expression in the exposed vines. Moreover, sensory evaluation of the wine indicated no significant differences in terms of astringency and colour intensity (both parameters related to the content of phenolic compounds). However, it is worth noting that wine derived from sheltered vines consistently received higher ratings for overall quality.

Vines in windy sites may develop asymmetrical canopy architecture depending upon the angle of attack of the prevailing wind, which can have implications for fruit quality, an attribute highly linked to the exposure of grape clusters to solar radiation. Tarara et al. (2005) conducted a field experiment in the windy wine region of Columbia River, USA, and measured irradiance at the fruiting zone and shoot geometry in two contiguous vineyards differing only in row orientation. The study revealed that wind-induced canopy asymmetry led to uneven berry ripening, particularly in areas with high solar irradiance. As a result, the authors suggest that growers in windy locations should consider establishing row orientation based on both sun-earth geometries to maximize radiation interception by the canopy and the potential consequences of radiation distribution at the fruiting zone due to wind-induced canopy asymmetry. In established vineyards, growers have the opportunity to mitigate the effects of non-uniform canopy architecture to some extent through modifications to the trellis system and standard training practices. These adjustments can help promote more balanced berry ripening and improve overall fruit quality.

12.3 Conclusions and Future Directions

In this article, a thorough review of the existing but limited literature on the influence of wind on grapevines has been undertaken. Our primary objective was to evaluate the potential implications for vineyards in the Patagonia region and propose future directions to foster the long-term sustainability of viticulture in this Argentine wine-growing area. From a physiological standpoint, wind exposure can harm grapevine growth and water relations. The reviewed studies concur that wind induces stomatal closure and reduces vegetative expression. However, research is scarce on the effects of wind on yield components and the enological quality of berries, and the results obtained with diverse cultivars and in different climates so far are contradictory. From an agronomic standpoint, it is essential to consider the production objective when assessing the impact of wind.

In the case of vineyards aiming for high yields, grapevine growers in windy regions, such as Patagonia, may need to consider implementing windbreaks that are maintained throughout the vineyard's lifecycle. On the other hand, vineyards to produce lower yields with high quality may benefit from moderate winds. A sparse canopy resulting from moderate winds can improve lighting conditions for the clusters, leading to an increase in compounds related to oenological quality. However, further studies are necessary to confirm this hypothesis. Another crucial area for future research is the study of how different varieties of *Vitis vinifera* respond to wind, as their responses to environmental factors can vary (Schultz, 2003). In addition, it is important to investigate the interaction between wind and water relations in grapevines. Assessing the combined effect of wind and different irrigation strategies on stomatal conductance, transpiration rates, water use efficiency, and overall vine water status would provide valuable insights into optimizing irrigation practices in windy regions and effectively managing water resources.

Finally, to comprehend the impact of wind in the specific climatic and soil conditions of the Argentine Patagonia, it will be crucial to conduct comprehensive and long-term field trials in the region. These trials should compare the performance of grapevines grown with and without windbreaks. It is important to assess various parameters, including vegetative growth, yield components, fruit composition, and wine quality, in order to determine the benefits and potential disadvantages of implementing windbreaks. By conducting these future studies, researchers can expand the knowledge base on the effects of wind on grapevines, provide practical recommendations for grape growers, and facilitate the successful expansion of grape cultivation in the Patagonia region of Argentina.

References

Alonso, R., Berli, F. J., Fontana, A., Piccoli, P., & Bottini, R. (2021). Abscisic acid's role in the modulation of compounds that contribute to wine quality. *Plants (Basel), 10*(5), 938. https://doi.org/10.3390/plants10050938

Arias, L. A., Berli, F., Fontana, A., Bottini, R., & Piccoli, P. (2022). Climate change effects on grapevine physiology and biochemistry: Benefits and challenges of high altitude as an adaptation strategy. *Frontiers in Plant Science, 13*, 835425.

Bettiga, L. J., Dokoozlian, N. K., & Williams, L. E. (1997). Response of Chardonnay grapevines to windbreaks in the Salinas Valley. *American Journal of Enology and Viticulture, 48*(3), 374–374.

Biddington, N. L. (1986). The effects of mechanically-induced stress in plants – A review. *Plant Growth Regulation, 4*(2), 103–123. https://doi.org/10.1007/BF00025193

Borges, A. A., Jiménez-Arias, D., Expósito-Rodríguez, M., Sandalio, L. M., & Pérez, J. A. (2014). Priming crops against biotic and abiotic stresses: MSB as a tool for studying mechanisms. *Frontiers in Plant Science, 5*, 642.

Bruce, T. J., Matthes, M. C., Napier, J. A., & Pickett, J. A. (2007). Stressful "memories" of plants: Evidence and possible mechanisms. *Plant Science, 173*(6), 603–608.

Cabré, F., & Nuñez, M. (2020). Impacts of climate change on viticulture in Argentina. *Regional Environmental Change, 20*(1), 12.

Cabré, M. F., Quénol, H., & Nuñez, M. (2016). Regional climate change scenarios applied to viticultural zoning in Mendoza, Argentina. *International Journal of Biometeorology, 60*, 1325–1340.

Caffarra, A., & Eccel, E. (2011). Projecting the impacts of climate change on the phenology of grapevine in a mountain area. *Australian Journal of Grape and Wine Research, 17*(1), 52–61.

Campbell-Clause, J. M. (1998). Stomatal response of grapevines to wind. *Australian Journal of Experimental Agriculture, 38*, 77–82. https://doi.org/10.1071/EA97144

Cleugh, H. A., Miller, J. M., & Böhm, M. (1998). Direct mechanical effects of wind on crops. *Agroforestry Systems, 41*(1), 85–112. https://doi.org/10.1023/A:1006067721039

De Langre, E. (2008). Effects of wind on plants. *The Annual Review of Fluid Mechanics, 40*, 141–168.

Dry, P. R., Reed, S., & Potter, G. (1989). The effect of wind on the performance of Cabernet Franc grapevines. *Acta Horticulturae, 240*, 143–146. https://doi.org/10.17660/actahortic.1989.240.24

Dunn, M., Lindesay, J., & Howden, M. (2015). Spatial and temporal scales of future climate information for climate change adaptation in viticulture: A case study of user needs in the Australian winegrape sector. *Australian Journal of Grape and Wine Research, 21*(2), 226–239.

Freeman, B. M., Kliewer, W. M., & Stern, P. (1982). Influence of windbreaks and climatic region on diurnal fluctuation of leaf water potential, stomatal conductance, and leaf temperature of grapevines. *American Journal of Enology and Viticulture, 33*(4), 233–236.

Gardiner, B., Berry, P., & Moulia, B. (2016). Review: Wind impacts on plant growth, mechanics and damage. *Plant Science, 245*, 94–118. https://doi.org/10.1016/j.plantsci.2016.01.006

Gokbayrak, Z., Dardeniz, A., & Bal, M. (2008). Stomatal density adaptation of grapevine to windy conditions. *Trakia Journal of Sciences, 6*(1), 41–60. https://doi.org/10.1016/j.ultrasmedbio.2008.01.022

Gutiérrez-Gamboa, G., Pszczólkowski, P., Cañón, P., Taquichiri, M., & Peñarrieta, J. M. (2021). UV-B radiation as a factor that deserves further research in Bolivian viticulture: A review. *South African Journal of Enology and Viticulture, 42*(2), 201–212. https://doi.org/10.21548/42-2-4706

Gutierrez-Gamboa, G., Zheng, W., & Martinez de Toda, F. (2021). Current viticultural techniques to mitigate the effects of global warming on grape and wine quality: A comprehensive review. *Food Research International, 139*, 109946. https://doi.org/10.1016/j.foodres.2020.109946

Hannah, L., Roehrdanz, P. R., Ikegami, M., Shepard, A. V., Shaw, M. R., Tabor, G., Zhi, L., Marquet, P. A., & Hijmans, R. J. (2013). Climate change, wine, and conservation. *Proceedings of the National Academy of Sciences, 110*(17), 6907–6912.

INV. (2018). *Informe vitivinícola de la región Sur de la Argentina*. Instituto Nacional de Vitivinicultura. https://www.argentina.gob.ar/inv/vinos/estadisticas/regiones-vitivinicolas
INV. (2021). *Informe anual de superficie 2021*. Instituto Nacional de Vitivinicultura. https://www.argentina.gob.ar/inv/vinos/estadisticas/regiones-vitivinicolas
Jaffe, M. J. (1973). Thigmomorphogenesis: The response of plant growth and development to mechanical stimulation: With special reference to Bryonia dioica. *Planta, 114*, 143–157.
Jones, G. V. (2006). Climate and terroir: Impacts of climate variability and change on wine. *Geoscience Canada Reprint Series, 9*, 203–217.
Kobriger, J. M., Kliewer, W. M., & Lagier, S. T. (1984). Effects of wind on water relations of several grapevine cultivars. *American Journal of Enology and Viticulture, 35*(3), 164–169.
Mira de Orduña, R. (2010). Climate change associated effects on grape and wine quality and production. *Food Research International, 43*(7), 1844–1855. https://doi.org/10.1016/j.foodres.2010.05.001
Mitchell, C. A. (1996). Recent advances in plant response to mechanical stress: Theory and application. *HortScience, 31*(1), 31–35. https://doi.org/10.21273/hortsci.31.1.31
Palese, C., Lässig, J. L., Cogliati, M. G., & Bastanski, M. A. (2000). Wind regime and wind power in North Patagonia, Argentina. *Wind Engineering, 24*(5), 361–377. https://doi.org/10.1260/0309524001495738
Pienaar, J. W. (2005). *The effect of wind on the performance of the grapevine* (Master's thesis). Thesis for the degree of Master of Agricultural and Forestry Sciences, Stellenbosch University. https://scholar.sun.ac.za/handle/10019.1/50486
Pomarici, E., & Seccia, A. (2016). Economic and social impacts of climate change on wine production. In *Reference module in food science* (pp. 1–8). Elsevier. https://doi.org/10.1016/b978-0-08-100596-5.03062-6
Santillán, D., Garrote, L., Iglesias, A., & Sotes, V. (2020). Climate change risks and adaptation: New indicators for Mediterranean viticulture. *Mitigation and Adaptation Strategies for Global Change, 25*(5), 881–899.
Schultz, H. R. (2000). Climate change and viticulture: A European perspective on climatology, carbon dioxide and UV-B effects. *Australian Journal of Grape and Wine Research, 6*, 2–12. https://doi.org/10.1111/j.1755-0238.2000.tb00156.x
Schultz, H. R. (2003). Differences in hydraulic architecture account for near-isohydric and anisohydric behaviour of two field-grown Vitis vinifera L. cultivars during drought. *Plant, Cell & Environment, 26*(8), 1393–1405.
Seymour, G. B., Taylor, J. E., & Tucker, G. A. (2012). *Biochemistry of fruit ripening*. Springer Science & Business Media.
Tarara, J., Ferguson, J., Hoheisel, G.-A., & Peña, J. P. (2005). Asymmetrical canopy architecture due to prevailing wind direction and row orientation creates an imbalance in irradiance at the fruiting zone of grapevines. *Agricultural and Forest Meteorology, 135*(1–4), 144–155.
Van Gardingen, P., & Grace, J. (1991). Plants and wind. In J. A. Callow (Ed.), *Advances in botanical research* (Vol. 18, pp. 189–253). Academic Press. https://doi.org/10.1016/S0065-2296(08)60023-3
Van Leeuwen, C., & Destrac-Irvine, A. (2017). Modified grape composition under climate change conditions requires adaptations in the vineyard. *OENO One, 51*(2), 147–154. https://doi.org/10.20870/oeno-one.2016.0.0.1647
Van Leeuwen, C., Destrac-Irvine, A., Dubernet, M., Duchêne, E., Gowdy, M., Marguerit, E., Pieri, P., Parker, A., De Resseguier, L., & Ollat, N. (2019). An update on the impact of climate change in viticulture and potential adaptations. *Agronomy, 9*(9), 514.
Villarreal, P., Romagnoli, S., & Llorente, A. (2007). *Pautas tecnológicas: Vid para vinificar*. ISBN: 9789875212299. https://inta.gob.ar/documentos/pautas-tecnologicas-vid-para-vinificar
Williams, L. (2000). Bud development and fruitfulness of grapevines. In L. P. Christensen (Ed.), *Raisin production manual* (pp. 24–29). University of California.

Chapter 13
Growing Vines in the Mapuche Heartland: The First Report About the Vitiviniculture of the Araucanía Region

Gastón Gutiérrez Gamboa, Cristóbal Palacios-Peralta, Rafael López-Olivari, Pamela Castillo, Milton Almonacid, Raúl Narváez, Luis Morales-Salinas, Nicolás Verdugo-Vásquez, Marcela Hidalgo, Alejandra Ribera-Fonseca, and Ignacio Serra

13.1 Introduction

Climate change has affected the distribution of grapevine varieties in different wine growing regions (Jones et al., 2005; Santos et al., 2020; Gutiérrez-Gamboa et al., 2021a). Climate trends have suggested that climate change considerably affected

---

The original version of the chapter has been revised. A correction to this chapter can be found at https://doi.org/10.1007/978-3-031-51325-1_16

G. Gutiérrez Gamboa (✉)
Instituto de Investigaciones Agropecuarias, INIA Carillanca, Vilcun, Chile

Escuela de Agronomía, Facultad de Ciencias, Ingeniería y Tecnología, Universidad Mayor, Temuco, Chile
e-mail: gaston.gutierrez@inia.cl

C. Palacios-Peralta
Escuela de Agronomía, Facultad de Ciencias, Ingeniería y Tecnología, Universidad Mayor, Temuco, Chile

R. López-Olivari
Instituto de Investigaciones Agropecuarias, INIA Carillanca, Vilcun, Chile

P. Castillo
Vicerrectoría de Investigación y Desarrollo, Universidad de Concepción, Concepcion, Chile

M. Almonacid
Centre for Transnational Studies, University of Copenhagen, Copenhagen, Denmark

R. Narváez
Viña FANOA, San Fernando, Chile

G. Gutiérrez Gamboa, M. Fourment (eds.), *Latin American Viticulture Adaptation to Climate Change*, https://doi.org/10.1007/978-3-031-51325-1_13

vine cultivation in Chilean central valleys, increasing their heat unit accumulation during growing season (Montes et al., 2012; Verdugo-Vásquez et al., 2023). Bioclimatic indices have reported that most of the Chilean viticultural valleys changed their classification towards warmer climates over the last three decades (Verdugo-Vásquez et al., 2023), favoring the viticultural aptitude of the Southern Chilean valleys. In addition, Central and Southern valleys of Chile considerably increased their risk for temperatures higher than 35 °C (Verdugo-Vásquez et al., 2023), which have negatively affected grape and wine production (Gutiérrez-Gamboa et al., 2021b). Based on these reasons, vineyard plantation in Chile is moving to the south and vineyard surface in the Araucanía Region (37°35′ to 39°37′ South Latitude) increased by 953% from 2003 to 2020 (Morales-Henríquez et al., 2022).

The Decree 464 of the Chilean law divide the Araucanía Region in two different viticultural regions (BCN, 2020): South Region that considers Malleco valley (Angol, Collipulli, Ercilla, Los Sauces, Lumaco, Purén, Renaico, Traiguén and Victoria communes); Austral Region that considers Cautín valley (Perquenco and Galvarino communes). Malleco valley was the chosen zone to begin the establishment of the new commercial vineyards in the Araucanía Region (Leiva, 2007). The vines were planted in the 1990s by using grapevine varieties adapted to cold climate conditions similar to the viticulture developed in the Southern valleys of Australia and New Zealand (de Solminihac, 2015). Nevertheless, the new vineyard plantations in the Araucanía Region have been established with scarce edaphoclimatic background linked to viticultural aptitude of each variety (Leiva, 2007). Nowadays, the Araucanía Region surface of planted vines accounts 107.3 ha (0.08% of the total national vineyard surface), in which Traiguén (46%), Purén (14%), and Victoria (13%) communes are the most important (SAG, 2021). Some the European famous grapevine varieties, such as Pinot Noir (55.5%), Chardonnay (31.0%), and Sauvignon Blanc (5.3%) are the most planted in this region, which are mostly established in the Malleco valley (SAG, 2021).

The Araucanía Region holds comparative advantages with respect to the viticulture developed in the central valleys for the production of high-quality white and sparkling wines due to its climatic conditions (Verdugo-Vásquez et al., 2023) and its

L. Morales-Salinas
Laboratorio de Investigación en Ciencias Ambientales (LARES), Facultad de Ciencias Agronómicas, Universidad de Chile, Santiago, Chile

N. Verdugo-Vásquez
Instituto de Investigaciones Agropecuarias, INIA Intihuasi, La Serena, Chile

M. Hidalgo
Departamento de Suelos y Recursos Naturales, Facultad de Agronomía, Universidad de Concepción, Concepcion, Chile

A. Ribera-Fonseca
Centro de Fruticultura, Facultad de Ciencias Agropecuarias y Forestales, Campus Andrés Bello, Universidad de La Frontera, Temuco, Chile

I. Serra
Departamento de Producción Vegetal, Facultad de Agronomía, Universidad de Concepción, Concepcion, Chile

soils of volcanic origin (de Solminihac, 2015). Climatic conditions of the Araucanía Region are widely different compared Malleco to Cautín valleys. Summer thermal regime in Malleco valley is warm, whereas Cautín valley is cool (Leiva, 2007). Growing degree days accumulation in the Araucanía Region ranged from 968 (Nueva Imperial corresponding to Cautín valley) to 1409 (Los Sauces corresponding to Malleco valley) heat units (Leiva, 2007). The viticulture in this zone is performed over volcanic-ash-derived soils that have light porosity, high water retention capacity, acid to slightly acid pH, and high organic matter content (Gonzalez et al., 1974). Soils of the region are mostly cultivated with annual crops, such as wheat (*Triticum aestivum*), barley (*Hordeum vulgare*), oat (*Avena sativa*), and rapeseed (*Brassica napus*), reason for which the Araucanía Region is commonly called "the granary of Chile" (Román-Figueroa et al., 2017; González et al., 2021).

The vitiviniculture of the Araucanía Region has interesting peculiarities that must be promoted given its landscapes and its symbolism, which have been used in a racist and commercial manner by the national big companies (Martin Bastidas, 2020). Small viticulturists of the Araucanía Region generally reach low production levels at harvest due to factors associated mostly to (i) climate: short vegetative cycle, hot summer temperatures, spring and summer frosts, and water scarcity and (ii) management decisions: scarce availability of plant material, as productive clones and rootstocks adapted to the zone, low adoption of viticultural technology, deficiencies in pest and diseases management, scarce number of available specialists, and high productive costs. Based on this, the goal of this chapter is to provide the first scientific information of the viticulture of the Araucanía Region and give to viticulturists some guidelines to enhance production under the edaphoclimatic conditions of the zone.

13.2 Productive Characterization of Viticulturists in the Araucanía Region

A survey was performed on 20 viticulturists (34% of the total) of the Araucanía Region to understand basic aspects of their productive system in order to define the training needs (Leiva & Soto, 2016). The viticulturists develop this activity in a complementary way with forestry (22.0%), livestock (12.2%), crops and cereals (17.1%), fruit trees (34.2%), vegetables (2.3%), and others (12.2%). Dryland viticulture is developed by 30% of the viticulturists, whereas the rest of the producers are located in areas with irrigation availability (Leiva & Soto, 2016). Some technical difficulties are defined by the viticulturists of the Araucanía Region as the most important to taken into account to enhance the wine industry in the region, such as lack of specialists workers (24%), high incidence of spring frost (22%), low production levels (19%), lack of water or access to irrigation (13%), shortage of labor (9%), access to inputs (7%), heterogeneous maturity (4%), and fruit theft (2%) (Fig. 13.1). The viticulturists defined some topics of priority the be trained, such as

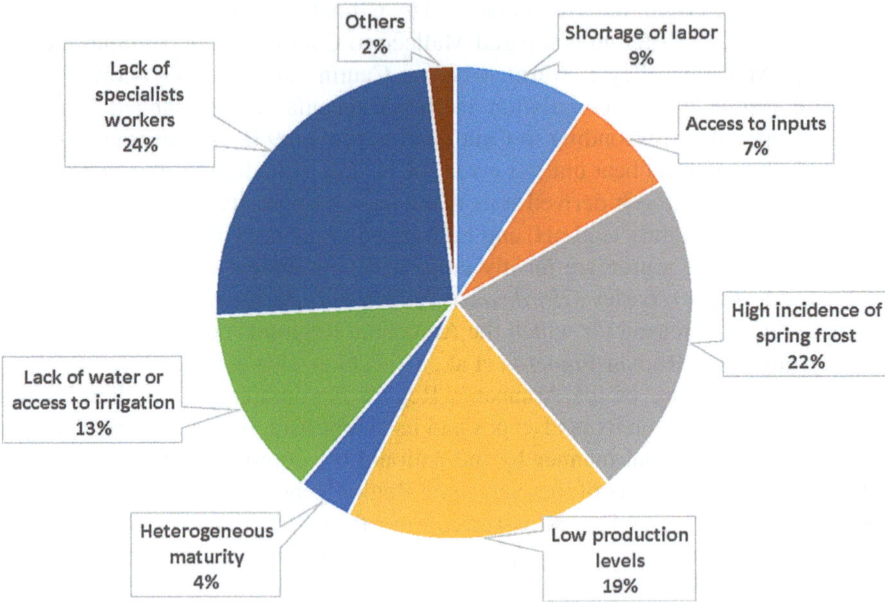

Fig. 13.1 Percentage (%) of the main productive problems defined by the viticulturists of the Araucanía Region

frost protection (16.7%), diseases control (14.6%), marketing of wines (10.4%), pruning strategies (10.4%), organic vineyard management (8.3%), irrigation systems (8.3%), vine nutrition (6.3%), pest management (4.2%), shoot management (4.2%), ripening monitoring (4.2%), planting (2.1%), vine physiology (2.1%), analysis interpretation (2.1%), among others (Leiva & Soto, 2016). The viticulturists defined some complementary topics the be trained, such as support in wine cellar equipment (19.2%), analysis of feasibility of wine production (15.4%), and physicochemical analysis of grapes and wines (11.5%) (Leiva & Soto, 2016).

13.3 Geomorphology of Soils of the Araucanía Vitiviniculture

13.3.1 Geology of the Araucanía Soils

The area landscape is determined by three main morpho-structural units: Coastal Range, Central Depression, and pre- and Andean Mountain Range (de Solminihac, 2015; Gutiérrez-Gamboa & Moreno-Simunovic, 2019). The morphologic elements most representative in the area correspond to the Central Depression and the Coastal Range, referred to as Nahuelbuta Cordillera at these latitudes (38°00′ to 38°30′ S). The hydrology is dominated by the fluvial basin of Cholchol river borne from the

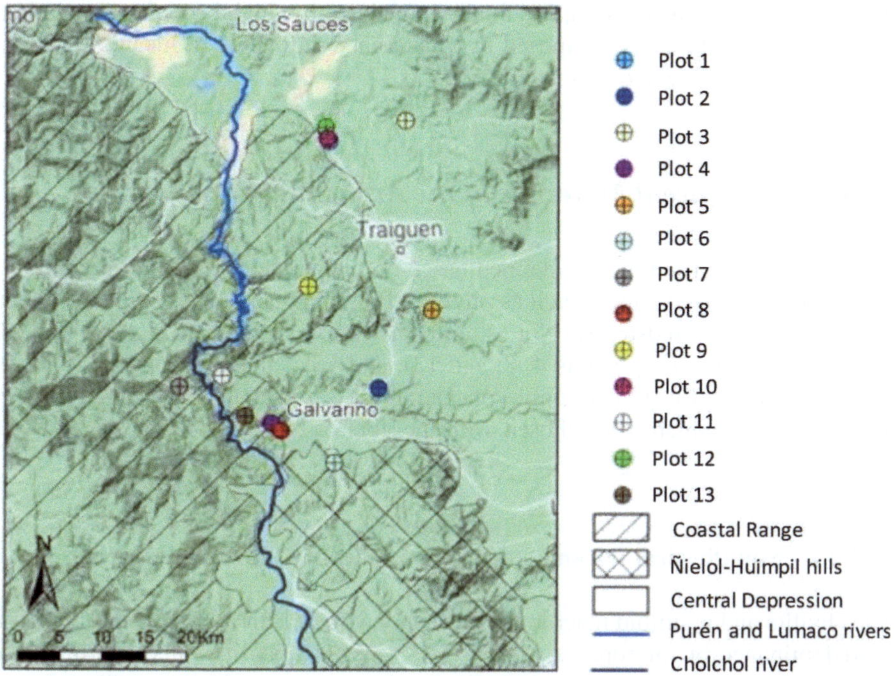

Fig. 13.2 Geomorphology of the Malleco valley. Note: Circles correspond to vineyards that are cultivated

Nahuelbuta Cordillera. That basin defines the limit between both morphological units, which can be observed to the west of Galvarino city (Fig. 13.2). The viticulture in the Araucanía Region is developed mostly in the oriental hills of the Nahuelbuta Cordillera and in the Intermediate Depression (Fig. 13.2), specifically in the viticultural zone of Malleco valley (de Solminihac, 2015).

Geology of the area is dominated by Paleozoic metamorphic and intrusive rocks that form the Coastal Range, aged between the Devonian and Triassic periods (~420 Ma; million years before the present). In the Central Depression, these Paleozoic rocks constitute the basement where younger volcanic and sedimentary formations were deposited (SERNAGEOMIN, 2004). The rocks in Nahuelbuta Cordillera correspond to the Bahía Mansa Metamorphic Complex, a group of metamorphic rocks developed in the subduction zone (Mella & Quiroz, 2010), under high pressure and temperature conditions. This metamorphic complex is in contact with granitic rocks of Coastal Batholith, which are distributed at the east of the Nahuelbuta Cordillera and at the western section of the Central Depression in the north of the area. Coastal Batholith Unit corresponds to a large mountain range elongated north–south. In the Central Depression appears (i) sedimentary lacustrine and fluvial rocks of Triassic age (237–200 Ma), (ii) volcanic and sedimentary rocks of 33 Ma (Mella & Quiroz, 2010), recognized in the Ñielol-Huimpil hills (Fig. 13.2), (iii) marine sedimentary rocks dated in 14 Ma, and (iv) glacio-fluvial deposits aged from 3 to 0.8 Ma. The landscape is finally modulated by fluvial activity, generating the current fluvial basins composed by Quaternary sediments in meandric rivers, alluvial and landslide deposits.

13.3.2 *Soil Parent Materials of the Araucanía Vitiviniculture*

Different soil macrounits were defined based on the parental soil material of vines cultivated in Malleco valley, such as phyllite (Plots 7, 9, and 11), granitoids (Plots 1, 3, 10, and 12), sandstones (Plot 6), alluvial with volcanic ash (Plot 5), alluvial with metamorphic rocks (Plots 4, 8, and 13), and volcaniclastic deposits (Plot 2) macrounits (Fig. 13.2). This only concentrates a sample of producers from the Malleco valley, which were chosen with the goal of covering a large part of the territory.

13.3.2.1 Soil Physical Parameters of the Araucanía Vitiviniculture

The phyllite and granitoid macrounits present clayey soils of high bulk density with a predominance of micropores of high soil water availability for root extraction close to 12% of soil volume (Table 13.1). Sandstone macrounit presents loamy texture in the first horizon and sandy in the second and lower horizons of high bulk density with a predominance of micropores in the first horizon and macropores in the lower horizons, which favors water drainage (Table 13.1). Sandstone macrounit has a water availability available for root extraction of close to 20% of the soil volume (Table 13.1). Soils of alluvial origin present clayey texture oh high bulk density and a predominance of micropores with a water availability for root extraction of close to 16% of the soil volume (Table 13.1). Soils of volcaniclastic deposits present a loamy to sandy texture of high bulk capacity with pores of variable size due to the presence of rocks or clasts of different sizes, which confers them a good oxygenation and soil drainage. This macrounit has a water availability for root extraction of close to 17% on average of the soil volume (Table 13.1).

13.3.2.2 Soil Chemical Parameters of the Araucanía Vitiviniculture

Phyllite and granitic soils present low pH and low levels of organic matter (OM), macro- and micro-nutrients, and cation exchange capacity (CEC) (Table 13.2). Based on soil fertility it could be beneficial for the development of the viticulture since these soils could limit growth of vines (Gutiérrez-Gamboa et al., 2018). Phyllite soils present high aluminum saturation due to their geochemical composition, which results in calcium (Ca) and magnesium (Mg) deficiencies (Keller, 2020).

Table 13.1 Soil physical parameters of different macrounits of the Araucanía Region

Physical soil parameter	Phyllite	Granitoids	Sandstones	Alluvial with volcanic ash	Alluvial with metamorphic rocks	Volcaniclastic deposits
Horizon 1						
Sand (%)	15.2	26.8	26.0	7.2	16.0	20.2
Silt (%)	32.5	21.4	35.4	41.7	33.9	28.5
Clay (%)	52.4	51.8	38.6	51.1	50.1	51.3
Bulk density (g/ml)	1.35	1.34	1.31	1.31	1.37	1.33
Soil water availability (%)	12.09	10.66	22.79	13.96	13.64	18.36
Horizon 2						
Sand (%)	16.4	23.8	46.8	10.7	10.4	48.3
Silt (%)	36.8	23.8	44.7	34.2	31.1	29.3
Clay (%)	46.8	52.4	8.4	55.1	58.5	22.4
Bulk density (g/ml)	1.33	1.38	1.18	1.43	1.41	1.22
Soil water availability (%)	12.80	12.43	14.76	14.76	14.63	19.97
Horizon 3						
Sand (%)	n.d	n.d	n.d	12.4	58.1	80.2
Silt (%)	n.d	n.d	n.d	38.6	18.8	16.1
Clay (%)	n.d	n.d	n.d	49.0	23.1	3.7
Bulk density (g/ml)	n.d	n.d	n.d	1.35	1.35	1.30
Soil water availability (%)	n.d	n.d	n.d	19.05	13.00	12.17

n.d not determined

Aluminum (Al) tends to bond with phosphorus (P) in a less available and insoluble form in soils and plant roots, thereby creating a P deficiency for plant growth (Bojórquez-Quintal et al., 2017). Under these conditions, high concentrations of Al results on inhibition of root elongation and plant growth through a diversity of biochemical and physiological mechanisms (Kochian et al., 2015). Sandstone macrounit has low levels of OM and macronutrients as nitrogen (N), P and potassium (K). Sandstone soils presents high levels of Ca, Mg, iron (Fe), manganese (Mn) and medium CEC, which could be attributed to the presence of montmorillonite-type clays (Macías-Quiroga et al., 2018). Low OM content was shown in alluvial soils with volcanic ash. These soils also presented medium levels of N and high levels of Ca, Mg, Fe, Mn, copper (Cu), and CEC. The mineralogy of these soils is not as consistent with the chemical results (Table 13.2). The medium levels of CEC and

Table 13.2 Soil chemical parameters of different macrounits of the Araucanía Region

Chemical soil parameter	Phyllite	Granitoids	Sandstones	Alluvial with volcanic ash	Alluvial with metamorphic rocks	Volcaniclastic deposits
pH in water	5.59	4.88	6.30	5.82	5.63	5.81
Organic matter (%)	0.31	0.70	0.56	0.69	0.21	0.02
Nitrate (mg/kg)	0.9	15.5	1.1	1.2	1.2	0.9
Ammonium (mg/kg)	2.3	11.5	4.8	28.0	3.0	6.4
Nitrogen (mg/kg)	3.2	27.0	5.8	29.2	4.3	7.3
Extractable P (mg/kg)	3.3	7.0	0.1	0.2	4.7	3.6
Extractable K (mg/kg)	12.0	94.8	83.9	99.8	174.7	219.6
Exchangeable K (cmol/kg)	0.03	0.24	0.22	0.26	0.45	0.56
Exchangeable Ca (cmol/kg)	1.89	2.55	11.32	8.47	5.73	11.81
Exchangeable Mg (cmol/kg)	1.36	0.80	4.64	3.57	2.18	3.77
Exchangeable Na (cmol/kg)	0.08	0.10	0.35	0.10	0.18	0.17
Sum of bases (cmol/kg)	3.36	3.72	16.52	12.39	8.53	16.32
Exchangeable Al (cmol/kg)	0.02	0.31	0.10	0.19	0.17	0.15
CEC (cmol/kg)	3.57	4.03	16.63	12.58	8.70	16.46
Al saturation (%)	5.69	7.65	0.62	1.49	1.91	0.89
K saturation (%)	0.86	6.04	1.29	2.04	5.15	3.42
Ca saturation (%)	52.99	63.31	68.11	67.29	65.86	71.77
Mg saturation (%)	38.26	20.41	27.88	28.36	25.04	22.92
Extractable S (%)	30.4	12.6	0.1	2.3	4.9	0.1
Fe (mg/kg)	3.0	16.2	36.6	10.4	17.8	27.8
Mn (mg/kg)	2.3	39.6	6.0	6.8	31.8	17.4
Zn (mg/kg)	0.1	0.2	0.2	0.1	0.2	0.3
Cu (mg/kg)	0.1	1.3	1.5	1.5	0.7	1.3
B (mg/kg)	0.4	0.2	0.2	0.2	0.2	0.1
EC (dS/m)	0.1	0.1	0.1	0.1	0.1	0.1

CEC cation exchange capacity, *EC* electrical conductivity, NO_3^- nitrate, NH_4^+ ammonium, *P* phosphorous, *K* potassium, *Ca* calcium, *Mg* magnesium, *Na* sodium, *Al* aluminum, *S* sulfur, *Fe* iron, *Mn* manganese, *Zn* zinc, *Cu* copper, *B* boron

Table 13.3 Geographical information of the weather station located in Malleco and Cautín valleys

Weather station name	Latitude (°South)	Longitude (°West)	Elevation (m)	Distance to the Pacific Ocean (km)
Traiguén	−38.26	−72.65	234	72
Carillanca	−38.68	−72.42	470	87
Maquehue	−38.77	−72.64	92	64

the high sum of bases of these soils could be explained by their high percentage of clay minerals (>85%) despite that kaolinite is a low CEC clay. The origin of these results could probably be due to the presence of volcanic ash in these alluvial deposits, which can produce amorphous minerals such as allophanes and imogolite (Parfitt, 2018). These minerals have high CEC due to the high superficial area. Alluvial soils with metamorphic rocks have low levels of OM, N, P, and CEC and high levels of Fe, Mn, and Cu. These microelements are transition metals that exist in various oxidation states but only Mn^{2+} and Cu^{2+} could be taken by roots and transported throughout the plant in both the xylem and phloem (Keller, 2020). Cu availability increases at low pH in soils and long-term application of Cu-based fungicides allows its accumulation in the soil surface (Kakutey et al., 2023). Mn toxicity is rare and occurs on acid soils (<5.5) with high Mn availability (Pittman, 2005). This soil macrounit presents low pH and high aluminum saturation (Table 13.2), which may affect vine root development and P availability (Bojórquez-Quintal et al., 2017). Volcaniclastic deposit macrounit presents low content of OM, N, and P, whereas it has high levels of K, Ca, Mg, Fe, Mn, Cu, sum of bases, leading to medium levels of CEC (Table 13.2). Similar to the Sandstone soil macrounit, the medium CEC levels of volcaniclastic deposit macrounit may be attributed to the presence of montmorillonite type clays (Macías-Quiroga et al., 2018).

13.4 Climatic Conditions in the Araucanía Region

13.4.1 Meteorological, Bioclimatic, and Risk Indices

The climatic information available in the Araucanía valley allows to obtain trends and general averages correspond to those located in the sub-valleys of Malleco (Traiguén station) and Cautín valley (Carillanca and Maquehue stations) due to the existence of data from more than three decades of data. The coordinates of these stations are shown in Table 13.3.

The above stations provide systematized information on temperatures (maximum and minimum) and precipitation, with a daily resolution, for the period 1985–2015, making it possible to obtain averages and trends for different types of indices (climatic, bioclimatic, and risk), according to the methodology described by Verdugo-Vásquez et al. (2023). Table 13.4 shows the mean values of different meteorological and risk indices for the period 1985–2015. Minimum temperature

Table 13.4 Mean values (1985–2015 period) obtained in each weather station for meteorological and risk indices

Weather station name	T_{min} (°C)	T_{max} (°C)	PP (mm)	Day $T > 30$ °C	Day $T > 35$ °C	Day $T < 0$ °C
Traiguén	7.0	17.8	1002.2	8.3	0.6	1.9
Carillanca	5.6	17.5	1310.7	6.0	0.6	12.9
Maquehue	6.3	18.0	1128.8	6.5	0.7	8.0

T_{min} minimum temperature, T_{max} maximum temperature, *PP* precipitation, *T* temperature

Table 13.5 Mean values (1985–2015 period) obtained in each weather station for bioclimatic indices

Weather station name	GST (°C)	GDD (heat units)	BEDD (heat units)	HI (heat units)	CI (°C)	SONMean (heat units)	SONMax (heat units)
Traiguén	15.1	1104.3	1009.2	1655.8	9.0	1069.7	1580.4
Carillanca	14.0	883.9	883.0	1479.3	7.3	1015.6	1556.3
Maquehue	14.5	970.3	944.0	1558.8	7.8	1053.7	1596.0

GST Growing Season Temperature, *GDD* Growing Degree Days, *BEDD* Biologically Effective Degree Days, *HI* Huglin index, *CI* cold night index, *SONMean* sum of the mean temperatures of September, October, and November, *SONMax* sum of the maximum temperatures of September, October, and November

average for the 1985–2015 period ranged from 5.6 to 7.0 °C (Carillanca and Traiguén, respectively), whereas maximum temperature average varied from 17.5 to 18.0 °C (Carillanca and Maquehue, respectively). The highest precipitation reached in Carillanca (1311 mm), followed by Maquehue (1128 mm) and Traiguén (1002 mm), respectively. The risk of temperature higher than 30 °C was higher in Traiguén (8 days) than in Maquehue (7 days) and Carillanca (6 days), whereas in the studied stations the risk of temperatures higher than 35 °C was lower than 1 day. The risk of frost temperatures (lower than 0 °C) is considerably lower in Traiguén (2 days) than in Maquehue (8 days) and Carillanca (13 days).

Table 13.5 shows the mean values of bioclimatic indices calculated for the 1985–2015 period with data obtained in Malleco (Traiguén station) and Cautín valleys (Carillanca and Maquehue stations). The Growing Season Temperature (GST) varied from 14.0 to 15.1 °C, which characterize Malleco valley as intermediate and Cautín valley as cool climate (Jones et al., 2005). The Growing Degree Days (GDD) varied from 884 to 1104 heat units, whereas Biologically Effective Degree Days ranged from 883 to 1009 heat units. GDD characterizes Malleco and Cautín valleys in the Region I of climatic classification in which the most suitable grapevine varieties for cultivation are the early ripening varieties (Amerine & Winkler, 1944). Huglin index varied from 1479 to 1656 heat units, which allows to classify Carillanca as very cool climate and Maquehue and Traiguén as cool climates (Huglin, 1978). Cool Night Index varied from 7.3 to 9.0 °C, which allows to classify Malleco and Cautín valleys with cold nights, which favors the accumulation of secondary metabolites in grapes (Tonietto & Carbonneau, 2004).

Table 13.6 Trends for the 1985–2015 period obtained in each weather station for meteorological and risk indices

Weather station name	T_{min} (°C/year)	T_{max} (°C/year)	PP (mm/year)	Days $T > 30$ °C (day/year)	Days $T > 35$ °C (day/year)	Days $T < 0$ °C (day/year)
Traiguén	**−0.025**	0.0273	2.656	0.200	0.000	0.000
Carillanca	−0.0167	0.0017	−6.888	0.091	**0.001**	0.118
Maquehue	−0.0171	**0.0213**	−2.352	**0.222**	**0.048**	**0.278**

Data in bold correspond to significant trends for the period described

13.4.2 Trends in the Meteorological, Bioclimatic and Risk Indices

Table 13.6 shows the trends in meteorological and risk indices calculated for the 1985–2015 period for Malleco (Traiguén station) and Cautín valleys (Carillanca and Maquehue stations).

Maquehue was the zone most influenced by climate change in the last three decades in the Araucanía Region, in which the trends in maximum temperature and risk indices varied significantly (Table 13.6, Fig. 13.3). Contrary to this, the trend in precipitations did not change in Traiguén, Carillanca, and Maquehue for the studied period (Table 13.6, Fig. 13.3).

In the 1985–2015 period, the minimum temperature decreased 0.03 °C by year in Traiguén, whereas the maximum temperature increased 0.02 °C by year in Maquehue (Table 13.6, Fig. 13.3). The number of days with temperatures higher than 30 °C increased 0.22 days per year in Maquehue (Table 13.6, Fig. 13.3). The number of days with temperatures higher than 35 °C increased 0.001 and 0.048 days per year in Carillanca and Maquehue (Table 13.6, Fig. 13.3), respectively. The number of days with frost temperatures (lower than 0 °C) increased 0.278 days per year in Maquehue (Table 13.6, Fig. 13.3).

13.5 Trends in Reference Evapotranspiration in the Araucanía Region

The reference evapotranspiration (ET_0) is the first step in estimating crop water requirements and irrigation water management, thereby a better understanding of its trends is crucial for scientific management of water resources (Yassen et al., 2020). The Araucanía Region has been undergoing a process of significant climatic variability, where a significant increase in maximum temperature, a significant decrease in minimum temperature, and a non-significant decrease of pluviometry is shown in the last three decades (Verdugo-Vásquez et al., 2023). The trends in the ET_0 values were compared between two-time intervals in 1960–1990 and 1985–2015 in different agroclimatic zones of the Araucanía Region (Fig. 13.4).

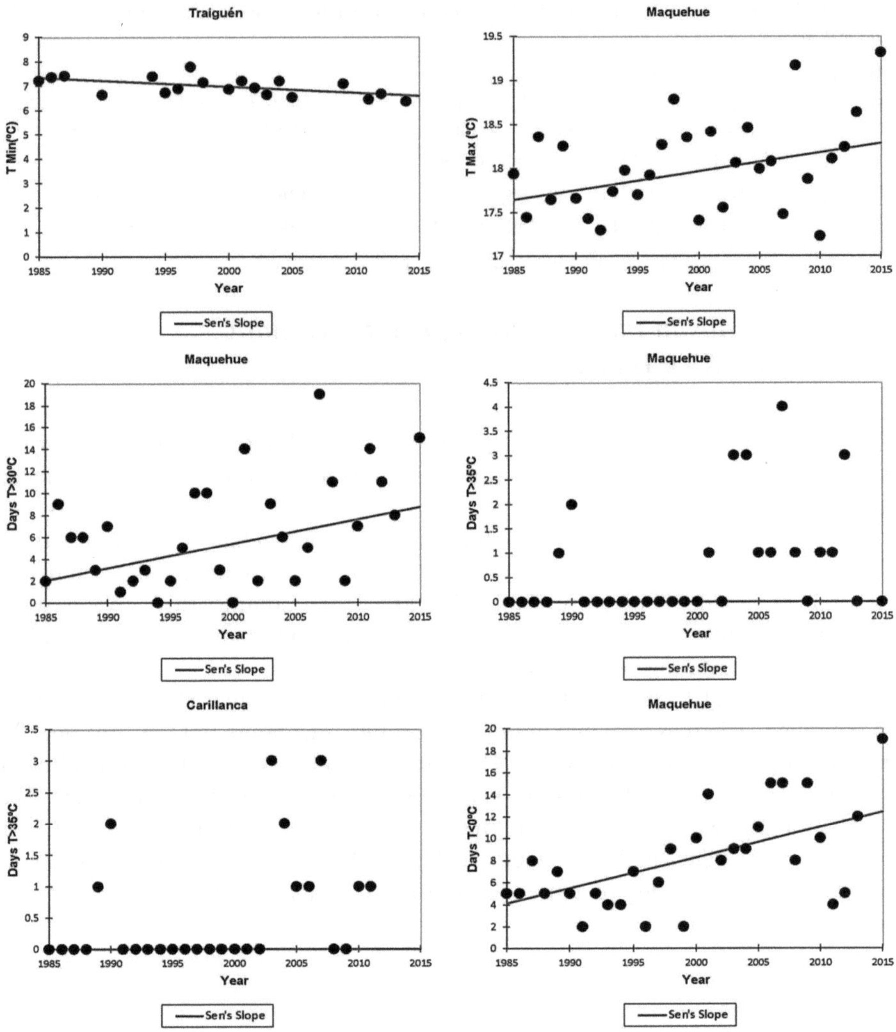

Fig. 13.3 Significant trends for the 1985–2015 period obtained in each weather station for meteorological and risk indices. Note: *T* Min: minimum temperature. *T* Max: maximum temperature. *T*: temperature

The average annual ET_0 value was 875 mm for the 1960–1990 period and 1045 mm for the 1985–2015 period, which means an average increase close to 22% (Fig. 13.4). The most affected areas by the increase in the ET_0 were the (i) Interior Drylands (Los Sauces, Purén, Lumaco, Galvarino, and Loncoche), (ii) the Intermediate Depression (Angol and Renaico), (iii) the Coastal Dryland (Carahue, Puerto Saavedra, Puerto Domínguez, and Toltén), and (iv) the Dryland Valley (Collipulli, Ercilla, Victoria, Traiguén, Perquenco, Lautaro, Vilcún, Temuco, Nueva Imperial, Freire, Pitrufquén, Gorbea, and Villarrica) with average increases of 26% (213 mm), 31% (281 mm),

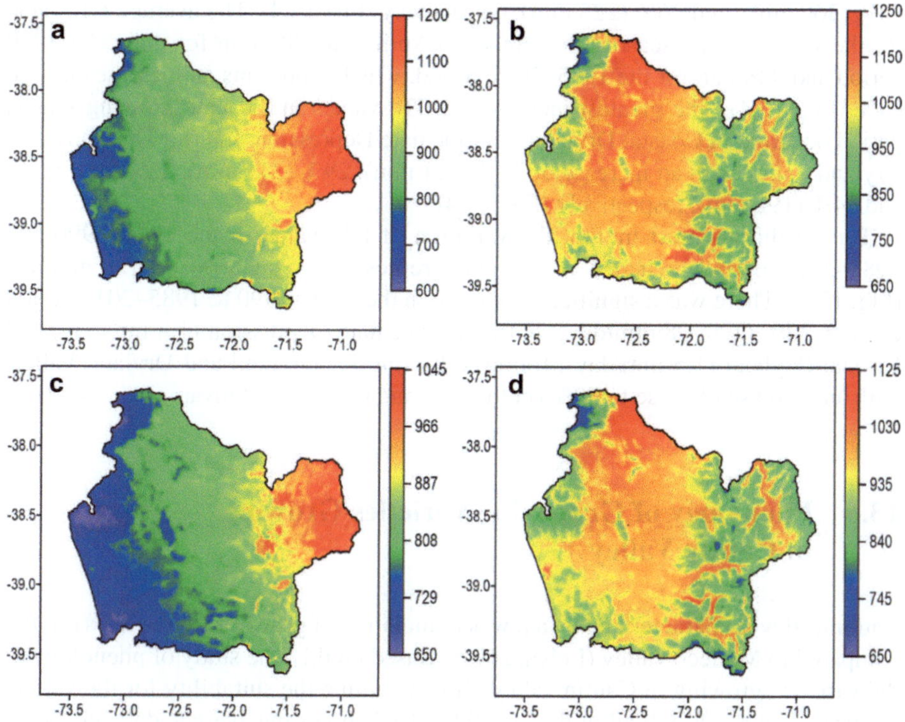

Fig. 13.4 Reference evapotranspiration (ET_0) average values of the Araucanía Region during the calendar year (January to December) and vine growing season (September to April). ET_0 values of 1960–1990 (**a, c**) and ET_0 of 1985–2015 (**b, d**). (Source: Figure obtained from the publication of López-Olivarí et al. (2022))

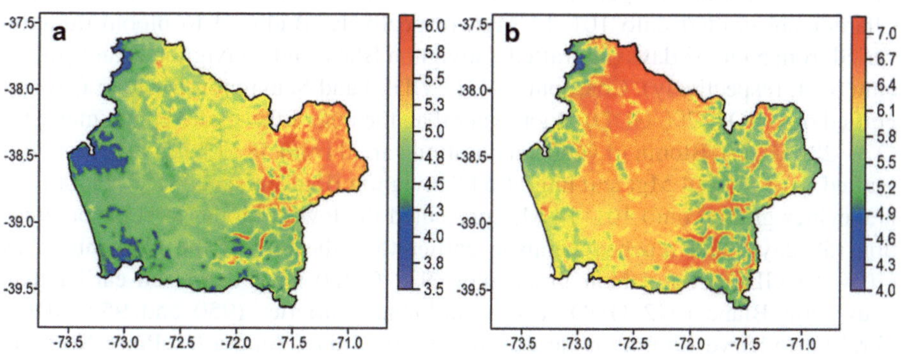

Fig. 13.5 Reference evapotranspiration (ET_0) average values of the Araucanía Region during the warmest month (January). ET_0 values of 1960–1990 (**a**) and ET_0 of 1985–2015 (**b**). (Source: Figure obtained from the publication of López-Olivarí et al. (2022))

35% (269 mm), and 26% (225 mm), respectively (Fig. 13.4). The average ET_0 value for the vine growing season (September to April) was 784 mm for the 1960–1990 period and 948 mm for the 1985–2015 period, which represents an increase close to 24% (Fig. 13.4). The greatest increases in ETo values in the vine growing season period occurred in Coastal Dryland, Intermediate Depression, Interior Drylands, and Dryland Valley with an increase of 31% (221 mm), 29% (243 mm), 25% (184 mm), and 24% (192 mm), respectively (Fig. 13.4).

During the warmest month, the variation of ET_0 value in the 1960–1990 and 1985–2015 periods reached a 28% that represents a quantity of 1.3 mm/day (Fig. 13.5). There was a significant increase in the 1960–1990 to 1985–2015 period close to 34% (1.5–1.8 mm/day), in the Intermediate Depression (1.8 mm/day) and Coastal Dryland (1.5 mm/day), followed by Interior Dryland and Dryland Valley, with an increase of close to 30% (both 1.4 mm/day) (López-Olivarí et al., 2022).

13.6 Phenology of Grapevine Varieties Growing in Cautín Valley

Cautín valley is characterized by a low accumulation of growing degree days (GDD) compared to Malleco valley (Leiva, 2007). Based on this, the study of phenology of the varieties growing in Cautín valley allow to define the suitability for the variety cultivation in a cold viticultural zone. Table 13.7 shows the average data and accumulated degree days of the phenological stages of ungrafted Chardonnay, Sauvignon Blanc, and Pinot Noir and grafted onto different rootstocks. The timing and duration of some phenological stages, such as bloom, veraison, and harvest, was significantly affected by the plant material.

Budburst timing and duration was not statically affected by the plant material. The date of budburst ranged from 31.9 to 50.3 GDD in ungrafted Chardonnay and Chardonnay grafted onto 101–14 Mgt, respectively. Budburst to bloom duration varied from 54 to 63 days (ungrafted Sauvignon Blanc and Sauvignon Blanc grafted onto SO4, respectively). Chardonnay (311 GDD) and Sauvignon Blanc grapevines grafted onto SO4 (298 GDD) bloom later than the rest of the studied plant materials (270–284 GDD). Bloom to veraison duration varied from 58 to 70 days (Pinot Noir grafted onto 101–14 Mgt and ungrafted Sauvignon Blanc, respectively). Pinot Noir grapevines grafted onto 101–14 Mgt presented the lowest bloom to veraison duration (58 days) and reached veraison earlier than the rest of the plant materials (701–758 GDD). Pinot Noir grapevines (932 GDD) were harvested earlier than Sauvignon Blanc (942 GDD) and Chardonnay varieties (950 and 956 GDD). Veraison to harvest duration varied from 56 to 65 days (ungrafted Pinot Noir and Chardonnay grafted onto 101–14 Mgt). The grapevines grafted onto 101–14 Mgt rootstocks presented higher veraison to harvest duration than ungrafted grapevines (Table 13.8).

Table 13.7 Phenological stages occurrence (date) in varieties growing in Cautín valley from 2021 to 2023 seasons

Plant material	Budburst	GDD	Bloom	GDD	Veraison	GDD	Harvest	GDD
Chardonnay onto 101–14 Mgt	20-Oct ± 1.71a	50.3	16-Dec ± 1.00b	284.4	21-Feb ± 2.28ab	752.5	25-Apr ± 8.00a	950.2
Chardonnay	13-Oct ± 2.25a	31.9	20-Dec ± 0.00a	311.8	22-Feb ± 1.99a	757.9	25-Apr ± 0.00a	955.6
Sauvignon Blanc	19-Oct ± 1.42a	50.3	15-Dec ± 0.00b	277.5	22-Feb ± 1.27ab	757.9	21-Apr ± 0.56a	942.2
Sauvignon Blanc onto SO4	18-Oct ± 1.12a	48.6	18-Dec ± 0.56a	298.0	18-Feb ± 1.28b	735.6	21-Apr ± 0.48a	942.2
Pinot Noir onto 101–14 Mgt	18-Oct ± 1.43a	48.6	15-Dec ± 0.79b	277.5	12-Feb ± 0.46c	701.4	17-Apr ± 0.98b	932.2
Pinot Noir	16-Oct ± 2.42a	39.6	14-Dec ± 0.00b	270.3	21-Feb ± 2.22ab	752.5	17-Apr ± 1.55b	932.2
Significance	n.s.		***		***		***	

For each parameter, different letters within a column represent significant differences (LSD Fisher < 0.05)
GDD Growing Degree Days base 10 °C, *n.s.* non statistical differences
**p-value < 0.05; ***p-value < 0.001

Table 13.8 Length between phenological phases (days) in varieties growing in Cautín valley from 2021 to 2023 seasons

Plant material	Bb to Bl	Bl to Ver	Ver to Har
Chardonnay onto 101–14 Mgt	57.55 ± 2.55a	66.33 ± 2.73ab	64.75 ± 3.35a
Chardonnay	60.50 ± 6.50a	70.00 ± 0.00ab	56.00 ± 0.00b
Sauvignon Blanc	54.53 ± 1.86a	70.33 ± 1.28a	57.17 ± 1.28b
Sauvignon Blanc onto SO4	63.11 ± 1.24a	64.00 ± 1.42b	59.89 ± 10.69ab
Pinot Noir onto 101–14 Mgt	57.50 ± 3.05a	58.13 ± 0.84c	64.25 ± 0.47a
Pinot Noir	62.50 ± 6.50a	68.00 ± 0.00ab	55.50 ± 0.50b
Significance	n.s.	***	**

For each parameter, different letters within a column represent significant differences (LSD Fisher < 0.05)
Bb budburst, *Bl* bloom, *Ver* veraison, *Har* harvest, *n.s.* non statistical differences
**p-value < 0.05; ***p-value < 0.001

13.7 Proposal Guidelines

The vitiviniculture in the Araucanía Region has comparative advantages in relation to its pedoclimatic conditions and cultural landscape compared to the rest of the viticultural valleys of Chile. Despite this, it is important to develop an own narrative that considers its potential in order to provide a specific identity. This proposal narrative could consider the following issues: (i) gathering information about the plant material brought by European conquerors in the Southern Chile; (ii) the development and sustainability of a wine tourist route taking advantage of the region's scenic beauty and typical products; and (iii) the development of sparkling wines given the climatic conditions using local varieties of the area.

References

Amerine, M. A., & Winkler, A. J. (1944). Composition and quality of musts and wines of California grapes. *Hilgardia, 15*, 493–675.
BCN. (2020). *Decreto 464. Establece zonificación vitícola y fija normas para su utilización.* Ministerio de Agricultura, Biblioteca del Congreso Nacional.
Bojórquez-Quintal, E., Escalante-Magaña, C., Echevarría-Machado, I., & Martínez-Estévez, M. (2017). Aluminum, a friend or foe of higher plants in acid soils. *Frontiers in Plant Science, 8*, 1767.
de Solminihac, F. (2015). Dinamismo de la vitivinicultura chilena: el valle de Malleco. In R. Aravena (Ed.), *Patrimonio vitivinícola. Aproximaciones a la cultura del vino en Chile* (pp. 58–64). Ediciones Biblioteca Nacional.
Gonzalez, R., Appelt, H., Schalscha, E. B., & Bingham, F. T. (1974). Molybdate adsorption characteristics of volcanic-ash-derived soils in Chile. *Soil Science Society of America Journal, 38*(6), 903–906.
González, A., Ellena, M., Neumann, G., & Romero, I. (2021). *Potencial frutícola de la Araucanía: Estado actual, oportunidades y desafíos* (Boletín INIA – Instituto de Investigaciones Agropecuarias, No. 449). INIA.

Gutiérrez-Gamboa, G., & Moreno-Simunovic, Y. (2019). Terroir and typicity of Carignan from Maule Valley (Chile): The resurgence of a minority variety: Terroir and typicity of Carignan. *OENO One, 53*(1), 75–93.
Gutiérrez-Gamboa, G., Carrasco-Quiroz, M., Martínez-Gil, A. M., Pérez-Álvarez, E. P., Garde-Cerdán, T., & Moreno-Simunovic, Y. (2018). Grape and wine amino acid composition from Carignan noir grapevines growing under rainfed conditions in the Maule Valley, Chile: Effects of location and rootstock. *Food Research International, 105*, 344–352.
Gutiérrez-Gamboa, G., Zheng, W., & Martínez de Toda, F. (2021a). Strategies in vineyard establishment to face global warming in viticulture: A mini review. *Journal of the Science of Food and Agriculture, 101*, 1261–1269.
Gutiérrez-Gamboa, G., Zheng, W., & Martínez de Toda, F. (2021b). Current viticultural techniques to mitigate the effects of global warming on grape and wine quality: A comprehensive review. *Food Research International, 139*, 109946.
Huglin, P. (1978). Nouveau mode d'evaluation des possibilités heliothermiques d'un milieu viticole. *Comptes Rendus de l'Académie d'agriculture de France, 64*, 117–1126.
Jones, G. V., White, M. A., Cooper, O. R., & Storchmann, K. (2005). Climate change and global wine quality. *Climatic Change, 73*, 319–343.
Kakutey, K., Sackey, L. N. A., & Akoto, O. (2023). Impact of accumulation of copper from application of copper-based fungicides on soil properties in Ghana. *Discover Environment, 1*, 1.
Keller, M. (2020). The Science of Grapevines. Third Edition. Academic Press, Elsevier. pp. 554.
Kochian, L. V., Piñeros, M. A., Liu, J., & Magalhaes, J. V. (2015). Plant adaptation to acid soils: The molecular basis for crop aluminum resistance. *Annual Review of Plant Biology, 66*, 571–598.
Leiva, J. (2007). *Viticultura en la Región de La Araucanía. Manual para el Productor Vitícola.* Agraria.
Leiva, J., & Soto, O. (2016). *Diagnóstico de la vitivinicultura región de la Araucanía.* Corporación Agencia Regional de Desarrollo Productivo.
López-Olivarí, R., Morales-Salinas, L., & Pauchard-Cuevas, H. (2022). Análisis histórico de la evapotranspiración en la Región de la Araucanía. La evapotranspiración de referencia (ETo) es una variable clave en la toma de decisiones para la gestión hídrica intrapredial. *Campo & Tecnología*, 32–35.
Macías-Quiroga, I. F., Giraldo-Gómez, G. I., & Sanabria-González, N. R. (2018). Characterization of Colombian clay and its potential use as adsorbent. *The Scientific World Journal, 2018*, 5969178.
Martin Bastidas, M. N. (2020). Identity and multiculturalism: Representation of the Mapuche people in Chilean wine labels. *Revista Iberoamericana de Viticultura, Agroindustria y Ruralidad, 9*(25), 173–190.
Mella, M., & Quiroz, D. (2010). *Geología del área Temuco-Nueva Imperial, Región de La Araucanía* (Carta Geológica de Chile, Serie Geología Básica 122, 46 p.). SERNAGEOMIN.
Montes, C., Perez-Quezada, J. F., Peña-Neira, A., & Tonietto, J. (2012). Climatic potential for viticulture in Central Chile. *Australian Journal of Grape and Wine Research, 18*, 20–28.
Morales-Henríquez, T., Gutiérrez-Gamboa, G., Zheng, W., & de Toda, M. (2022). Principles of vineyard establishment and strategies to delay ripening under a warming climate. *IVES Technical Reviews, 6*, 1–2.
Parfitt, R. L. (2018). Allophane and imogolite: Role in soil biogeochemical processes. *Clay Minerals, 44*(1), 135–155.
Pittman, J. K. (2005). Managing the manganese: Molecular mechanisms of manganese transport and homeostasis. *New Phytologist, 167*, 733–742.
Román-Figueroa, C., Montenegro, N., & Paneque, M. (2017). Bioenergy potential from crop residue biomass in Araucania Region of Chile. *Renewable Energy, 102*, 170–177.
SAG. (2021). *Catastro vitícola nacional 2021.* División de Protección Agrícola-Forestal y Semillas Subdepartamento de Viñas, Vinos y Bebidas Alcohólicas. Servicio Agrícola y Ganadero.
Santos, J. A., Fraga, H., Malheiro, A. C., Moutinho-Pereira, J., Dinis, L.-T., Correia, C., Moriondo, M., Leolini, L., Dibari, C., Costafreda-Aumedes, S., Kartschall, T., Menz, C., Molitor, D., Junk, J., Beyer, M., & Schultz, H. R. (2020). A review of the potential climate change impacts and adaptation options for European viticulture. *Applied Sciences, 10*(9), 3092.

SERNAGEOMIN. (2004). *Mapa Geológico de Chile*. Servicio Nacional de Geología y Minería. Digital Publication. Available on: https://www.sernageomin.cl/

Tonietto, J., & Carbonneau, A. (2004). A multicriteria climatic classification system for grape-growing regions worldwide. *Agricultural and Forest Meteorology, 124*, 81–97.

Verdugo-Vásquez, N., Orrego, R., Gutiérrez-Gamboa, G., Reyes, M., Zurita Silva, A., Balbontín, C., Gaete, N., & Salazar-Parra, C. (2023). Climate trends and variability in the Chilean viticultural production zones during 1985–2015. *OENO One, 57*(1), 345–362.

Yassen, A. N., Nam, W. H., & Hong, E. M. (2020). Impact of climate change on reference evapotranspiration in Egypt. *Catena, 194*, 104711.

Chapter 14
Heroic Viticulture in Itata Valley, Chile: Characteristics and Challenges for the Development of Unique Wines in Southern Chilean Vineyards

Ignacio Serra, Arturo Calderón-Orellana, and Marcela Hidalgo

14.1 Characteristics of the Heroic Viticulture of the Itata Valley

The Itata Valley is located in the Ñuble region, Chile, between latitudes −36 to −37° South and longitudes −72.8°O to −71.6°E, and includes 13 municipalities (Quirihue, Ninhue, San Carlos, San Nicolas, Portezuelo, Treguaco, Coelemu, Ranquil, Chillán, Chillán Viejo, Quillón, Florida, and Bulnes) (Fig. 14.1).

The formation of the Itata Valley dates back to the Upper Carboniferous, about 320 Myr B.P. (million years before present). It began when subduction was activated at the margin of the tectonic plates, and convergence began to frontally deform the present-day Eastern Series protolith (Willner et al., 2005). Gradually, basal accretion was generated by a high accumulation of material, forming the Eastern Series (Willner et al., 2005), which vertically deformed the subduction wedge (Hervé et al., 2013). Near the Carboniferous-Permian boundary, at 305 Myr B.P., subduction generated large masses of magmatic material, which formed the metamorphic complex. The successive magmatic pulses continued until the Upper Triassic (237 Myr B.P.), when the tectonic subduction ceased and convergence stopped, ending the magmatic generation and the formation of the metamorphic complex series. Progressively, the high accumulation of magmatic pulses generated a large batholith-like intrusive body, which in the Coastal Cordillera occurred in a large, elongated, north-south trending intrusive rock mass with hilly complex topography (known as the Coastal Batholith) (Hervé et al., 2013). The tectonic cycle that

I. Serra (✉) · A. Calderón-Orellana
Departamento de Producción Vegetal, Facultad de Agronomía, Chillán, Chile

M. Hidalgo
Departamento de Suelos y Recursos Naturales, Facultad de Agronomía, Chillán, Chile

G. Gutiérrez Gamboa, M. Fourment (eds.), *Latin American Viticulture Adaptation to Climate Change*, https://doi.org/10.1007/978-3-031-51325-1_14

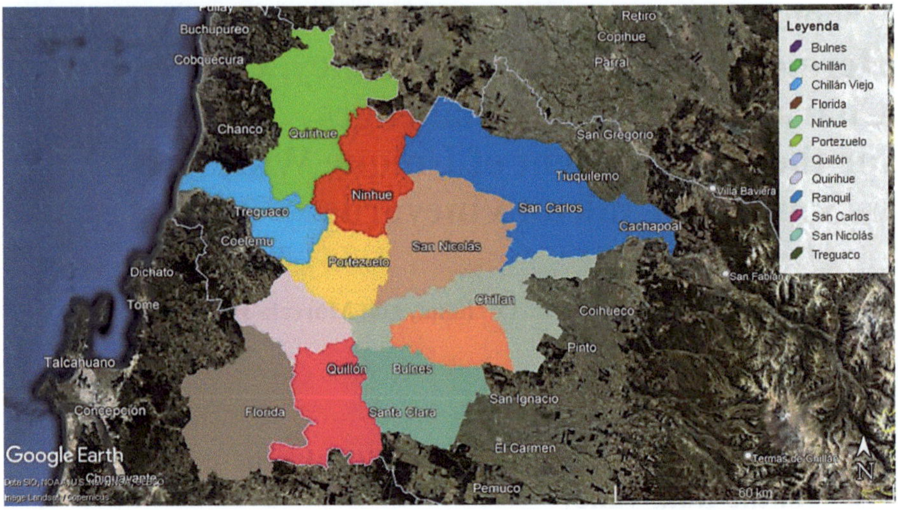

Fig. 14.1 Municipalities of the Itata valley, scale 1:1.000.000. (Own elaboration using the available shapes from the "Infraestructura de datos geoespaciales de Chile, IDE_{Chile}" for the Google Earth platform)

gave rise to the Andes began in the Early Jurassic (less than 200 Myr B.P.) when new subduction activity was reactivated, generating continuous and uninterrupted arc magmatism (to the present) (Charrier et al., 2007). Throughout the Andean Cycle, the volcanic arc has moved eastward to its present position, with the youngest rocks being close to the volcanoes of the Andes (Charrier et al., 2007). This process has modeled the main geomorphology of Chile and the evolution of the soil parent material. The Itata valley covers five geomorphological zones: (1) Coastal mountain range, (2) marginal granitic basins, (3) central fluvial-glacial-volcanic plain, (4) fluvial or alluvial sedimentation plains, and (5) marine or fluvial-marine plain (Cedeus, 2015) (Fig. 14.2) and a variety of parent materials, described in the Geological Map of Chile (scale 1:1.000.000), such as CPg: Plutonic ambient with granites, granodiorites, tonalites and diorites, hornblende and biotite, locally of muscovite rocks; Q1: Continental sedimentary ambient presenting parent material of alluvial, colluvial, and landslide deposits; Qf: Continental-sedimentary environment with parent material of fluvial deposits such as gravels, sands, and silts from the current course of major rivers or their sub-current terraces and floodplains; Pz4b: Metamorphic environment representing slates, phyllites, and meta-sandstones with low-gradient metamorphism; and Jig: Plutonic environment representing pyroxene diorites, gabbros and monzodiorites, quartz diorites and granodiorites, and hornblende and biotite tonalites (SERNAGEOMIN, 2003). The soil found in the valley is mainly derived from the geomorphological formations of the coastal mountain range, conglomerates, volcanic tuffs, and alluvial plain, as mentioned before, from where parent material can be convened in three main groups: (i) weathered metamorphic rock, (ii) granitic origin, and (iii) coarse to fine alluvial sediment.

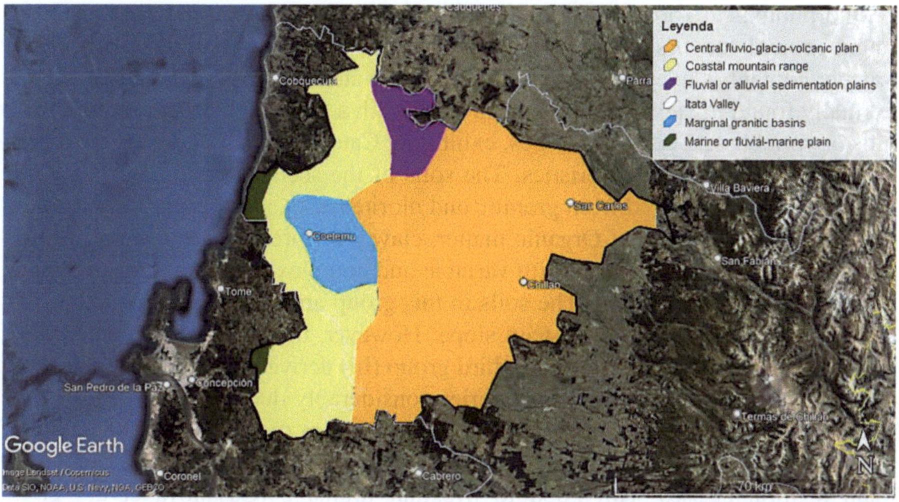

Fig. 14.2 Geomorphology of the Itata valley, scale 1:1.000.000. (Own elaboration using the available shapes from the "CEDEUS - Centre for Sustainable Urban Development UC - UdeC" for the Google Earth platform)

There are different soil groupings according to their origin. The Metamorphic soils, formed from the formation of the coastal batholith of the Cordillera de la Costa, are composed by the weathering of metamorphic rocks such as schists, micas, and gneiss (Stolpe, 2006). The Granitic soils are formed mainly by the weathering of granitic rocks or quartz diorites, which form a large part of the coastal mountain range (Casanova et al., 2013). A group of sandy soil, coming from sediments originated by alluvium in the last period of glaciation, being the alluvial cone of the Laja River, the formation present in part of the Itata Valley extending from the Itata River in the north to the Rarinco River in the south, and from the coastal mountain range in the west to the quarries in the east with Andes mountain range (Casanova et al., 2013; Stolpe, 2006). Finally, Sedimentary alluvial fine sedimentary corresponds to soil of similar origin to the previous group but with different distributed characteristics originated by the removal and redistribution of sediments product of erosive processes of the rivers Biobío and Laja that have removed great extensions of material Flavio glacial of the last glacial epoch originating sediments of fine particles (Stolpe, 2006). In this soil, aggrupation described by Stolpe (2006) classified soils as the following soil orders such as Mollisols, Vertisols, and Entisols predominate in groupings of sandy soils and fine alluvial sediments, while soil orders such as Ultisols, Alfisols, Inceptisols predominate in groupings of granitic and metamorphic soils. From the interaction of soil forming factors, parent material, climate, weather, soil micro and macro biota, and topography, today we find a variety of soils depending on their origin, evolution, different degrees of weathering, and composition of secondary minerals, which confer different attributes and properties to the soils. In the case of Itata Valley, as discussed above, the soil is grouped about its geomorphological formation and geological formation ambient, where the soils found around the first metamorphic group (i) exhibit low natural fertility, low organic content, clayey soil texture, and slow water infiltration. This soil type is normally found in elevated positions in areas with a topography that presents steep hills or variable and complex slopes, exhibiting Catena formation due to the topography and drainage characteristics. The soils of the second group (ii), of granitic origin, have been derived from granitic and diorite rocks, also presenting low natural fertility, low content of Organic matter, clayey textures, and low water infiltration in the topography of hills with variable and complex slopes that make the soil susceptible to water erosion. The soils in this group are normally classified as forestry soils, mainly due to the steep slope. However, these soils have exceptional qualities for viticulture. Finally, the third group (iii) derived from fine alluvial sediments. The thickness of the deposit varies considerably due to the influence of the rivers present in the area that transport large quantities of sediments, especially of the fine material that has formed the soils of this group with sandy loam and clayey loam textural classes with good to irregular drainage (Casanova et al., 2013; Stolpe, 2006).

The classification of soils is based on taxonomic characteristics based on their physical and biochemical properties, such as texture, bulk density, water holding capacity, and arrangement of soil aggregates. As well as chemical characteristics such as pH, cation exchange capacity, nutrient availability, mineral content, among others. The mineralogical and climatic conditions of the area mainly determine these.

The valley has an elevation from 70 to 700 m.a.s.l., (meters above sea level) where the highest altitude of the valley is in the coastal range mountain, marginal granitic basins, and fluvial or alluvial sediment plains geomorphology (CEDEUS, 2015) (Fig. 14.3) representing almost the half of the valley, where there are also found catena's formations due to the topography from strongly undulating to loins with a complex slope from 15° to 30° and a Mediterranean climate with rainy winters (IDE, 2016). Catena formation describes a repeating sequence of soils from the top of a hillslope to the adjacent valley bottom. A consequence of soil erosion processes is the redistribution of soil material downslope. The catena is defined as a sequence of soils of about the same age, derived from similar parent material, and occurring under similar climatic conditions, but have different characteristics due to variation in relief and drainage (Grunwald, 2016). Characteristics that give more variability to the soil of the valley, even in a single hill, showing differences in soil properties such as soil texture, nutrient availability and soil water content.

The land use capacity is a technical classification system which classifies soils in terms of the limiting constraints they may have, which hamper or impede their use for crop production. Based on the classification system, almost the whole valley presents a land use capacity of four (arable land with severe use limitations for agricultural crop production requiring special soil management and conservation practices for its conditions and limitations) to seven (no arable land with very severe limitations, not suitable for crops, mainly forestry use) (Stolpe, 2006). Nevertheless, due to the soil limitations in the valley, either for low natural fertility or steep slope of the hills. The viticulture for wine production is favored due to the restriction of

Fig. 14.3 Contour lines 100 m of the Itata valley, scale 1:1.000.000. (Own elaboration using the available shapes from the "CEDEUS - Centre for Sustainable Urban Development UC - UdeC" for the Google Earth platform)

conditions for vine growth and development, resulting in naturally stressed vines that generate a higher concentration of secondary metabolites favoring the production of higher quality grapes with unique wines.

Considering the whole geographic area of Itata Valley, approximately 32% is covered by forests (mostly pine and eucalyptus plantations), 41% is used in agriculture, and 24% are grasslands and scrubs (Serra et al., 2017). Compared to the other Chilean wine valleys, the Itata Valley has the highest number of registered vineyards. The average vineyard size in the Itata Valley is 2.2 ha, while in the Maipo Valley, it is 32 ha (Serra et al., 2021). As the vineyards are very small, growers use the number of plants rather than the area planted to describe the size of the vineyard. The vineyards of the Itata Valley are among the oldest in Chile, with vines that can be several decades old since planting, up to 100 to 200 years old, cv. País (Mora-Penrose et al., 2020). The predominant cultivars in the Itata Valley, País, and Muscat of Alexandria, among others, are considered "heritage cultivars" since they played a pivotal role in the wine history of Chile. Wine production in Chile began with the País cultivar, brought by the Spanish conquerors.

The Itata Valley was very relevant for the early Chilean wine production during the colonial period. However, in the nineteenth century, french cultivars were introduced to Chile, which moved the production of high-quality wine to the Central Valley, the modern center for wine exportation in Chile. Currently, Chile is the fourth largest wine exporter in the world (OIV, 2022), generating US$1.69 billion (The Growth Lab at Harvard University, 2020). In this context, for the last 100 years, the Itata Valley has mostly produced fruit oriented to elaborate table wines for the local market. However, during the last decades, the Itata Valley has experienced a resurgence due to its potential to produce high-quality wines with a unique identity.

Over the past century, the development of viticulture and oenology, and the introduction of new technologies like fertilization, commercial yeast, and oak casks, have allowed "wine production" to become safer and more stable, and feasible, at the expense of the product's distinctiveness (Lukacs, 2012). Some Itata winemakers have found "natural winemaking" an interesting commercial niche that benefits from the characteristics of traditional agroecosystems. Although there is no single definition of a "natural wine," there is consensus that it promotes the inherent naturalness of the product by diversifying the winemaking processes in the vineyard and winery, focusing on social equity and environmental and economic sustainability (Fabbrizzi et al., 2021). In general, natural wines use native, unconventional yeasts that favor the synthesis of new volatile aromatic compounds in the fermentation process. In this context, the meteorological conditions of Itata would be favorable for the production of natural wines since the high pluviometry and relative humidity of air during harvest seems to stimulate the growth of native yeasts (*Metschnikowia*), capable of synthesizing new volatile aromatic compounds, and can help to produce sweet, tropical, and aromatic wines (Jara et al., 2016).

The vineyards of the Itata Valley have characteristics that make them very different from those of the Central Zone, where a large part of Chile's exported wines come from (Serra et al., 2021). Many of the cultural practices and technologies currently used in the Itata Valley are the same as those developed by the Spanish conquerors during the colonial period 200 years ago (Mora-Penrose et al., 2020). These practices include (1) the head training system, (2) dry farming, (3) manual harvesting, (4) low use of agrochemicals, (5) the use of indigenous yeasts for fermentations, and (6) traditional weed control by digging the soil under the vines (i.e., "cava" and "recava" in August and September, respectively). For example, the Itata Valley is dominated by rainfed viticulture (84%) and a head-trained vine training system (76%). At the same time, in the Central Valley (Valley of Maipo, Rapel, Curicó and Maule), vineyards are irrigated (92%) and mainly trained on trellises (>90%) (SAG, 2021).

Heroic viticulture, also known as Mountain Viticulture, is defined as a type of viticulture that occurs on land with more than 30% decline; altitude higher than 500 meters above sea level; on terraced land and viticulture or small islands in a structural and socioeconomic context penalized from the point of view of business profitability (CERVIM, 2004). The heroic viticulture of the Itata Valley represents a traditional type of agroecosystem characterized by internal regulation since the application of external inputs (irrigation, fertilizers, herbicides, fungicides, insecticides) is low. This type of agriculture has a high level of nitrogen and carbon recycling; weeds are usually controlled based on competition; and it is a suitable habitat for beneficial insects, among others (Peterson et al., 2018). Although traditional agroecosystems are often less productive than conventional ones, they are much more resilient and diverse. While Itata vineyards are traditionally rainfed, the availability of irrigation water is low at higher elevations. There is a strong relationship between water access difficulties and population poverty (WWAP, 2009). This is particularly important for the Itata Valley, as this viticultural area is considered a highly vulnerable zone ("Zona de rezago") due to its geographical isolation, low presence and coverage of public services, and high poverty indices (SUBDERE, 2019). The relationship between water availability and community development depends on (i) physical water scarcity, (ii) lack of access to water due to infrastructure or institutional frameworks, (iii) exposure to hazards related to water availability, such as drought or floods, and (iv) water productivity, which is low for dry-farmed agriculture (Cook et al., 2011).

In dry-farmed vineyards, water supplementation depends exclusively on rainfall, the water storage capacity of the soil and the presence of groundwater tables. At these latitudes, rainfall is concentrated in the winter months (June to September) and averages between 500 and 800 mm per year. The peculiar hilly topography of this valley has led many vineyards to be planted in areas with complex slopes (>15%). Under these topographic characteristics, surface, and subsurface movements of soil particles, by migration or leaching, in favor of the slope, have generated the presence of a sequence of soil profiles, with progressive variations in their physical properties from the interfluve to the valley floor (i.e., catena). In addition, applying cultural practices that involve tillage, such as weed control with draft animal ploughs, and their effect on creating the "plow foot" can create preferential water flows in favor of the slope (Dorner & Dec, 2008). The interaction of these edaphic and cultural factors has a major impact on the ease of movement and storage of water, gases, and nutrients in Itata vineyards established in a catena. Also, the high geomorphological variability, the ambient diversity in geological soil formation and the diversity of parent materials of the Itata valley have induced a considerable heterogeneity in pedogenic characteristics. These soils usually have clayey textures with slow water infiltration and topography of hills with complex slopes between 20% and 30%. It predisposes the soil to water erosion, loss of surface horizons and gully formation. Consequently, it is unsurprising that vineyards planted in a catena from the Itata valley exhibit high variability in vegetative and reproductive development. The variability of these vineyards is partially explained by a heterogeneous water status among vines during the season. The plant water status of the vineyard determines the cellular turgor and photosynthetic capacity that sustains the growth and development of each plant organ. Comparing the severity of water stress among vines planted across an Inceptisol catena in the Itata Valley (Footslope–summit difference in elevation = 11 m) (Fig. 14.4), Calderón-Orellana et al. (2022) showed no differences in plant water status, measured as midday stem water potential (Ψ_{SWP}), and between vigorous vines planted on the footslope and weak vines planted on the summit (Ψ_{SWP} of −0.9 MPa). However, the variability in Ψ_{SWP} of the summit was almost twice that of the footslope. Taylor et al. (2010) indicated that as soils dry, variability in plant water status increases, related to heterogeneity in soil physical properties and their ability to store water.

It is often assumed by the wine industry that higher variability in the vineyard can lead to lower wine quality. However, studies in California have shown that vineyards with high variability in maturity do not produce grapes of lower commercial value (Calderón-Orellana et al., 2014). In fact, modern winemaking practices consider blending fruit from vineyards with very different characteristics to produce "premium" wines. This practice aims to obtain a greater organoleptic complexity of

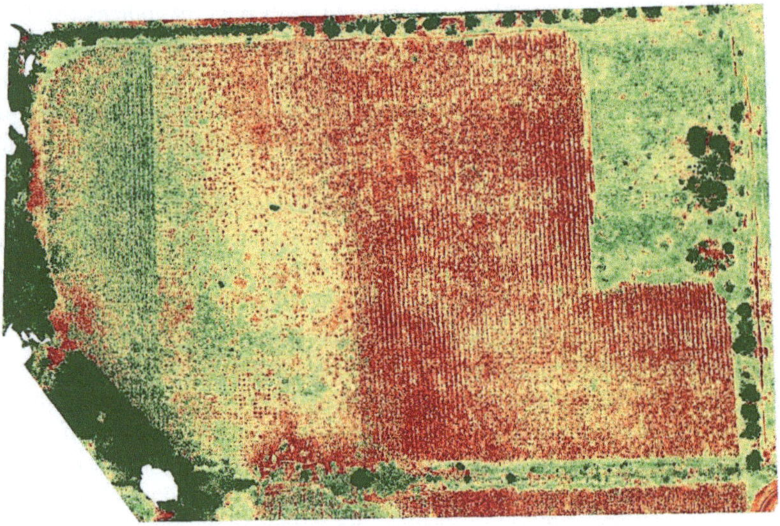

Fig. 14.4 Image of cv. País vineyard established in a catena in the Itata Valley with an illustration of the Visible Atmospheric Resistance Index (VARI) obtained from the drone deploy software. On the left side of the image, the green shades indicate higher vigor and a greater number of plants (Footslope); in the center of the image, the yellow shades indicate moderate-low vigor and a smaller number of plants (Backslope), on the right side of the image, the red shades represent very low vigor and a low number of plants (Summit) (Ortiz, 2023)

the wines, which suggests that the artificial introduction of different sources of variability could improve the quality of the wines. Therefore, the high variability of the Itata vineyards may be an advantage for obtaining wines with a distinctive character.

The head training is the oldest and most economical system used in viticulture to define plant architecture, where the vines support their weight and are kept at a short distance from the ground (Freeman et al., 1992). This training system, also known as the bush training or goblet, consists of a central trunk from which the spurs (two to three buds) or canes (more than four buds) are born. In the Itata Valley, grape growers often place the fruit zone at the ground level or between 0.5 and 1.0 m above the ground (Lobato et al., 2003) (Fig. 14.5).

Head training tends to be less productive than other training systems (Reynolds & Vanden Heuvel, 2009), mainly because the leaf area does not cover a large soil area. The yield per plant in head-trained Muscat of Alexandria vines under dry-farmed conditions in the Itata valley was 50% lower than those of the same cultivar located under irrigated conditions and with the pergola-type training (Puentes, 2019). In this context, the number of clusters and the individual cluster weight contributed similarly to the low yield per plant. The dryness of the vineyards, where the plants compete for the moisture available in the soil, is one of the factors that could explain the low yield of the Muscat of Alexandria vines in Itata. The soil water is likely exhausted a few weeks before harvest due to the low water-holding capacity of the soils of the Itata Valley and the high evaporation demand of the area, favored

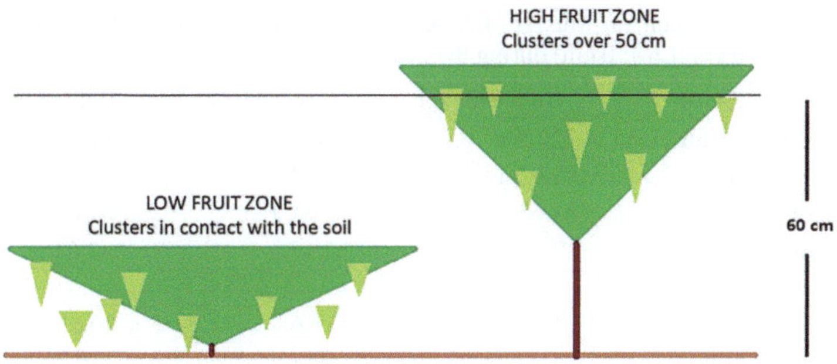

Fig. 14.5 Head training or Gobelet in vine plants conducted at two heights from the trunk (High: 50 cm; Low: 0 cm) (Ibarra et al., 2023)

Table 14.1 Yield components in cultivar Muscat of Alexandria vine plants conducted at two heights from the trunk (High: 50 cm; Low: 0 cm) during the 2017/2018 season

Yield estimates	High fruit zone	Low fruit zone	Reference values
Yield per vine (kg)	3.82 a	3.74 a	7.8*
Cluster number	16.1 a	15.7 a	34–50**
Cluster weight (g)	236.1 a	212.9 a	475***

*Muscat of Alexandria in overhead pergola (Agraria Sur)
**Flame Seedless (Lavin et al., 2003)
***Flame Seedless (Salazar, 2012)

by the high wind speeds under the canopy (van Zyl & van Huyssteen, 1980). Stem water potential values evaluated on País vines grown in a catena (Fig. 14.4) confirmed that rainfed vines were moderately water-stressed after harvest, which may affect berry differentiation (Matthews & Anderson, 1988). Although Muscat of Alexandria vines from the Itata Valley had a low cluster number per plant, the cluster weight was also low (Table 14.1). These results suggest that vine capacity, which determines the number of grapes that can adequately ripen on each plant, is probably low under the growing conditions of the Itata Valley.

The head training system was used by ancient grape growers to produce wines made with mature berries in cold climates, taking advantage of the proximity of clusters to the warm soil to accelerate berry ripening (Freeman et al., 1992). In South Africa, van Zyl and van Huyssteen (1980) compared different training systems and recorded the highest grape and soil temperatures on head-trained plants. In the Itata Valley, a study of head-trained Muscat of Alexandria vines under rainfed conditions showed that during the phenological stage of veraison, which initiates the grape ripening stage, the air temperature in the fruiting zone was between 30 and 35 °C for about 45% of the day (Pascual et al., 2017). Although harvest often occurs in late summer or fall, the technical maturity of berries exposed to warm air and soil temperatures usually ranges between 21 and 24 Brix. The fact that a variety such as Muscat of Alexandria does not reach technical over-ripeness values (>25 °Brix) when grown head-on and without irrigation would indicate that the lack of water and the high air temperatures would not be sufficient to cause dehydration of the

Table 14.2 Quality of cultivar Muscat of Alexandria vine plants conducted at two heights from the trunk (High: 50 cm; Low: 0 cm) during the 2017/2018 season

Berry quality attributes	High fruit zone	Low fruit zone	Reference values
Soluble solids concentration	20.5 a	20.7 a	17–20*
Polar diameter (mm)	1.93 a	1.73 a	>1.6*
Equatorial diameter (mm)	1.65 a	1.54 a	>1.6*
Berry volume (mm)	2.92 a	2.42 a	>2.6*
Color			
Chroma	15.5 a	13.6 a	16.3**
Hue	119.7 a	90.1 a	110.2**
CIRG	2.2 a	1.5 a	1.2*- 1.3**

*Export Standard Del Monte Fresh Pink Muscat
**Muscat of Alexandria (Carreño et al., 1996)
Adapted from Puentes (2019).

grapes and a significant increase in the concentration of soluble solids before harvest.

Results of this study showed that skin coloration for Muscat of Alexandria berries collected in the Itata Valley (CIRG between 1.5 and 2.2) was higher than the skin color standard for that cultivar (CIRG of 1.2) (Carreño et al., 1996). The CIRG values measured in the Itata valley were closer to those obtained in grape cultivars with pink pigmentation or more color, probably indicating the damage of "sunburn" in berry skins (Table 14.2).

14.2 Challenges of the Heroic Itata Valley

One of the most relevant silvicultural and livestock farming activities in the Ñuble Region is forestry plantations, which have some 277,653 ha (INFOR, 2020), of which 148,568 ha are in the communes that make up the Itata Valley. According to the Chilean Ecological Society (SOCECOL, 2023), the occurrence of forest fires is due to the combination of three main factors: (i) favorable weather conditions, (ii) ignition source, and (iii) fuel accumulation. Regarding this last point, a large part of the 10,736 ha of vineyards in this valley (SAG, 2021) are enclosed by forest plantations (Serra et al., 2017), which represents a high fire risk since forest plantations are made up of high flammable invasive species (SOCECOL, 2023). Forest fires have increased by 50% over the last 10 years (CONAF, 2018). In the summer of 2017, the largest fire in Chile's history occurred, affecting more than 518,000 ha (CONAF, 2017). These fires occurred mainly in the South Central Zone of the country (Fig. 14.6).

Six years later (summer of 2023), fires occurred in the South Central Zone of the country, covering an area of approximately 363 thousand ha (SENAPRED, 2023) and mainly affecting the Itata and Bio Bio Valleys. Numerous publications have shown that smoke from fires negatively affects wine quality (Kennison et al., 2009;

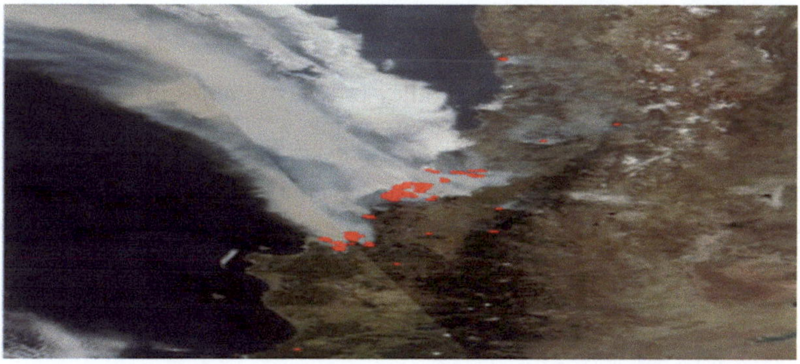

Fig. 14.6 NASA satellite image of the outbreaks of the mega-fire* (January 28, 2017) in south-central Chile (Pineda, 2020). *Red dots indicate the outbreaks of the mega wildfire

Table 14.3 Hedonic test of Cinsault wines before and after malolactic fermentation (MLF) from the Ránquil and Pinihue locations during the 2016–2017 season (Pineda, 2020)

Treatments	Acceptability Before MFL	Acceptability After MLF
Ránquil	6.17 b	6.33 b
Pinihue	4.67 a	4.61 a

Different letters within a column indicate a significant difference according to the Kruskal Wallis test ($P < 0.05$). The Ránquil wines came from vineyards far from the fires while the Pinihue vineyards were located next to one of the fire outbreaks

Krstic et al., 2015). Due to wood combustion, several volatile compounds are generated, including vinyl phenols (Fine et al., 2001), which deliver smoke aromas. Studies conducted on wines from the Itata Valley after the mega-fire of 2017 confirmed the presence of the vinyl phenols guaiacol, syringol, and 4-ethyl guaiacol and the negative effect on wine acceptance (Table 14.3) (Sabugo, 2019; Pineda, 2020).

A study of small producers in the Itata and Bio Bio Valleys showed that production activities are generally carried out individually (65%) rather than associatively (25%), which harms their ability to obtain certifications, form strategic alliances, and generate innovation. This contrasts with the situation of the medium and large wineries in the Central Valley, which are united through "Wines of Chile," an entity financed by the producers themselves and with support from the Chilean government that allows them to face challenges ranging from market access to certifications and R&D (Innovation and Development). In general, small producers in the Itata Valley have serious deficiencies in the marketing and/or commercialization of their products, and most of their wine production is destined for the domestic market (Serra et al., 2021).

14.3 Conclusions

The unique conditions that the Itata Valley presents in terms of geology, soil, climate, and wine management, as well as its people who have been making wine for generations, give it great potential to produce quality wines with identity. However, several challenges range from aspects related to climate change, such as forest fires, to socioeconomic, and associative gaps that make wine production and marketing difficult.

References

Calderon-Orellana, A., Matthews, M. A., Drayton, W. M., & Shackel, K. A. (2014). Uniformity of ripeness and size in Cabernet Sauvignon berries from vineyards with contrasting crop price. *American Journal of Enology and Viticulture, 65*(1), 81–88.

Calderón-Orellana, A., Serra-Stepke, I., Ortiz, I., Bustos, C., Correa, C., Cuevas, J., & Hermosilla, N. (2022). Vine capacity, abiotic stress severity, and wine composition of a non-irrigated vineyard along an Inceptisol catena in the Itata Valley. In 29th International applied geochemistry symposium IAGS2022. Linking geology and geochemistry to viticulture and wine, October 2022, Viña del Mar. p. 83.

Carreño, J., Martínez, A., Almela, L., & Fernández-López, J. A. (1996). Measuring the color of table grapes. *Color Research & Application, 21*(1), 50–54.

Casanova, M., Salazar, O., Seguel, O., & Luzio, W. (2013). *The soils of Chile* (Vol. 10, p. 185). Springer.

CEDEUS. (2015). *Centro de Desarrollo Urbano Sustentable*. Retrieved from http://datos.cedeus.cl/layers/

CERVIM. (2004). *Centro de Investigación, Estudio, Salvaguarda, Coordinación y Valorización de la Viticultura de Montaña CERVIM STATUTE*. Retrieved from https://www.cervim.org/d/146/statute-en.pdf

Charrier, R., Pinto, L., & Rodríguez, M. (2007). Tectonostratigraphic evolution of the Andean Orogen in Chile. In T. Moreno & W. Gibbons (Eds.), *The Geology of Chile* (The geological society) (pp. 21–114).

CONAF. (2017). *Análisis de la afectación y severidad de los incendios forestales, ocurridos en enero y febrero 2017 sobre los usos de suelo y los ecosistemas naturales presentes entre las regiones de Coquimbo y la Araucanía de Chile*. Informe Técnico. CONAF.

CONAF. (2018). *Estadísticas históricas*. Corporación Nacional Forestal, Chile. Retrieved from http://www.conaf.cl/incendios-forestales/incendios-forestales-en-chile/estadisticas-historicas.

Cook, S., Fisher, M., Tiemann, T., & Vidal, A. (2011). Water, food and poverty: Global- and basin-scale analysis. *Water International, 36*, 1–16. https://doi.org/10.1080/02508060.2011.541018

Dörner, J., & Dec, D. (2008). Efecto de la estructura sobre el movimiento de agua en una catena de suelos. *Agro Sur, 36*(2), 93–100.

Fabbrizzi, S., Alampi Sottini, V., Cipollaro, M., & Menghini, S. (2021). Sustainability and natural wines: An exploratory analysis on consumers. *Sustainability, 13*(14), 7645.

Fine, P. M., Cass, G. R., & Simoneit, B. R. T. (2001). Chemical characterization of fine particle emissions from fireplace combustion of woods grown in the northeastern United States. *Environmental Science & Technology, 35*(13), 2665–2675.

Freeman, B. M., Tassie, E., & Rebbechi, M. D. (1992). Training and trellising. *Viticulture, 2*, 42–65.

Grunwald, S. (Ed.). (2016). *Environmental soil-landscape modeling: Geographic information technologies and pedometrics*. CRC Press.

Hervé, F., Calderon, M., Fanning, C. M., Godoy, E., & Pankhurst, R. J. (2013). Provenance variations in the Late Paleozoic accretionary complex of Central Chile as indicated by detrital zircons. *Gondwana Research, 3*(2), 1122–1135.

Ibarra, K., Serra, I. M., Peña-Neira, Á., Bambach, N., Puentes, P., & Calderón-Orellana, A. (2023). Sunburn and its Relation to Maturity and Concentration of Aromatic Compounds in Bush-Trained Muscat of Alexandria Vines. *American Journal of Enology and Viticulture, 74*(2).

IDE. (2016). *Infraestructura de datos Geoespaciales*. Retrieved from https://www.ide.cl/index.php/informacion-territorial/descargar-informacion-territorial

INFOR. (2020). *Superficie de plantaciones forestales Región de Ñuble*. Retrieved from https://ifn.infor.cl/index.php/informacion-regional/region-de-nuble

Jara, C., Laurie, V. F., Mas, A., & Romero, J. (2016). Microbial terroir in Chilean valleys: Diversity of non-conventional yeast. *Frontiers in Microbiology, 7*, 663.

Kennison, K. R., Wilkinson, K. L., Pollnitz, A. P., Williams, H. G., & Gibberd, M. R. (2009). Effect of timing and duration of grapevine exposure to smoke on the composition and sensory properties of wine. *Australian Journal of Grape and Wine Research, 15*(3), 228–237.

Krstic, M. P., Johnson, D. L., & Herderich, M. J. (2015). Review of smoke taint in wine: Smoke-derived volatile phenols and their glycosidic metabolites in grapes and vines as biomarkers for smoke exposure and their role in the sensory perception of smoke taint. *Australian Journal of Grape and Wine Research, 21*, 537–553.

Lavín, A., Lobato., A., Muñoz., y Valenzuela, J. (2003). Viticultura. Poda de la vid Boletín INIA°99. INIA Cauquenes, Cauquenes, Chile.

Lobato, A., Lavín, A., Muñoz, I., & Valenzuela, B. (2003). Viticultura: Poda de la Vid. Boletín INIA - Instituto de Investigaciones Agropecuarias. N° 99. Retrieved from: https://hdl.handle.net/20.500.14001/6975

Lukacs, P. (2012). *Inventing wine: A new history of one of the world's most ancient pleasures*. WW Norton & Company.

Matthews, M. A., & Anderson, M. W. (1988). Fruit ripening in *Vitis vinifera* L.: Responses to seasonal water deficits. *American Journal of Enology and Viticulture, 39*, 313–320.

Mora-Penrose, Z., Arriagada-Cortés, F., & Serra-Stepke, I. (2020). *Chile se hace Vino en el Itata: La enología ancestral de un valle único*. Book. 278 p.

OIV. (2022). *State of the world vine and wine sector in 2022*. Retrieved from https://www.oiv.int/sites/default/files/documents/OIV_State_of_the_world_Vine_and_Wine_sector_in_2022_2.pdf

Ortiz, I. (2023). Comparación de parámetros de desarrollo vegetativos en plantas y variables como predictores del rendimiento en vides de secano conducidas en cabeza a lo largo de una catena. Undergraduate Thesis, Ingeniero Agrónomo. Universidad de Concepción, Facultad de Agronomía. Chillán, Chile.

Pascual, G. A., Serra, I., Calderón-Orellana, A., Laurie, V. F., & Lopéz, M. D. (2017). Changes in concentration of volatile compounds in response to defoliation of Muscat of Alexandria grapevines grown under a traditional farming system. *Chilean Journal of Agricultural Research, 77*(4), 373–381.

Peterson, C. A., Eviner, V. T., & Gaudin, A. C. (2018). Ways forward for resilience research in agroecosystems. *Agricultural Systems, 162*, 19–27.

Pineda, N. (2020). Efectos del mega incendio forestal del 2017 en la composición química y características sensoriales del vino cv. Cinsault. Undergraduate Thesis, Ingeniero Agrónomo. Universidad de Concepción, Facultad de Agronomía. Chillán, Chile.

Puentes, P. (2019). Aptitud del cultivo Moscatel de Alejandría del Valle del Itata para la producción comercial de uva de mesa. Undergraduate Thesis, Ingeniero Agrónomo. Universidad de Concepción, Facultad de Agronomía. Chillán, Chile.

Reynolds, A. G., & Vanden Heuvel, J. E. (2009). Influence of grapevine training systems on vine growth and fruit composition: A review. *American Journal of Enology and Viticulture, 60*(3), 251–268.

Sabugo, D. F. (2019). Efecto de la quema de distintas maderas (pino, eucalipto y hualle) en las características organolépticas y químicas del vino cv. Cinsault. Undergraduate Thesis, Ingeniero Agrónomo. Universidad de Concepción, Facultad de Agronomía. Chillán, Chile.

SAG. (2021). Catastro vitícola nacional 2021. División de Protección Agrícola-Forestal y Semillas Subdepartamento de Viñas, Vinos y Bebidas Alcohólicas. Servicio Agrícola y Ganadero.

Salazar, D. (2012). Descripción de los componentes del rendimiento de uva de mesa variedad Flame Seedless. Memoria de título, Ingeniera Agrónoma, Mención Fruticultura. Universidad de Chile, Facultad de Ciencias Agronómicas. Santiago, Chile.

SENAPRED. (2023). *Informe de Emergencia "Consolidado de incendios forestales relevantes"*. Servicio Nacional de Prevención y Respuesta ante Desastres. Retrieved from https://senapred.cl/incendios-forestales-relevantes/

SERNAGEOMIN. (2003). *Mapa Geológico de Chile: Versión digital*. Servicio Nacional de Geología y Minería, Publicación Geológica Digital, N°. 004. Santiago, Chile. Retrieved from: https://www.sernageomin.cl/visor-de-mapas/

Serra, I., Castillo, P., Hidalgo, M., Fuentealba, O., Cerda, R., & Salazar, F. (2017). Itata Valley: How traditional farming and geomorphological characteristics can became from a limiting factor to key factor in competitiveness. In Fifth International Congress on Mountain and Steep Slope Viticulture. March 31–April 01, 2017. Centre for Research, Environmental Sustainability and Advancement of Mountain Viticulture. Conegliano, Italy. pp. 19–27.

Serra, I., Cerda, R., Aguilera, S., & Díaz, N. (2021). Economical characterization of small wine producers from three wine regions from Chile: Challenges for the future. In R. Vasconcelos Botelho & A. M. Jordao (Eds.), *Vitis products composition, health benefits and economic valorization* (p. 440). Nova Science Publishers, Inc.

SOCECOL. (2023). *Declaración pública de la Sociedad de Ecología de Chile frente a los incendios forestales catastróficos que nuevamente afectan al sur de Chile*. Sociedad de Ecología de Chile. Retrieved from: https://www.socecol.cl/wp-content/uploads/2023/02/declaracion-socecol-2023.docx-2.pdf

Stolpe, N. (2006). *Descripciones de los principales suelos de la VIII Región de Chile*. Publicaciones del Departamento de Suelos y Recursos Naturales 1. Trama Impresores S.A. Publicaciones - Departamento de Suelos y Recursos Naturales. N° 01. Retrieved from: https://hdl.handle.net/20.500.14001/56740

SUBDERE. (2019). *Reglamento que fija la Política Nacional sobre Zonas Rezagadas en Materia Social*. Subsecretaría de Desarrollo Regional y Administrativo. Retrieved from: http://territoriosdeconvergencia.subdere.gov.cl/files/doc_zonas_rezagadas/Presentaci%C3%B3n%20Reglamento%20Pol%C3%ADtica%20Nacional%20Zonas%20Rezagadas%20en%20Materia%20Social.pdf

Taylor, J. A., Acevedo-Opazo, C., Ojeda, H., & Tisseyre, B. (2010). Identification and significance of sources of spatial variation in grapevine water status. *Australian Journal of Grape and Wine Research, 16*(1), 218–226.

The Growth Lab at Harvard University. (2020). *The Atlas of Economic Complexity*. Retrieved from: http://www.atlas.cid.harvard.edu

Van Zyl, J. L., & Van Huyssteen, L. (1980). Comparative studies on wine grapes on different trellising systems: I. Consumptive water use. *South African Journal of Enology and Viticulture, 1*, 7–14.

Willner, A. P., Hervé, F., Kröner, A., Thomson, S. N., Wartho, J. A., & Wijbrans, J. R. (2005). Time markers for the evolution and exhumation history of a late Paleozoic paired metamorphic belt in north-Central Chile (34°– 35°30'S). *Journal of Petrology, 46*, 1835–1858.

WWAP. (2009). *The United Nations World Water Development Report 3 (WWDR3). Water in a changing world*. World Water Assessment Programme, United Nations Educational, Scientific and Cultural Organization (UNESCO). Retrieved from: http://unesdoc.unesco.org/images/0018/001819/181993e.pdf#page=5 (Accessed July 2023).

Chapter 15
Concluding Remarks and Future Directions of Latino America Vitiviniculture

Mercedes Fourment and Gastón Gutierrez Gamboa

15.1 Overview of the Book's Content

A diversity of knowledge characterizes and determines Latin American viticulture. Latin American vitiviniculture is currently taking into account social and historical aspects in its development to create a true identity and to diversify the range of products for the consumer. These efforts have been generated from the association of different actors, such as the public sector, the private sector, producers, universities, and the academic world, who seek to promote initiatives that promote equity in the sector. Based on the show, it was evidenced that the Anglo-Saxon European scientific journals related to vine and wine sciences encourage a high degree of discrimination related to the origin of their editorial board members compared to Revista Iberoamericana de Viticultura, Agroindustria y Ruralidad (RIVAR). This Chilean journal presented the highest diversity in the number of editors, reaching 53.8% of Latino Americans and 46.2% of the rest of the world members, comparing the most important journals related to vine and wine sciences. Notably, the editorial board of RIVAR presents an equitable incorporation of gender in the editorial board, reaching the same number of male and female editors. The scientific world does not widely discuss this discrimination related to genre and origin in access to science. Still, they are now a matter of concern not only for science but also for international guidelines for promoting development and the medical sector (Eliason et al., 2010; Amrein et al., 2011; Higa et al., 2014; Flanagin et al., 2021). Similar to those observed in the vine and wine sciences journals, Shim et al. (2021) confirmed that

M. Fourment (✉)
Facultad de Agronomía, Universidad de la República, Montevideo, Uruguay
e-mail: mfourment@fagro.edu.uy

G. Gutierrez Gamboa
Instituto de Investigaciones Agropecuarias, INIA Carillanca, Vilcún, Chile

G. Gutiérrez Gamboa, M. Fourment (eds.), *Latin American Viticulture Adaptation to Climate Change*, https://doi.org/10.1007/978-3-031-51325-1_15

the race and ethnicity of editorial board members and editors indicate structural racism in Psychiatry and Neuroscience journals.

The book's chapters reflect and exalt Latin American viticulture from various angles. On the one hand, an approach to the historical development and local paradigms is shown, exploring Latin American viticulture through different paradigms and highlighting the influence of colonial, French, and Anglo-Saxon models on the wine industry (Villanueva, 2017; Lacoste, 2019). Through it, the historical trajectory and the evolution of the identity of Latin American wines can be elucidated, which determines the market positioning and consumption patterns. Since the introduction of the first cuttings in the American colonial period, viticulture in Latin America has greatly evolved. In less than two centuries, the wine industry went from the pre-industrial towards the industrial model and later to the technological, scientific model. These revolutions in the wine industry were carried out under the idea of progress and development, which allowed to marginalize the knowledge acquired during the first stage of the history of Latin American wine. The ceremonial character of colonial viticulture was also highlighted, which was in the hands of women. Colonization processes resulted in the process of invisibility and dispossession of knowledge of women that were important in pre-hispanic societies.

Sustainability and global changes are another issue explored throughout the book. The sustainability of wine production systems and the current challenges of environmental problems and the context of climate change generate the need to focus on national sustainability frameworks. Countries such as Argentina, Chile, Brazil, and Uruguay have a long way to go in this regard. This requires investing in sustainable practices, offering insights into achieving a sustainable competitive advantage and improving overall business performance. Examples of the seal of certification valid for the Sustainability Code of Chile, Argentina, and Uruguay are presented in Fig. 15.1. An interesting case is the one that has been developed in Chile and has united together most of the Chilean wine companies (Marola et al., 2020; Aguilera et al., 2022; Valenzuela et al., 2022). The Chilean Sustainability Code is a voluntary instrument that aims to guide the Chilean wine sector towards sustainable wine production based on high social, environmental, and quality standards and to motivate grape growers and winemakers to improve their management

Fig. 15.1 Seal of certification valid for the Sustainability Code of Chile (left), Argentina (middle), and Uruguay (right)

by complying with the requirements stipulated in the standard (Wines of Chile, 2015). To comply with the requirements of the Code, wineries must have an environmental and social management system that complies at least with current national legislation, regardless of the complexity of their operations (Wines of Chile, 2015). There are different areas, such as green (vineyard management), orange (facilities), red (winemaking), and purple (wine tourism), which have defined roles and act independently. The governance, updating and administration of the Code is the responsibility of "Wines of Chile," through the Wines of Chile R&D Consortium. Eighty-four wine companies (more than 80% of exported wine) have adhered to this code, substantially improving their winemaking practices. Certified wineries can sell their wine using a distinctive seal that lets consumers know that the bottle of wine was produced under the development of environmentally sustainable practices.

However, this is a critical issue for small and medium-sized producers, which usually promote internal initiatives for the development and positioning of their wines based on the participation of experts and political authorities (Fig. 15.2).

The pesticide applications in plant production cause an increase of heavy metal content in soils (Schwalbert et al., 2021). In humid subtropical climates, such as in the South of Brazil and Uruguay, fruit crops receive several fungicide applications during the growing season that can increase heavy metals accumulation in soils, mostly those belonging to the chemical group of dithiocarbamates and the inorganic Cu-based ones, which are used for the control of downy mildew (*Plasmopara viticola*) in orchards (Brunetto et al., 2017). High contents of heavy metals in soils induce great damage in the roots, leading to important changes in morphology and anatomy (Ambrosini et al., 2018). Some plant species activate specific strategies to shrink the acquisition and the allocation of the metals to the shoot (Ambrosini et al.,

Fig. 15.2 Sensory evaluation of the organic and conventional wines produced by the "Asociación Gremial Viñateros del Valle del Itata A.G." (**a**) Recognition to Viña Männle. (**b**) Sensory evaluation performed by academics, journalists, and specialists in the sector. (**c**) Closing photography with producers, specialists, and regional authorities

2018), which could be a useful input for future plant breeding. Several approaches have been proposed to heavy metal management, such as limestone, organic compost, and use of cover crops, among others (Cesco et al., 2021), which could improve the sustainability of grape production in these regions. Based on the difficulties above-mentioned, Brazil has developed a serious breeding program to provide the industry plant material adapted to subtropical and tropical viticulture, and the future perspective is to continue its breeding program to provide high-quality plant material for the international market (Gazolla et al., 2020; de Carvalho et al., 2023; de Leao & de Carvalho, 2023; de Oliveira et al., 2023; Verslype et al., 2023).

Book's chapters also discussed climate change and adaptation using sustainable practices. There are a variety of responses to climate change adaptation strategies that winegrowers can implement in the small, long, and medium term. Some viticultural strategies to face global warming effects on the wine industry were provided in terms of plant material selection (varieties, rootstocks and clones), vineyard design (row orientation, training systems, and pruning strategies), canopy management (late winter pruning, double cropping and forcing bud growth), strategies to decrease canopy temperature (shading nets, agrivoltaism, and water sprinkling), and strategies to limits sink to source ratio (apical leaf removal and severe shoot trimming). Besides, the study of enological strategies for this goal was also discussed in terms of must dilution (blending musts with juice from low-sugar grapes early harvested), microbiological strategies (use of unconventional or native yeast strains that reduce the efficiency of ethanol production), and post-fermentation technologies (blending high and low alcohol wines, and membrane-based techniques with physical removal of alcohol after alcoholic fermentation processes). Over the last few decades, the introduction and spread of recognized European grapevine varieties have caused a massive loss of minority and autochthonous grapevine varieties traditionally grown in several wine-growing regions (Balda & Martínez de Toda, 2017; Gutiérrez-Gamboa & Moreno-Simunovic, 2019; Sargolzaei et al., 2021). The exception to the rule is Georgia, which has been able to maintain its ancestral practices and preserve its genetic material for centuries (Imazio et al., 2013; De Lorenzis et al., 2015; Maghradze et al., 2014, 2015, 2020). The disappearance of a large number of old grapevine varieties and the varietal homogenization of the vineyards entail an increase in genetic vulnerability to biotic and abiotic stress against which the cultivated varieties are not resistant (Balda & Martínez de Toda, 2017). South American viticulture holds a valuable diversity of grapevine genetic material adapted to extreme desert conditions, including unknown vine genotypes, minority, and criolla varieties (Milla-Tapia et al., 2013; Franck et al., 2020). Some of them are Uva Anís (Listán Prieto × Moscatel de Alejandría), Picanta (Moscatel de Alejandría × unknown), Moscatel Amarillo (Listán Prieto × Moscatel de Alejandría), Canela (Listán Prieto × Muscat à Petit Grains), Moscatel Rosada (Moscatel de Alejandría × unknown), Blanca Ovoide (Moscatel de Alejandría × unknown), among others that are under study by Argentinean and Chilean researchers.

Climate change has brought a series of changes in viticulture, considerably affecting the distribution of grapevine varieties in different wine growing regions of the world (Jones et al., 2010; Nesbitt et al., 2022). The rise in annual air temperature has significantly expanded the distribution of grapevines to the poles and accelerated the growth cycle of grapevines in cool climate zones, such as Finland, Poland, Romania, Hungary, Serbia, Slovenia, Slovakia as well as in Latino America as can be observed in south zones of Argentina and Chile (Lisek, 2008; Baduca Campeanu et al., 2012; Vršič & Vodovnik, 2012; Vršič et al., 2014; Ruml et al., 2016; Kovács et al., 2018; Ivanišević et al., 2019; Maciejczak & Mikiciuk, 2019; Karvonen, 2020; Cabré & Nuñez, 2020; Bernáth et al., 2021; Verdugo-Vásquez et al., 2023). The vineyard plantation in Chile is moving to the south and vineyard surface in the Araucanía Region (37°35′ to 39°37′ South Latitude) increased by 953% from 2003 to 2020 (Gutiérrez-Gamboa et al., 2021), whereas Argentina's viticulture is expanding towards Southern latitudes in which Patagonia emerged as the southernmost wine-growing region in the world. However, Patagonia is one of the windiest regions and some guidelines about this subject were addressed in the book.

A little-known aspect technocrats consider is the cultural heritage and regional development of Latin American viticulture. The historical exploration of wine-growing regions in Chile, such as Pica, Maule, and Itata emphasizes the importance of preserving cultural heritage and the role it plays in territorial development (Pascual et al., 2017; Reyes Muñoz & Lavín Acevedo, 2022; Gutiérrez-Gamboa et al., 2023) (Fig. 15.3). The valorization of local plant material such as *criolla* varieties and the identification of collective *savoir-faire* in terroir delimitations linked to wine tourism are also fundamental pieces in Latin American regional development (Gabardo & Valduga, 2019; García-Rodea et al., 2022), which is usually poorly understood in other continents. The French paradigm promoted by some national elites in South America led to the disappearance of protected products and the import of European products with origin appellation (Castro et al., 2016; Lacoste et al., 2016). The technocrats at national industry service, admirers of the French paradigm, questioned the quality of the vineyards and wines produced in these traditional areas (Jerković et al., 2022). A strong narrative was built to denigrate the viticulturists to keep their traditional methods of grape and wine production from the prestige of universities, business associations, newspapers, and specialized

Fig. 15.3 Vestiges of winemaking materials preserved from the eighteenth century in Chile. **(a)** Barrels preserved in "Hacienda Cucha Cucha" which was property of the Society of Jesus (Itata Valley). **(b)** Wine press "Lagar de Matilla", whose construction is a national monument of Chile (Matilla Oasis)

journals (Hernández & Moreno, 2011; Jerković et al., 2022). Based on this dominant discourse, the wine industry should favor decisions based on market criteria in order to increase sales and profits, leading to a decrease of the surface of traditional vineyards from 60,000 to 15,000 ha (Jerković et al., 2022). Despite this pressure, some producers have maintained their vineyards and continue to make their own wines, such as the "asoleado" in Maule and Itata valleys, the "pajarete" in Huasco and Elqui valleys, the "pintatani" in Codpa valley, and the "pipeño" from the drylands of Maule, Ñuble, and Biobío regions (Castro et al., 2016; Lacoste et al., 2016; Gutiérrez-Gamboa et al., 2023).

15.2 Future Directions

Based on a deep understanding of the reality of Latin American vitiviniculture, it is important to consider certain key points to promote the sector's development in the region that is currently in crisis due to the economic consequences derived from the COVID-2019 pandemic, mainly by the decrease in exports to China. In this fashion, it is of wide importance to promote sustainable viticultural practices in grape and wine production in Latin America with the exception of Bolivia due to its rational particularities to develop this industry. Sensitizing viticulturists and winemakers can be achieved through the promotion of the current and future advantages of sustainable viticulture systems. To adopt these techniques, the population must be educated, resources must be available, and incentives must be provided by public and private institutions. Certification programs and labeling initiatives that are recognizably environmentally friendly and socially responsible for the wine sector must be developed and provided with equity. Recently, the "Asociación Nacional de Ingenieros Agrónomos Enólogos de Chile" promoted an online seminar to take the subject of regenerative viticulture. In this event, Chilean, French, Spanish, and American specialists will talk about their experiences.

An elemental issue to favor the vitiviniculture sector in Latin America is to foster innovation and local research through public and private investments for the specific challenges of each country in the region. In this sense, the in-depth study of the impacts of climate change and variability on viticulture, the development of adaptation, and mitigation responses to these changes in specific regions as well as exploration of the potential of criolla plant material could be lines of research that would have a cultural and economic impact on the development of the sector. To this end, it is key to foster collaborations between academics, industry professionals, and public policymakers to lead the exchange of research and innovation from a horizontal viewpoint of knowledge. In addition to the above, it is important to emphasize strengthening regional collaborations to facilitate knowledge sharing. Thus, promoting regional cooperation in areas such as marketing, sustainable practices, and research are key points in this regard. Basically, working together leverages collective knowledge and can help to enhance the resiliation of Latin American vitiviniculture to the current socioeconomic and environmental changes.

To conclude, the most valuable treasure is our cultural wine heritage that can be seen throughout the continent. The recognition and preservation of this material and immaterial heritage associated with traditional winemaking methods, local grape varieties, and historic wine regions make up a proper identity. Thus, initiatives that promote the protection and valorization of this heritage should be promoted, providing better opportunities for the ancestral and local communities.

15.3 Concluding Remarks

The book serves as a valuable contribution to the understanding and appreciation of Latin American vitiviniculture, highlighting its rich diversity, its ongoing efforts towards sustainability, and the importance of preserving its cultural heritage. By acknowledging these factors and working together to address challenges, Latin American viticulture can continue to thrive and gain recognition on the global stage.

References

Aguilera, P., Ortiz, N., Becerra, N., Turrini, A., Gaínza-Cortés, F., Silva-Flores, P., Aguilar-Paredes, A., Romero, J. K., Jorquera-Fontena, E., de La Luz Mora, M., & Borie, F. (2022). Application of arbuscular mycorrhizal fungi in vineyards: Water and biotic stress under a climate change scenario: New challenge for Chilean grapevine crop. *Frontiers in Microbiology, 13*, 826571.

Ambrosini, V. G., Rosa, D. J., Bastos de Melo, G. W., Zalamena, J., Cella, C., Simão, D. G., da Silva, L. S., dos Santos, H. P., & Brunetto, G. (2018). High copper content in vineyard soils promotes modifications in photosynthetic parameters and morphological changes in the root system of "Red Niagara" plantlets. *Plant Physiology and Biochemistry, 128*, 89–98.

Amrein, K., Langmann, A., Fahrleitner-Pammer, A., Pieber, T. R., & Zollner-Schwetz, I. (2011). Women underrepresented on editorial boards of 60 major medical journals. *Gender Medicine, 8*(6), 378–387.

Baduca Campeanu, C., Beleniuc, G., Simionescu, V., Panaitescu, L., & Grigorica, L. (2012). Climate change effects on ripening process and wine composition in Oltenia's vineyards from Romania. *Acta Horticulturae, 931*, 47–54.

Balda, P., & Martínez de Toda, F. (2017). Variedades minoritarias de vid en La Rioja. Consejería de Agricultura, Ganadería y Medio Ambiente. Gobierno de La Rioja.

Bernáth, S., Paulen, O., Šiška, B., Kusá, Z., & Tóth, F. (2021). Influence of climate warming on grapevine (*Vitis vinifera* L.) phenology in conditions of Central Europe (Slovakia). *Plants, 10*, 1020.

Brunetto, G., Ferreira, P. A. A., Melo, G. W., Ceretta, C. A., & Toselli, M. (2017). Heavy metals in vineyards and orchard soils. *Revista Brasileira de Fruticultura, 39*, 263–274.

Cabré, F., & Nuñez, M. (2020). Impacts of climate change on viticulture in Argentina. *Regional Environmental Change, 20*, 12.

Castro, A., Pszczólkowski, P., Mujica, F., Lacoste, P., Núñez, E., Cofré, C., Adunka, M. L., & Soto, N. (2016). Pajarete of Huasco and Elqui (Chile). Story of a selected wine. *Idesia, 34*(4), 23–35.

Cesco, S., Pii, Y., Borruso, L., Orzes, G., Lugli, P., Mazzetto, F., Genova, G., Signorini, M., Brunetto, G., Terzano, R., Vigani, G., & Mimmo, T. (2021). A smart and sustainable future for viticulture is rooted in soil: How to face Cu toxicity. *Applied Sciences, 11*(3), 907.

de Carvalho, J. N., de Carvalho, P. A., Pio, R., Barbosa, M. A. G., & de Souza Leão, P. C. (2023). Multitrait selection in seedless grape hybrids in semiarid regions of Brazil. *Crop Science, 63*(4), 2091–2102.

de Leao, P. C. S., & de Carvalho, J. N. (2023). Assessment of table grape progenies and correlation between seedlessness and other agronomic traits. *Revista Brasileira de Fruticultura, 45*, e-225.

De Lorenzis, G., Chipashvili, R., Failla, O., & Maghradze, D. (2015). Study of genetic variability in Vitis vinifera L. germplasm by high-throughput Vitis18kSNP array: The case of Georgian genetic resources. *BMC Plant Biology, 15*, 154.

de Oliveira, C. R. S., da Silva, F. B., Felinto Filho, E. F., de Mendonça Júnior, A. F., Ulisses, C., & de Leao, P. C. S. (2023). The influence of rootstock on vigor and bud fertility of BRS Tainá grape in the São Francisco Valley. *Revista Brasileira de Fruticultura, 45*, e-103.

Eliason, M. J., Dibble, S., & DeJoseph, J. (2010). Nursing's silence on lesbian, gay, bisexual, and transgender issues: The need for emancipatory efforts. *Advances in Nursing Science, 33*(3), 206–218.

Flanagin, A., Frey, T., & Christiansen, S. L. (2021). Manual of style committee. Updated guidance on the reporting of race and ethnicity in medical and science journals. *Journal of the American Medical Association, 326*(7), 621–627.

Franck, N., Zamorano, D., Wallberg, B., Hardy, C., Ahumada, M., Rivera, N., et al. (2020). Contrasting grapevines grafted into naturalized rootstock suggest scion-driven transcriptomic changes in response to water deficit. *Scientia Horticulturae, 262*, 109031.

Gabardo, W. O., & Valduga, V. (2019). Cultural systems and the Brazilian winescapes: Resources for wine tourism. *Estudios y Perspectivas en Turismo, 28*(3), 759–779.

García-Rodea, L. F., Thomé-Ortiz, H., & de Alcântara Bittencourt, P. C. (2022). Viticulture and wine tourism in Vale dos Vinhedos, Brazil: A governance study. *Revista Iberoamericana de Viticultura, Agroindustria y Ruralidad, 9*(27), 39–56.

Gazolla, R., Gründling, R. D. P., & Aragao, A. A. (2020). A produção e o comércio internacional de uva. *Revista Brasileira de Agrotecnologia, 10*(3), 68–74.

Gutiérrez-Gamboa, G., & Moreno-Simunovic, Y. (2019). Terroir and typicity of Carignan from Maule Valley (Chile): The resurgence of a minority variety: Terroir and typicity of Carignan. *OENO one, 53*(1), 75–93.

Gutiérrez-Gamboa, G., Zheng, W., & Martínez de Toda, F. (2021). Strategies in vineyard establishment to face global warming in viticulture: A mini review. *Journal of the Science of Food and Agriculture, 101*, 1261–1269.

Gutiérrez-Gamboa, G., Contreras Cortez, V., Jara, S., Verdugo-Vásquez, N., & Pszczólkowski, P. (2023). Andean viticulture: Edaphoclimatic characterization of vineyards located in the Codpa Valley. *Revista Iberoamericana de Viticultura, Agroindustria y Ruralidad, 10*(28), 26–40.

Hernández, A., & Moreno, Y. (2011). *The origins of Chilean wine*. Curicó, Maule, Itata, Biobío. Origo Ediciones.

Higa, D., Hoppe, M. J., Lindhorst, T., Mincer, S., Beadnell, B., Morrison, D. M., Wells, E. A., Todd, A., & Mountz, S. (2014). Negative and positive factors associated with the well-being of lesbian, gay, bisexual, transgender, queer, and questioning (LGBTQ) Youth. *Youth & Society, 46*(5), 663–687.

Imazio, S., Maghradze, D., Lorenzis, G., Bacilieri, R., Laucou, V., This, P., et al. (2013). From the cradle of grapevine domestication: Molecular overview and description of Georgian grapevine (*Vitis vinifera* L.) germplasm. *Tree Genetics and Genomes, 9*, 641–658.

Ivanišević, D., Kalajdžić, M., Alesandar, T., & Jakšić, D. (2019). Phenological stages of some grapevine cultivars in North Serbia: Historical data and current state. *BIO Web of Conference, 15*, 01023.

Jerković, M., Skewes, J. C., Duhart, F., Mujica, F., & Lacoste, P. (2022). Chilean heritage vineyards and natural wines: The mestizo Gothic-Mapuche legacy. *Revista Austral De Ciencias Sociales, 42*, 349–371.

Jones, G. V., Hall, A., Duff, A., & Myers, J. (2010). Spatial analysis of climate in winegrape growing regions in the western United States. *American Journal of Enology and Viticulture, 61*(3), 313–326.

Karvonen, J. (2020). Changes in the grapevine's growth cycle in Southern Finland in the 2000s—Comparison between two first decades. *Climate Change, 6*, 94–99.

Kovács, E., Puskás, J., Pozsgal, A., & Kozma, K. (2018). Shift in the annual growth cycle of grapevines (*Vitis vinifera* L.) in West Hungary. *Applied Ecology and Environmental Research, 16*, 2029–2042.

Lacoste, P. (2019). El Tratado de la vid de Louis Dussieux y el Tratado del Vino de Jean Antoine Chaptal de 1796 conservados en la Biblioteca de San Millán. *Revista Iberoamericana de Viticultura, Agroindustria y Ruralidad, 6*(17), 129–131.

Lacoste, P., Castro, A., Rendón, B., Pszczólkowski, P., Soto, N., Adunka, M. L., Jeffs, J., Salas, A., Cofré, C., & Núñez, E. (2016). "Asoleado" ("Sunny") of Cauquenes and Conception: The rise and fall of a Chilean wine with Appellation of Origin. *Idesia, 34*(1), 85–99.

Lisek, J. (2008). Climatic factors affecting development and yielding of grapevine in Central Poland. *Journal of Fruit and Ornamental Plant Research, 16*, 285–293.

Maciejczak, M., & Mikiciuk, J. (2019). Climate change impact on viticulture in Poland. *International Journal of Climate Change Strategies and Management, 11*(2), 254–264.

Maghradze, D., Vashakidze, L., Abashidze, E., Chipashvili, R., Mdinaradze, I., Failla, O., et al. (2014). Multidisciplinary study of traditional grape cultivars from Kartli province of Georgia (the Caucasus region) and activities for their preservation. *Acta Horticulturae, 1032*, 235–241.

Maghradze, D., Salimov, V., Melyan, G., Musayev, M., Ocete, C. A., Chipashvili, R., et al. (2015). Sanitary status of the Eurasian wild grapevine in the South Caucasian region. *Vitis, 54*, 203–205.

Maghradze, D., Melyan, G., Salimov, V., Chipashvili, R., Iñiguez, M., Puras, P., et al. (2020). Wild grapevine (*Vitis sylvestris* C.C.Gmel.) wines from the Southern Caucasus region. *OENO one, 54*, 809–822.

Marola, E., Schöpfner, J., Gallemore, C., & Jespersen, K. (2020). The bandwidth problem in telecoupled systems governance: Certifying sustainable winemaking in Australia and Chile. *Ecological Economics, 171*, 106592.

Milla-Tapia, A., Gómez, S., Moncada, X., León, P., Ibacache, A., Rosas, M., et al. (2013). Naturalised grapevines collected from arid regions in Northern Chile exhibit a high level of genetic diversity. *Australian Journal of Grape and Wine Research, 19*, 299–310.

Nesbitt, A., Dorling, S., Jones, R., Smith, D. K., Krumins, M., Gannon, K. E., Dorling, L., Johnson, Z., & Conway, D. (2022). Climate change projections for UK viticulture to 2040: A focus on improving suitability for Pinot noir. *OENO one, 56*(3), 69–87.

Pascual, G. A., Serra, I., Calderón-Orellana, A., Laurie, V. F., & Lopéz, M. D. (2017). Changes in concentration of volatile compounds in response to defoliation of Muscat of Alexandria grapevines grown under a traditional farming system. *Chilean Journal of Agricultural Research, 77*(4), 373–381.

Reyes Muñoz, M. M., & Lavín Acevedo, J. A. (2022). Estación Experimental Cauquenes. History and contributions to national viticulture. *Revista Iberoamericana de Viticultura, Agroindustria y Ruralidad, 9*(27), 129–142.

Ruml, M., Korać, N., Vujadinović, M., Vuković, A., & Ivanišević, D. (2016). Response of grapevine phenology to recent temperature change and variability in the wine-producing area of Sremski Karlovci, Serbia. *The Journal of Agricultural Science, 154*, 186–206.

Sargolzaei, M., Rustioni, L., Cola, G., Ricciardi, V., Bianco, P. A., Maghradze, D., Failla, O., Quaglino, F., Toffolatti, S. L., & De Lorenzis, G. (2021). Georgian grapevine cultivars: Ancient biodiversity for future viticulture. *Frontiers in Plant Science, 12*, 630122.

Schwalbert, R., Stefanello, L. O., Schwalbert, R. A., Tarouco, C. P., Drescher, G. L., Trentin, E., Tassinari, A., da Silva, U. B., Brunetto, G., & Nicoloso, F. T. (2021). Soil tillage affects soybean growth and promotes heavy metal accumulation in seeds. *Ecotoxicology and Environmental Safety, 216*, 112191.

Shim, R. S., Tully, L. M., Yu, G., Monterozza, E. C., & Blendermann, M. (2021). Race and ethnicity of editorial board members and editors as an indicator of structural racism in Psychiatry and Neuroscience journals. *JAMA Psychiatry, 78*(10), 1161–1163.

Valenzuela, L., Ortega, R., Moscovici, D., Gow, J., Alonso Ugaglia, A., & Mihailescu, R. (2022). Consumer willingness to pay for sustainable wine—The Chilean case. *Sustainability, 14*(17), 10910.

Verdugo-Vásquez, N., Orrego, R., Gutiérrez-Gamboa, G., Reyes, M., Zurita Silva, A., Balbontín, C., Gaete, N., & Salazar-Parra, C. (2023). Climate trends and variability in the Chilean viticultural production zones during 1985–2015. *OENO one, 57*(1), 345–362.

Verslype, N. I., do Nascimento, A. C. A., dos Santos Musser, R., de Souza Caldas, R. M., Martins, L. S. S., & de Souza Leão, P. C. (2023). Drought tolerance classification of grapevine rootstock by machine learning for the São Francisco Valley. *Smart Agricultural Technology, 4*, 100192.

Villanueva, E. (2017). Innovation as a cause of the Anglo-Saxon New World wine exports boom. *Revista Iberoamericana de Viticultura, Agroindustria y Ruralidad, 1*(1), 1–13.

Vršič, S., & Vodovnik, T. (2012). Reactions of grape varieties to climate changes in North East Slovenia. *Plant Soil and Environment, 58*, 34–41.

Vršič, S., Šuštar, V., Pulko, B., & Šumenjak, T. K. (2014). Trends in climate parameters affecting winegrape ripening in Northeastern Slovenia. *Climate Research, 58*, 257–266.

Wines of Chile. (2015). Código de Sustentabilidad de la Industria Vitivinícola Chilena. Reglamento General | Versión 3.1/04.201. Wines of Chile. Retrieved from: https://www.sustentavid.org/

Correction to: Growing Vines in the Mapuche Heartland: The First Report About the Vitiviniculture of the Araucanía Region

Gastón Gutiérrez Gamboa, Cristóbal Palacios-Peralta, Rafael López-Olivari, Pamela Castillo, Milton Almonacid, Raúl Narváez, Luis Morales-Salinas, Nicolás Verdugo-Vásquez, Marcela Hidalgo, Alejandra Ribera-Fonseca, and Ignacio Serra

Correction to:
Chapter 13 in: G. Gutiérrez Gamboa, M. Fourment (eds.), ***Latin American Viticulture Adaptation to Climate Change*****,**
https://doi.org/10.1007/978-3-031-51325-1_13

The original version of Figure 13.2 and Section 13.3.2 was inadvertently published with errors. The chapter has now been corrected and approved by the author.

The updated version of this chapter can be found at
https://doi.org/10.1007/978-3-031-51325-1_13

G. Gutiérrez Gamboa, M. Fourment (eds.), *Latin American Viticulture Adaptation to Climate Change*, https://doi.org/10.1007/978-3-031-51325-1_16

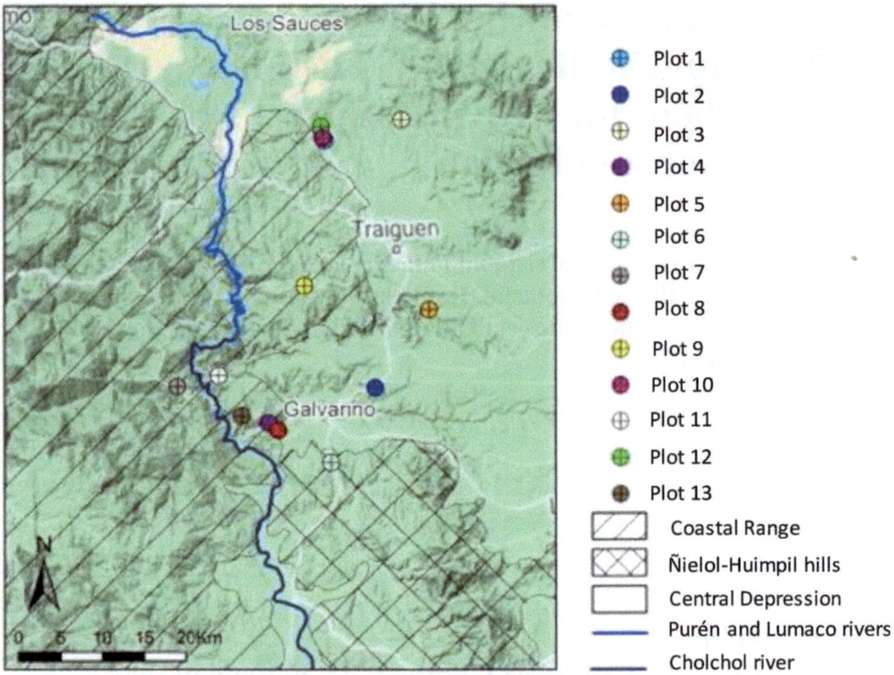

Fig. 13.2 Geomorphology of the Malleco valley. Note: Circles correspond to vineyards that are cultivated

Index

G. Gutiérrez Gamboa, M. Fourment (eds.), *Latin American Viticulture Adaptation to Climate Change*, https://doi.org/10.1007/978-3-031-51325-1

MIX
Papier aus verantwortungsvollen Quellen
Paper from responsible sources
FSC® C105338

If you have any concerns about our products,
you can contact us on
ProductSafety@springernature.com

In case Publisher is established outside the EU,
the EU authorized representative is:
Springer Nature Customer Service Center GmbH
Europaplatz 3, 69115 Heidelberg, Germany

Printed by Libri Plureos GmbH
in Hamburg, Germany